Advances in Experimental Medicine and Biology

Volume 988

More information about this series at http://www.springer.com/series/5584

Panayiotis Vlamos
Editor

GeNeDis 2016

Computational Biology and Bioinformatics

Springer

Editor
Panayiotis Vlamos
Department of Informatics
Bioinformatics and Human
Electrophysiology Laboratory
Ionian University
Corfu, Greece

ISSN 0065-2598 ISSN 2214-8019 (electronic)
Advances in Experimental Medicine and Biology
ISBN 978-3-319-56245-2 ISBN 978-3-319-56246-9 (eBook)
DOI 10.1007/978-3-319-56246-9

Library of Congress Control Number: 2017943973

Printed on acid-free paper

This Springer imprint is published by Springer Nature
The registered company is Springer International Publishing AG
The registered company address is: Gewerbestrasse 11, 6330 Cham, Switzerland

Contents

Biological Relevance of Network Architecture

Ioannis Gkigkitzis, Ioannis Haranas, and Ilias Kotsireas

Abstract Mathematical representations of brain networks in neuroscience through the use of graph theory may be very useful for the understanding of neurological diseases and disorders and such an explanatory power is currently under intense investigation. Graph metrics are expected to vary across subjects and are likely to reflect behavioural and cognitive performances. The challenge is to set up a framework that can explain how behaviour, cognition, memory, and other brain properties can emerge through the combined interactions of neurons, ensembles of neurons, and larger-scale brain regions that make information transfer possible. "Hidden" graph theoretic properties in the construction of brain networks may limit or enhance brain functionality and may be representative of aspects of human psychology. As theorems emerge from simple mathematical properties of graphs, similarly, cognition and behaviour may emerge from the molecular, cellular and brain region substrate interactions. In this review report, we identify some studies in the current literature that have used graph theoretical metrics to extract neurobiological conclusions, we briefly discuss the link with the human connectome project as an effort to integrate human data that may aid the study of emergent patterns and we suggest a way to start categorizing diseases according to their brain network pathologies as these are measured by graph theory.

Keywords Networks • Graphs • Cognition • Behavior • Memory • Connectome • Connectivity

---

I. Gkigkitzis (✉)
Department of Mathematics, East Carolina University, 124 Austin Building, East Fifth Street, Greenville, NC 27858-4353, USA
e-mail: gkigkitzisi@ecu.edu

I. Haranas • I. Kotsireas
Department of Physics and Computer Science, Wilfrid Laurier University, Science Building, Room N2078, 75 University Ave. W., Waterloo, ON, Canada N2L 3C5
e-mail: iharanas@wlu.ca; ikotsireas@wlu.ca

P. Vlamos (ed.), *GeNeDis 2016*, Advances in Experimental Medicine and Biology 988, DOI 10.1007/978-3-319-56246-9_1

1 Graph Theory

Graph theory may be very useful in neuroscience as a framework for the non-invasive exploration of the topological organization of brain connectivity both locally and globally [1]. Brain network graphs are composed of nodes representing regions or voxels and edges representing structural or functional connectivity among the nodes. Graphs can be undirected or directed, unweighted or weighted, and there exist several network metrics that can be used to study the structure and dynamics of networks [2] and to reveal brain network patterns intrinsically represented in the brain, and identify neural substrate relationships between intranetwork and internetwork activity during rest or/and task. The use of imaging modalities, such as structural MRI, diffusion MRI, functional MRI, and EEG/MEG, etc. He and Evans [3] is a requirement for the extraction of brain networks on which graph theoretical analysis can be applied. Graph metrics are expected to vary across subjects and are likely to reflect behavioral and cognitive performances. Statistical measures of dispersion (standard deviation, coefficient of variation, etc.) and measures of central tendency and may serve as markers of neurological dysfunctions, such as ADHD [4], aging and traumatic brain injury [5, 6]. Other important statistical/graph theoretical properties, including the degree distribution, cost of connection, clustering coefficients, measures of centrality, modularity, small-worldness, shortest path length, transitivity, local and global efficiency, highly connected network hubs, similarity measures, correlations, coherence, transfer entropy, etc., have been found to change during normal development, aging, and various neurological and neuropsychiatric diseases.

Modular (community) structures can be detected through the optimization of a function called modularity [7] which requires approximation algorithms such as in Clauset et al. [8], where communities merge to optimize the production of modularity, or in Guimera et al. [9] where the authors used simulated annealing, or more recently in Blondel et al. [10] where the algorithm finds hierarchical community structure for different resolutions of community detection. The last algorithm seems to circumvent the "resolution limit problem" [11].

An optimal parcellation scheme for neuroimaging that will support graph theory analysis is still to be found. There are three common approaches to parcellation schemes [1]: (1) The voxel wise approach with the number of nodes ranging from 3400 to 140,000 nodes. (2) The anatomical atlases approach with ROIs (nodes) ranging from 70 to 250 nodes. (3) The functional activation meta-analytic approach, which utilizes ROIs for all subsequent studies, usually modeled by spheres 3–6 mm radii and number of nodes for this parcellation strategy ranging from 10 to 264 nodes. Stanley et al. [1] recommend a non-biased network science dynamic view instead of "static functional unit" approach, because it allows the modeling of a brain performing complex functions which are emergent features of the brain system. The voxel based analysis does not require an a priori seed region ROI definition but, among other issues, it suffers from the need for arbitrary thresholds to define network properties [12]. A solution based on the intrinsic connectivity distribution has been suggested in Scheinost et al. [13]. However, mapping the human connectome starts with parcellation of grey matter on the basis of rfMRI time series and variations of connectivity across a parcel is assumed to be smaller than the differences in connectivity patterns between two parcels [14]. A "parcellated connectome" (parcels $\times$ parcels matrix) may be more preferable than a much larger original "dense connectome" (voxels $\times$ voxels matrix) [14]. The Human Connectome Project uses combinations of different imaging modalities and produces more than one parcellation. In general, current efforts focus one building brain atlases of minimal functional subunits based on BOLD signals rather than cytoarchitecture or other anatomic distinctions. Recently, a data driven approach was presented in Shen et al. [12] using groupwise clustering algorithm (groupwise multigraph K way clustering algorithm) to jointly optimize the group and the individual parcellation of the human subjects, with regions of high and low reproducibility. The groupwise parcellation acts as an implicit constraint that drives the optimization scheme for the parcellation of each graph towards a common configuration.

2 Graph Theory and Neuroimaging

Graph theoretical analysis of neuroimaging data is utilized for the investigation of structural and functional brain connectivity [3]. Structural MRI is based on correlated changes in gray matter morphology (cortical thickness and volume) between various anatomically or functionally linked areas [15, 16]. Structural MRI and graph theoretical methods combined have been used to the study of brain diseases such as Alzheimer's disease, schizophrenia, and multiple sclerosis [3]. He et al. [17] showed that Alzheimer's disease patients had increased clustering and path lengths and reduced betweenness centrality in temporal and parietal regions, suggesting a shift to more local processing and a disrupted structural integrity of the larger-scale brain systems. Bassett et al. [18] studied schizophrenia patients and showed a reduced hierarchy, an increased connection distance, and the loss of frontal hubs. He et al. [19] studied multiple sclerosis patients and showed disrupted small-world efficiency proportional to the extent of total white matter lesions mainly in the insula, precentral gyrus and prefrontal and temporal association cortical regions.

In the field of diffusion MRI, deterministic 'streamline' tractography infers the continuity of fiber bundles from voxel to voxel [20]. Probabilistic tractography allows to compute the connectivity probabilities rather than the actual white matter pathways between voxels [21]. Li et al. [22] used DTI tractography, and showed that higher intelligence quotient scores are associated with larger global efficiency in the brain networks. Shu et al. [23] used DTI tractography and showed that early blind subjects are associated with decreased connectivity degree and global efficiency in the visual cortex and increased connections in the motor or somatosensory areas, evident of a topological re-organization of structural brain connectivity. Gong et al. [24] showed a decrease in the local efficiency of structural brain networks constructed from diffusion-weighted MRI with age and a shift of regional efficiency from the parietal and occipital to the frontal and temporal neocortex in older adults. Also, female brains have greater overall connectivity and higher efficiencies than male brains, indicative of possible age-related and sex-related differences in cognition and behavior.

FMRI utilizes changes in cerebral blood flow and oxygen consumption in order to detect neuronal activity, with intermediate temporal (seconds) and spatial (mm) resolutions. van den Heuvel et al. [25] used fMRI to show a significantly negative correlation between the path length of brain networks (voxel level) and intelligence quotient, with more observable effects in the frontal and parietal regions, thus relating cognitive ability to the topology of the brain functional network. Wang et al. [26] showed that older adult brains were associated with an increased shortest path length and a reduction in the long-range connections during the performance of memory tasks, indicative of age-related declines in cognitive functions. Fair et al. [27] showed that, in children, modules are consistent with anatomical regions but in adults they reflect functional relationships. Supekar et al. [28] reported differences in hierarchical organization and interregional connectivity and a reduction in short-range connectivity and a strengthening of long-range connectivity from childhood to young adulthood. Achard and Bullmore [29] used resting state fMRI and showed reduced efficiency in older adults. Supekar et al. [30] reported that Alzheimer's disease patients had reduced clustering in the brain functional networks, indicative of disrupted local neighboring connectivity. Liu et al. [31] showed that various topological measurements, such as the clustering coefficient and the global and local efficiency, were reduced in the brain networks of schizophrenia patients as compared with controls, and the reduction was negatively correlated with the illness duration. Wang et al. [32] showed that boys with ADHD had increased local efficiency in the brain functional networks, with nodal efficiency changes in the prefrontal and temporal regions. These changes could reflect a compensatory recruitment or a developmental delay in brain topological organization in this disorder. Liao et al. [33] showed that patients with mesial temporal lobe epilepsy were associated with smaller clustering coefficients and shorter path lengths, indicating a more random-like configuration in the brain functional networks of the patients. In Nakamura et al. [34], patients with traumatic brain injury had reduced connectivity strength and increased small-world attributes from 3 to 6 months post injury, suggestive of a network recovery following severe brain injury.

EEG or MEG measure the changes in the electromagnetic field related to neuronal activity at a high temporal resolution (milliseconds) but a poor spatial resolution (cm). Bassett et al. [35] reported that the cost efficiency of a brain functional network in the left lateral frontal and parietal regions correlated positively with task performance (working memory). Micheloyannis et al. [36] showed that young adults had decreased clustering and increased path lengths in the brain functional networks as compared with children. In Alzheimer's disease, resting-state MEG network [37] revealed decreased clustering coefficients and path lengths, with a preferential decrease in connections between highly connected network hubs. A resting-state EEG study [38] reported similar changes in the brain functional network topology in Alzheimer's disease patients but observed changes in the opposite direction (increased clustering and path lengths) in patients with frontotemporal dementia, suggesting a different pathophysiology. These findings are inconsistent with those shown in a previous structural MRI study demonstrating increased clustering and shortest path lengths in structural brain networks in Alzheimer's disease [17]. The discrepancies could be attributable to different imaging modality, population size, network node, and edge definitions applied in these studies. Graph theoretical analysis based on EEG/MEG data has been also applied to other diseases such as schizophrenia [39], epilepsy [40], and depression [41].

Functional connectivity (in the resting state) has been positively correlated with structural connectivity, but functional connectivity has also been observed between regions where there is little or no structural connectivity [42]. Functional connectivity is based on seed-based analysis and independent component analysis (ICA) [1] while structural connectivity analysis is based on methods such as diffusion tensor imaging (DTI). Previous structural, functional, and diffusion MRI studies have used both anatomical and/or functional brain atlases and image voxels, but the resultant networks exhibited significantly different topological properties [43–46]. Stability and reproducibility of graph metrics are highly desirable for brain network analysis and some studies have structural [47] and functional (derived from MEG and fMRI) [48–50] reproducibility of graph metrics across different task states and subjects. Honey et al. [51] showed results of topological correlations of structural and functional networks. A disagreement between functional and structural measurements may be a reflection of false positives in resting state fMRI (physiological noise induced by cardiac pulse and respiration, from the BOLD signal), or false negatives in DTI (difficulty resolving small tracts as they cross larger tracts, incorrect measure of principal diffusion direction for a particular tract, etc.) [42]. A combined approach of different imaging modalities that yield convergent results [52–54] may have diagnostic and prognostic capacities [55]. Functional networks are constrained by structural connectivity, which they modulate over long timescales [56]. *Behavior* can thought of as an extension of brain networks into the world, an association that selects inputs that perturb the interplay between brain functional networks, producing changes that result in modulation of subsequent behavior. Behavioral variability may be mediated by competition between two networks (such as DMN and Attention Network [4]) and the combination of multimodal imaging techniques may aid the study of the brain connectivity underlying cognition and behaviors in humans.

According to Byrge et al. [56]: (1) Connectivity goes beyond information channeling among brain regions, and generates complex system-wide dynamics that enable local regions to participate to cognitive and behavioral tasks. (2) External inputs do not just trigger or activate of specific subroutines of neural processing that are encapsulated in local regions, but they rather perturb ongoing activity and become integrated with the system's current dynamic state. (3) The cumulative history of perturbations as recorded in changing patterns of connectivity—in various timescales—defines the system's capacity to respond to input and to generate internal dynamics. In parallel, in the field of developmental robotics [56], the compositionality reflected in higher-order behaviors emerges in systems with multiscale temporal dynamics that are grounded in sensorimotor processing [57], and self-generated movements are critical to the development of pattern-based object recognition supported by changes in connectivity in higher-order visual circuits [58].

3 Behavior, Cognition and Disease Example Applications of Graph Theory

The study of the structure and functional interactions in the brain may aid the understanding of behavior, cognition, clinical disorders and disease [12]. Cognition may be defined as a behavioral outcome of the neuronal integrative processing that links sensation and action and includes the manifestations of memory, emotion, attention, language, thought and consciousness [59]. The neural pathways, processing centers and synapses provide the biological constraint that shapes cognition and behavior. For instance, limbic processing of rapid learning of behaviorally relevant relationships may help new memories to enter associative readjustment [59]. The interaction between environmental events and internal representations is represented in neural ensembles, resonant with the goals and constraints of the context. The neuronal paths that link sensory information to perception, recognition, recollection and behavior can be thought of as noisy informational channels. The replication of real life events this information represents is unlikely to be faithful and also, at the higher synaptic levels, emotion, motivation and selective attention introduce subjective biases. Consciousness is embedded within an illusory reality, where a person's mnemonic abilities optimize the adaptive value of behavior [59]. We discuss some studies on behavior, cognition, disorders and diseases that used graph theory methods on brain networks.

Gong et al. [60] examined the relation between active learning and neural plasticity by analyzing Action Video Game (AVG) experience related to integration between Salience Network (attention) and Central Executive Network (working memory), in AVG experts' and amateurs' resting-state brain functions through graph theoretical analyses (global efficiency, cost, mean clustering coefficient and nodal characteristics) and functional connectivity. They showed enhanced intra- and internetwork functional integrations in AVG experts compared to amateurs.

Stanley et al. [61] used local and global efficiency of information transfer for predicting individual variability in working memory performance on an n-back task in both young and older adults. Decreases in local efficiency during the working memory task were associated with better working memory performance in both age cohorts. Increases in global efficiency were associated with much better working performance for young adults and with a slight decrease in working memory performance for older adults. Significant group whole-brain functional network decreases in local efficiency also were observed during the working memory task compared to rest.

Tang et al. [62] obtained intelligence factors (general, speed of reasoning, spatial, memory, and numerical) from tests completed by young adults, who also completed DTI and fMRI during a working memory n-back task. Clusters of activation during the n-back task were found in Anterior Cingulate and bilateral Prefrontal, Parietal, Insular, visual cortices. For males, activation in the right prefrontal and right parietal cortices was inversely correlated with spatial intelligence and activation at the right prefrontal cortex was inversely correlated with general intelligence. For females, general intelligence was inversely correlated with activation in left and right parietal cortex and numerical intelligence was inversely related to activation in the anterior cingulate, left and right parietal cortex. Speed of reasoning was inversely correlated with activation in the left parietal cortex. All correlations were negative (higher intelligence was associated with less brain activation). Memory showed no correlations with activation.

Rosenberg et al. [63] identified functional brain networks whose strength during a sustained attention task predicted individual differences in performance. These networks were shown to predict behavioral performance of novel individuals from their task based or resting state connectivity, and can also predict symptoms of attention deficit hyperactivity (ADHD). This indicated that functional connectivity is a robust neuromarker of sustained attentional abilities. Connections between motor cortex, occipital lobes and the cerebellum were predictors of better sustained attention. Connections between temporal and parietal regions as well as intratemporal and intracerebellar connections predicted worse attention between subjects. Connections within the subcortical-cerebellum network and the frontoparietal network appeared more frequently in a low than in a high attention network. Connections between subcortical-cerebellum network and the medial frontal, motor, visual I and visual association networks appeared more frequently in the high attention network. The test of gradCPT test and performance was assessed with sensitivity *d*.

Whitfield-Gabrieli et al. [64] used a priori bilateral anatomical amygdala seed-driven resting connectivity and probabilistic tractography of the right inferior longitudinal fasciculus together with a data-driven multivoxel pattern analysis of whole-brain resting-state connectivity before treatment to predict improvement in social anxiety after cognitive behavioral therapy in patients with social anxiety disorder (SAD). Each connectomic measure improved the prediction of individuals' treatment outcomes significantly better than a clinical measure of initial severity, and combination of multimodal connectomics yielded large improvement in predicting treatment response, indicating that imaging methods, may provide brain-based neuromarkers for neuropsychiatric diseases.

Davis et al. [65] studied the relationship between impulsivity and the modularity derived from resting-state fMRI data. In highly impulsive individuals, regulatory structures including medial and lateral regions of the prefrontal cortex were isolated from subcortical structures associated with appetitive drive, whereas these brain areas clustered together within the same module in less impulsive individuals. Changes in the functional connectivity were observed between visual, sensorimotor, cortical, and subcortical structures across the impulsivity spectrum.

Cao et al. [66] used resting-state functional MRI and graph-theory to chart the lifespan trajectory of the topological organization of brain functional networks in healthy individuals ranging from 7 to 85 years and observed highly preserved non-random modular and rich club organization over the entire age range. They found linear decreases in modularity and inverted-U shaped trajectories of local efficiency and rich club architecture. Regionally heterogeneous age effects were mainly located in several hubs (e.g., default network, dorsal attention regions). Inverse trajectories of long- and short-distance functional connections indicated that the reorganization of connectivity concentrates across the entire human lifespan, perhaps underlying variations in behavior and cognition.

Dosenbach et al. [67] investigated brain's moment-to-moment information processing by studying interactions of regions that carry signals related to task-control initiation, maintenance, and adjustment by applying graph theory to resting state fMRI data. They identified two distinct task-control networks. A frontoparietal network included the dorsolateral prefrontal cortex and intraparietal sulcus. This network emphasized start-cue and error-related activity and may initiate and adapt control on a trial-by-trial basis. The second network included dorsal anterior cingulate/medial superior frontal cortex, anterior insula/frontal operculum, and anterior prefrontal cortex. This network may control goal-directed behavior through the stable maintenance of task sets. These two independent networks appear to operate on different time scales and affect downstream processing via dissociable mechanisms.

Miraglia et al. [68] studied functional networks in males and females in resting state EEGs. They studied the Attentional Network, Frontal Network, Sensorimotor Network, Default Mode Network. Gender small-worldness differences in some of resting state networks indicated that there are specific brain differences in the EEG rhythms when the brain is in the resting-state condition. These specific regions could be considered related to the functions of behavior and cognition.

DeSalvo et al. [69] used graph theory and compared functional connectivity in healthy subjects between the resting state and when performing a clinically used semantic decision task. A language-related module emerged during the semantic decision task. Overall and within-module connectivity were greater in default mode network (DMN) and language areas during semantic decision making compared to rest, while between-module connectivity was diffusely greater at rest. Semantic decision making was associated with a reduction in distributed connectivity through hub areas of the DMN and an increase in connectivity within default and language networks.

Muller and Lindenberger [70] analyzed EEG of heterosexual couples during romantic kissing to kissing one's own hand, and to kissing one another while performing silent arithmetic. Using graph-theory methods, they identified theta-alpha hyper-brain networks. Network strengths were higher and characteristic path lengths shorter when individuals were kissing each other than when they were kissing their own hand. In both partner-oriented kissing conditions, greater strength and shorter path length for 5-Hz oscillation nodes correlated reliably with greater partner-oriented kissing satisfaction. This correlation was especially strong for inter-brain connections in both partner-oriented kissing conditions but not during kissing one's own hand. They concluded that hyper-brain networks may capture neural mechanisms that support interpersonally coordinated voluntary action and bonding behavior.

Liu et al. [71] used an attention network test (ANT) as behavioral measure of the efficiency of the three attention networks (alerting, orienting and executive networks) within a single task. They investigated the effect of passive hyperthermia on the attention network with event-related fMRI. Passive hyperthermia of 50 °C and 40% relative humidity impaired the executive function, but showed no effect on the alerting and orienting networks. Passive hyperthermia enhanced the activity in the right superior frontal gyrus and depressed the activity in the right middle occipital gyrus, left inferior parietal lobule and left culmen in the alerting network. It also enhanced the activity in the temporal lobe and depressed the activity in the frontal, parietal and occipital lobes in the orienting network, and enhanced the activity in the dorsolateral prefrontal cortex but did not affect the activity in the anterior cingulate. Passive hyperthermia impaired executive function, especially the efficiency of resolving conflict and the negative effects of passive hyperthermia on alerting and orienting were overcome through variant regional brain activation.

Qian et al. [72] simulated environmental heat exposure to 18 participants, and obtained resting state fMRI data. Brain networks in both normal and hyperthermia conditions exhibited economical small-world property, but significant alterations in both global and nodal network metrics were demonstrated during hyperthermia: a lower clustering coefficient, maintained shortest path length, a lower small-worldness, a lower mean local efficiency were found, indicating a tendency shift to a randomized network. Significant alterations in nodal efficiency were found in bilateral gyrus rectus, bilateral parahippocampal gyrus, bilateral insula, right caudate nucleus, bilateral putamen, left temporal pole of middle temporal gyrus, right inferior temporal gyrus. Alterations of normalized clustering coefficient, small-worldness, mean normalized local efficiency were significantly correlated with the rectal temperature alteration. Behavioral attention network test (ANT) results were altered and correlated with the alterations of some global metrics (normalized shortest path length and normalized global efficiency) and prefrontal nodal efficiency (right dorsolateral superior frontal gyrus, right middle frontal gyrus and left orbital inferior frontal gyrus), implying behavioral deficits in executive control effects and maintained alerting and orienting effects during passive hyperthermia. This study indicates brain functional disorder during passive hyperthermia.

Manelis et al. [73] studied win/loss anticipation in depressed individuals with bipolar disorder (BDD) versus depressed individuals with major depressive disorder (MDD) versus healthy control subjects. Participants were scanned while performing a number guessing reward task that included the periods of win and loss anticipation. Density of connections, path length, and the global connectivity direction ('top-down' versus 'bottom-up') were compared across groups and conditions. Loss anticipation was characterized by denser top-down fronto-striatal and fronto-parietal connectivity in healthy control subjects, by bottom-up striatal-frontal connectivity in

MDD, and by sparse connectivity lacking fronto-striatal connections in BDD. Win anticipation was characterized by dense connectivity of medial frontal with striatal and lateral frontal cortical regions in BDD, by sparser bottom-up striatum-medial frontal cortex connectivity in MDD, and by sparse connectivity in healthy control subjects.

Luft et al. [74] reviewed transcranial current brain stimulation (tCS) as a way to provide a causal link between a function or behavior and a specific brain region (e.g., primary motor cortex) and they argued in favor of combining tCS method with other neuroimaging techniques (e.g., fMRI, EEG) and employing graph theory to obtain a deeper understanding of the underlying spatiotemporal dynamics of functional connectivity patterns and cognitive performance. Finally, the authors investigated the neural correlates of human creativity.

Ash and Rapp [75] reviewed the literature on how alterations in structural and functional networks may contribute to individual differences in cognitive phenotypes in advanced aging (healthy and pathological aging phenotypes) and outlined analytic strategies that attempt to quantify graph theory metrics more precisely, with the goal of improving diagnostic sensitivity and predictive accuracy for differential trajectories in neurocognitive aging.

Sadaghiani et al. [76] investigated whether ongoing changes in baseline functional connectivity correlate with perception. They used a continuous auditory detection task and time intervals that permitted investigation of baseline connectivity unaffected by preceding evoked responses. They showed that functional connectivity before the target predicted whether it was heard or missed and graph theoretical measures characterized the difference in functional connectivity between states that lead to hits vs. misses. Before misses compared with hits and task-free rest, connectivity showed reduced modularity in the default mode and visual networks and was caused by both reduced within-network connectivity and enhanced across-network connections before misses. Therefore, dynamic changes in baseline functional connectivity may shape subsequent behavioral performance.

Tang et al. [77] studied silent lacunar infarcts and their association with subtle deficits in cognition. They used DTI tractography and graph theory to examine the topology of white matter networks in patients with silent lacunar infarcts in the basal ganglia. Compared with controls, the patients exhibited a significant reduction in local and global efficiency and significantly reduced nodal efficiency, indicating that small lesions in subcortical brain regions may affect large-scale cortical white matter network which may lead to associated cognitive impairments.

Lin et al. [78] studied the relationship between the DMN functional connectivity and cognitive behavior during resting-state and task performance. Nodal degree of the DMN was calculated as a metric of network topology. They found that the static and dynamic posterior cingulate cortex (PCC) nodal degree within the DMN was associated with task performance (Reaction Time), suggesting that the PCC plays a key role in cognitive function.

Godwin et al. [79] examined functional brain changes associated with awareness using graph theoretical analysis applied to functional connectivity data acquired at ultra-high field while subjects performed a simple masked target detection task.

They found that awareness of a visual target is associated with a degradation of the modularity of the brain's functional networks brought about by an increase in intermodular functional connectivity.

Paolini et al. [80] studied obesity and used global efficiency to study and the possibility that integration across the Hot-State Brain Network of Appetite (HBN-A) predicts WL after 6-months of treatment in older adults. Imaging involved a baseline food-cue visualization fMRI scan following an overnight fast. They used principal component analysis to show that insula, anterior cingulate cortex (ACC), superior temporal pole (STP), amygdala and the parahippocampal gyrus were highly integrated. The HBN-A is comprised of limbic regions important in emotional processing and visceral sensations and the ACC is key for translating such processing into behavior. Improved integration of these regions may leading to more successful self-regulation and to greater weight loss.

Khazaee et al. [81] aimed to identify the changes in brain networks in patients with AD and mild cognitive impairment (MCI), and used a graph theoretical approach and advanced machine learning methods. Multivariate Granger causality analysis was performed on rs-fMRI data and found that patients with AD experience disturbance of critical communication areas (hubs) in their brain network as AD progresses. Earlier, Khazaee et al. [82] combined graph theory with Support vector machine (SVM) to explore the ability of graph measures in diagnosis of AD. In addition to the machine learning approach, statistical analysis was performed on connectivity matrices to find altered connectivity patterns in patients with AD. The authors were able to accurately classify patients with AD from healthy subjects with accuracy of 100%.

Miraglia et al. [68] investigated the neuronal network characteristics in Alzheimer's disease (AD), mild cognitive impairment (MCI), and normal elderly subjects. They used EEG recordings and graph theory parameters (Characteristic Path Length, Clustering coefficient, and small-world network) on weighted networks. EEG analysis showed significant differences in delta, theta, and alpha 1 bands. For the normalized Characteristic Path Length the pattern differences between normal cognition and dementia were observed in the theta band (MCI subjects are find similar to healthy subjects), while for the normalized Clustering coefficient a significant increment was found for AD group in delta, theta, and alpha 1 bands; The small world parameter showed a theta increase in MCI. The fact that AD patients with respect the MCI subjects were significantly impaired in theta but not in alpha bands connectivity are in line with the hypothesis of an intermediate status of MCI between normal condition and overt dementia.

Taya et al. [83] tried to interpret EEG neurophysiological training signal patterns by assuming that the brain first works hard to learn how to use task-relevant areas, followed by improvement in the efficiency derived from disuse of irrelevant brain areas for good task performance during the training on a piloting. Global information transfer efficiency of the functional network in a high frequency band first decreased and then increased during the training while other measures such as local information transfer efficiency and small-worldness showed opposite patterns.

Additionally, the centrality of nodes changed due to the training at frontal and temporal sites. They concluded that network metrics can be used as biomarkers for quantifying the training progress.

Barttfeld et al. [84] studied adults with attention deficit/hyperactivity disorder (ADHD) or euthymic bipolar disorder (BD) relative to a control group. Resting EEG was measured and connectivity and graph theory metrics were computed. The ADHD group showed an enhancement of functional connectivity (fronto-occipital connections), intrinsic variability compared to control temporal variability of functional connections association with executive function and memory deficits and depression, hyperactivity and impulsivity levels association with abnormal intrinsic connectivity. The BD showed diffuse connections, reduced intrinsic connectivity, levels of anxiety and depression were associated to abnormal frontotemporal connectivity. ADHD in children was also studied in Cao et al. [85]. Children with ADHD have abnormal small-world properties in both functional and structural brain networks characterized by higher local clustering and lower global integrity, redistribution of regional nodes and connectivity involving the default-mode, attention, and sensorimotor systems. ADHD-associated alterations have significantly correlated with behavior disturbances and imaging-based biomarkers may be sufficient for clinical diagnosis and treatment evaluation in ADHD. Cao et al. [86] used diffusion MRI and probabilistic tractography method to examine whole-brain white matter structural connectivity in boys with ADHD and healthy controls. Small-world and network efficiency were analyzed using graph theoretical approaches. Both the ADHD and control groups showed an efficient small-world organization in the whole-brain white matter networks. ADHD exhibited decreased global efficiency and increased shortest path length, with the most pronounced efficiency decreases in the left parietal, frontal, and occipital cortices. The ADHD group showed decreased structural connectivity in the prefrontal-dominant circuitry and increased connectivity in the orbitofrontal-striatal circuitry, and these changes significantly correlated with the inattention and hyperactivity/impulsivity symptoms, respectively.

Elton et al. [87] investigated the inhibitory control of behavior as a target of childhood maltreatment (abuse and neglect) using a Childhood Trauma Questionnaire (CTQ) and fMRI while performing a stop-signal task. Independent component analysis identified a brain inhibitory control network. Graph theoretical analyses and structural equation modeling investigated the impact of childhood maltreatment on this neural processing network. Graph theory outcomes revealed sex differences in the relationship between network functional connectivity and inhibitory control which were dependent on the severity of childhood maltreatment exposure. A maltreatment dose-related negative modulation of dorsal anterior cingulate (dACC) activity by the left inferior frontal cortex (IFC) predicted better response inhibition and lesser ADHD symptoms in females, but poorer response inhibition and greater ADHD symptoms in males. Less inhibition of the right IFC by dACC in males with higher CTQ scores improved inhibitory control ability. The childhood maltreatment-related reorganization of a brain inhibitory control network provides sex-dependent mechanisms by which childhood adversity may confer greater risk for drug use and related disorders and by which adaptive brain responses protect individuals from this risk factor.

Brown et al. [88] performed diffusion tensor imaging (DTI), T1 structural imaging, and cognitive testing on aging APOE-4 allele (risk factor for AD) noncarriers and APOE-4 carriers. Fiber tractography was used to derive whole brain structural graphs, and graph theory was applied to assess structural network properties. APOE-4 carriers demonstrated an accelerated age-related loss of mean local interconnectivity and regional local interconnectivity decreases in the precuneus, medial orbitofrontal and lateral parietal cortices, significant age-related loss in mean cortical thickness and significant negative correlations of age and performance on two episodic memory tasks.

Jin et al. [89] showed abnormal reconfiguration of brain networks in focal hand dystonia (FHD) patients during motor task, using graph theoretic measures to assess efficiency and high-resolution EEGs at rest and during a simple sequential finger tapping task. Mutual information (MI) values of wavelet coefficients were estimated to produce adjacency matrices or graphs, by thresholding with network cost. They found that FHD patients have economical small-world properties in the alpha and beta bands. During a motor task, in the beta band network, FHD patients have decreased efficiency of small-world networks, whereas healthy volunteers increase efficiency. Reduced efficient beta band network in FHD patients during the task was consistently observed in global efficiency, cost-efficiency, and maximum cost-efficiency, representing a loss of long-range communication and abnormal functional integration in large-scale brain functional cortical networks. Moreover, negative correlations between efficiency measures and duration of disease were found, indicating that the longer duration of disease, the less efficient the beta band network in FHD patients. FHD patients at rest have high regional efficiency at supplementary motor cortex (SMA) compared with healthy volunteers; however, it is diminished during the motor task, possibly reflecting abnormal inhibition in FHD patients.

Smit et al. [90] examined overall connectivity (synchronization likelihood-SL), local clustering coefficient and average path length of connectivity, suing resting state EEG in 1438 subjects and showed that connectivity is more random at adolescence and old age, and more structured in middle-aged adulthood. These parameters, in the alpha band, showed high phenotypic and genetic stability from 16 to 25 years. Heritability for parameters in the beta band was lower, and less stable across ages. The investigators concluded that these connectivity parameters in the alpha band are good endophenotype for behavior and developmental disorders.

Schroter et al. [91] showed that propofol-induced loss of consciousness is associated with a breakdown of subcortical-cortical and cortico-cortical connectivity, with decrease of connectivity pronounced in thalamocortical connections. Compared with a random network, normalized clustering was significantly increased, small-worldness and long-range connections were significantly reduced. A breakdown of connectivity within higher-order association cortices and between higher-order association and primary sensory cortices was observed. Previous studies showed a linear association between functional connectivity of frontoparietal networks and levels of consciousness [92] where waning of consciousness was associated with a loss of cross-modal interactions between visual and auditory networks, and a reorganization of key nodes of the DMN [93] where connectivity of the posterior cingulate, during sedation, includes the motor/somatosensory cortices, the anterior thalamic nuclei, and the reticular activating system.

Liao et al. [33] used resting state fMRI data to study functional connectivity in mesial temporal lobe epilepsy during interictal periods and observed that DMN regions have a significant decreased number of connections to other regions. They found no change in normalized clustering coefficient and a decrease in normalized characteristic path length. This study also found a negative correlation of the functional connectivity between the right inferior frontal gyrus (opercular) and the left inferior frontal gyrus (triangular) with the epilepsy duration, suggesting a relationship between the decreased connectivity and the functional impairment associated with epilepsy duration.

Vlooswijk et al. [94] used fMRI with a word generation paradigm and an intelligence test to investigate brain network properties and their association with intellectual decline in patients with frontal and temporal lobe epilepsy. They observed lower values of the normalized cluster coefficient and global and local efficiency in patients when compared with healthy controls. The authors showed a diffuse disruption of small-worldness in the patient group. They observed that topology changes in patients with epilepsy were associated with a decline in intellectual abilities, although it is difficult to distinguish the effect of antiepileptic drugs from epilepsy itself.

Zhang et al. [95] and Onias et al. [96] used DTI and resting state fMRI to study whole-brain networks of healthy subjects vs. patients with generalized tonic–clonic seizures (GTCS). A less-optimized network organization in the second group was observed during interictal activity, represented by a decrease in "small-worldness" due to a decrease in the normalized clustering coefficient, with no change in normalized characteristic path length. GTCS patients showed increased total connection strength in functional connectivity. They also found a decrease in functional nodal topological properties in the DMN. The amygdala showed an increase in functional nodal degree, efficiency and centrality. These changes were positively correlated with the duration of epilepsy.

In Askren et al. [97], the authors used fMRI to examine pre-treatment predictors of post-treatment fatigue and cognitive dysfunction in women treated with adjuvant chemotherapy for breast cancer. Patients treated with or without chemotherapy and healthy controls were scanned coincident with pre- and 1-month post-chemotherapy during a verbal working memory task (VWMT) and assessed for fatigue, worry, and cognitive dysfunction. The chemotherapy group reported greater pre-treatment fatigue than controls and showed compromised neural response, characterized by higher spatial variance in executive network activity, than the non-chemotherapy group. Pre-treatment neural inefficiency (indexed by high spatial variance) in the executive network, which supports attention and working memory, was a better predictor of post-treatment cognitive and fatigue complaints than exposure to chemotherapy, indicating that this executive network compromise could be a pre-treatment neuromarker for patients most likely to benefit from early intervention for fatigue and cognitive dysfunction. Along these lines, executive network inefficiency and neurocognitive performance deficits pre-adjuvant treatment were predictors of cognitive dysfunction 1-year post-baseline, particularly in chemotherapy-treated patients [98].

In Wixted and Mickes [99], Rugg and Vilberg [100] a retrieval cue (such as a recognition memory test item) can elicit two qualitatively distinct kinds of mnemonic information: a multi-dimensional recollection signal that provides information about qualitative aspects of a prior event, including its context, and a scalar familiarity signal that can support simple judgments of prior occurrence [100]. The medialeff temporal lobe MTL (hippocampus and surrounding perirhinal, entorhinal and parahippocampal cortices) supports episodic memory and perirhinal cortex plays an important role in familiarity-driven recognition. Hippocampal regions manifest recollection-related enhancement. Recent studies have identified a network of cortical regions (each interconnected with the MTL) that are consistently engaged during successful episodic retrieval, and that constitute a content-independent network that acts in concert with cortical regions representing the contents of retrieval to support consciously accessible representations of prior experiences. The retrosplenial cortex and the mPFC (which connections with the hippocampus, parahippocampal and retrosplenial cortices) may play a role in the processing of contextual information [100]. Recollection-sensitive fMRI effects have consistently been identified in the hippocampus, parahippocampal, retrosplenial/posterior cingulate and lateral parietal cortices, and mPFC.

Qiu et al. [101] studied diabetes mellitus (DM) related changes in the topology in functional brain networks by using fluoro-D-glucose positron emission tomography (FDG-PET) data to construct functional brain networks of DM patients and sex- and age-matched normal controls, followed by a graph theoretical analysis. Both DM patients and control had a small-world topology. The DM group was found to have significantly lower small-world index, lower normalized clustering coefficients and higher normalized characteristic path length. Moreover, for diabetic patients, the nodal centrality was significantly reduced in the right rectus, the right cuneus, the left middle occipital gyrus, and the left postcentral gyrus, and it was significantly increased in the orbitofrontal region of the left middle frontal gyrus, the left olfactory region, and the right paracentral lobule. The diabetic brain was associated with disrupted topological organization in the functional PET network.

Park et al. [102] identified functional brain networks associated with obesity and examined how the networks were related to eating behaviors. They used fMRI scans from people of healthy weight and non-healthy weight, ICA, and a variety of centrality measures to associate connectivity with group-wise differences and correlate connectivity with scores on a three-factor eating questionnaire (TFEQ) describing restraint, disinhibition, and hunger eating behaviors. Frontoparietal and cerebellum networks showed group-wise differences between groups. Frontoparietal network showed a high correlation with TFEQ disinhibition scores. Both frontoparietal and cerebellum networks showed a high correlation with body mass index (BMI) scores. Brain networks with significant group-wise differences between showed a high correlation with eating behavior scores.

Laird et al. [103] Generated network images and associated metadata that may aid the study neuroimaging ontology development relevant to human cognition and behavior and the interpretation of the functional significance of future resting fMRI.

Tewarie et al. [104] showed that the minimum spanning tree (MST) characteristics were equally sensitive to alterations in network topology as the conventional graph theoretical measures. The MST parameters were very strongly related to changes in the characteristic path length when the network changed from a regular to a random configuration and to the degree of scale-free networks that were rewired to random networks. Moreover, MST avoids certain methodological biases and can describe network topology as well as conventional graph measures.

De Vico Fallani et al. [105] studied human social interactions by introducing the concept of hyper-brain network, a connectivity pattern representing at once the information flow among the cortical regions of a single brain as well as the relations among the areas of two distinct brains. Graph analysis of hyper-brain networks constructed from the EEG scanning of 26 couples of individuals playing the Iterated Prisoner's Dilemma reveals the possibility to predict non-cooperative interactions during the decision-making phase. The hyper-brain networks of two-defector couples have significantly less inter-brain links and overall higher modularity—i.e., the tendency to form two separate subgraphs—than couples playing cooperative or tit-for-tat strategies. The decision to defect can be "read" in advance by evaluating the changes of connectivity pattern in the hyper-brain network.

Lv et al. [106] found small-world properties of brain functional networks altered when human sleeps without EEG synchronization. Park et al. [107] discussed that heart failure association with resting-state spontaneous brain dysfunction between multiple sites, and the possibility that autonomic, cognitive, and affective deficits may stem from the altered functional connectivity, and used graph metrics (degree, betweenness, efficiency, clustering coefficient) to study the lateralization of changes to the right hemisphere.

Wu et al. [108] used resting fMRI to show positive correlations between IQ and the regional nodal properties in brain regions related to the attention system and negative correlations were found in various brain regions involved in the default mode, emotion, and language systems. Li et al. [22] used DTI and graph theory to show that higher intelligence scores are associated with a shorter characteristic path length and a higher global efficiency of the networks, indicating a structural aspect of information transfer in the brain. van den Heuvel et al. [25] showed strong negative association between the normalized characteristic path length of the resting fMRI brain network and IQ. They associated human intellectual performance to the efficiency of functional information integration between multiple brain regions especially in frontal and parietal regions. Langer et al. [109] used clustering coefficient and path length to show that the functional network is strongly related to intelligence. Higher intelligence resembles a small-world network and the parietal cortex may be a main hub as indicated by increased degree centrality that is associated with higher intelligence. Crossley et al. [110] used a meta-analysis of the literature that used fMRI or PET to measure task-related activation (1985–2010). They used similarity (Jaccard index) and weighted graphs to show that occipital, central, and default-mode modules were co-activated by specific cognitive domains (perception, action, and emotion). Rich club of hub nodes were located in parietal and prefrontal cortex and often connected over long distances, which were coactivated by a diverse range of experimental tasks. In a further detailed study, they concluded that the community structure of human brain networks is relevant to cognitive function. Deactivations may play a role in the integration between modules, periphery and central rich club properties according to cognitive demand. In general, Stam [111] says that brain network organization enables optimal cognitive function at a low wiring cost which can be used as cognitive background to consider what happens in cognitive and neurological diseases.

Doucet et al. [112] used neuropsychological testing pre- and post-surgery in verbal and nonverbal episodic memory, language, working memory, and attention domains to patients who underwent anterior temporal lobectomy and resting fMRI connectivity graph-theory measures such as local efficiency, distance and centrality to provide evidence that pre-surgery resting fMRI measures may help determine neurocognitive outcomes following ATL.

Pandit et al. [113] used resting-state fMRI on traumatic brain injury patients and graph analysis (average path length, reduced network efficiency) to show that a network critical to cognitive function (including PCC) shows a shift away from small-world characteristics and that cognitive function become less small-world after TBI, perhaps due to diffuse white matter damage.

Finally, Pessoa [114], based on the existence of hubs that regulate the flow and integration of information between regions, proposes that cognition and emotion are effectively integrated in the brain, since cognitive and emotional contributions to executive control cannot be separated (based on a control circuit contains traditional control areas (anterior cingulate lateral prefrontal cortices), affect (amygdala) and motivation (nucleus accumbens) and the ventral tegmental area.

4 Human Connectome Project

The human connectome is “a comprehensive structural description of the network of elements and connections forming the human brain” [115–117]. Connectomics is the study of the connectome and it refers to mapping of the brain’s functional architecture and the annotating it with the cognitive or behavioral functions they subtend [118]. The Human Connectome Project is sponsored by the National Institutes of Health, and its focus is to build a network map of the human brain in healthy adults that will assist the study of how functional brain states emerge from their underlying structural substrate, and how brain function is affected if this structural substrate is disrupted [116, 117]. The goal is that connectomic maps at different scales will be joined together into a single hierarchical map from neurons (microscale) to cortical areas (macroscale). Given the methodological uncertainties and the large differences in the connectomes of different subjects, any unified map will likely rely on probabilistic representations of connectivity data [116, 117]. A macroscale connectome (millimeter resolution) requires accurate anatomical or functional parcellation of the brain into functionally distinct brain regions with distinct architectonics, connectivity and function [119] that can be compared to disease states. In an effort to generate group level brain parcellations and associated network matrices and combine these with other data modalities and high field strengths, HCP has focused on the optimization of data acquisition methods and protocols and the development of robust data preprocessing pipelines [14]. However, some of the important issues will need further investigation [14] are: (1) Residual effects of head motion can create complex temporal patterns that the denoising processes (ICA-FIX) reduce but do not eliminate. "Motion scrubbing" identifies and excise time points that are irreversibly damaged by head motion [120]. (2) Global signal regression (regressing the mean time series over all brain voxels out of every individual voxel/grayordinate time series): it helps remove artefacts shared across all voxels and may improve the specificity of cortical-subcortical functional connectivity [121] but it negatively biases all computed correlations. The mean grey time course MGT may be an alternative. (3) The number of parcels: HCP uses ICA myelin maps, task fMRI constraints, and T1-w image intensity and cortical folding to achieve multimodal intersubject alignment for comparisons of subject specific parcellated connectome. An optimal core parcellation that will survive the noise of averaging (present in small parcels) is still to be found. (4) Optimal interpretability: Correlations are the least meaningful and biophysical models. Direct vs. indirect connections, dominant direction of information flow, temporal nonstationariries, etc. need to be analyzed.

The human connectome project uses data from more than 1000 subjects, drawn from families with twins and non-twin siblings, with 1 h of 3T rfMRI data for each subject (3T is considered to be the field strength most suitable for high quality data acquisition from a large cohort of subjects −200 subjects will also be scanned at higher strength), with an isometric spatial resolution of 2 mm and a temporal resolution of 0.7 s based on recent developments in multiband accelerated echo-planar imaging [14]. Each subject's 15 min rfMRI is spatially and temporally preprocessed, resulting in two versions of the preprocessed time series data (volumetric and grayordinates (surface vertices plus subcortical and cerebellar grey matter voxels)). Grayordinate version is more compact and should provide better alignment across subjects than the volumetric version. The dense connectome of average correlations could be estimated. This can be fed into a parcellation (through ICA or clustering algorithms), and from this, via the parcels' associated mean time series estimated from the group time series data) one can estimate the "parcellated connectome". HCP also uses accelerating fMRI acquisitions, which are highly advantageous due to: denser temporal sampling of physiological confounds, importance of temporal degree of freedom for many analysis techniques (such as high dimensional ICA or the use of partial correlations in network modeling), the value of temporal characteristics of resting state fluctuations [14]. However, high accelerations are limited in SNR by noise introduced by the-un aliasing of simultaneously acquired slices ("g factor noise") and incomplete slice separation ("L factor slice leakage") [14, 122]. Moreover, higher spatial resolution is feasible at higher strength fields such as 7T because of increases in image SNR [123] and in BOLD based susceptibility contrast [124]. At higher fields the microvascular BOLD signal becomes more significant (see Smith et al. [14] for more references). T2* blurring can become an increasing concern at higher field strength.

With respect to behavior, the objectives of the HCP are [125]: (1) to use behavioral measures that covary with brain connectivity and function, (2) to use standardized behavioral tests that will aid future projects to compare brain and behavior. The HCP uses various physical and mental state tests during each participant's visit in addition to a series of NIH Toolbox behavioral measures of cognition, emotion, language processing, motor and sensory function and other Non-Toolbox behavioral and individual difference measures of visual processing, personality, cognition, emotion, Psychiatric state, substance abuse and life function, physical function, sleep, etc. [126]. Behavioral studies will expand on the HCP. For example, Jia et al. [127] used a 3-level clustering approach (across space, time and subjects) has been used to allow for group inferences in dynamic functional connectivity and to explain behaviors (alertness, cognition, emotion and personality) in resting fMRI behavioral dataset from the HCP. Greater dynamics was associated with higher performance in behavioral tasks.

5 Discussion

Autonomic, cognitive, and affective deficits may stem from the altered functional connectivity that is directed from a need for functional non-lesion nodes that are more influential than in healthy subjects, as this is evident from altered (usually increased) connectivity degree in blindness [23], task performance [78], generalized tonic–clonic seizures [95, 96], and other studies. Also, variations in integration (measured by metrics such as efficiency) seem to be present in disease such as multiple sclerosis [19], in higher intelligence [22], aging [29] and sex-related differences in cognition and behavior [23], in schizophrenia [31], in ADHD (increase in local) [32], the relation between active learning and neural plasticity by analyzing Action Video Game (AVG) [60], in working memory performance (decrease in local, increase in global) [32], in hyperthermia (decrease local) [72], in silent lacunar infarcts and associated cognitive impairments (decrease local and global), [77], in weight loss [80], in training progress (first decrease then increase for global) [83], in focal hand dystonia (FHD) [89], intellectual abilities in in patients with frontal and temporal lobe epilepsy (decrease in local and global) [94], and in intelligence [22].

Finally variations in segregation through clustering which may indicate disruption of structural integrity of large scale networks or local neighborhood connectivity, are present in Alzheimer's disease (increased and decreased) [17, 30, 36, 38] in schizophrenia (decrease) [31], in mesial temporal lobe epilepsy (reduced) [33], in active learning and neural plasticity by analyzing Action Video Game (AVG) experience [60], in hyperthermia [72], mild cognitive impairment and aging [68], in ADHD (higher) [85], in the lateralization of changes in heart failure [107], in diabetes mellitus (lower) [101]. In generalized tonic–clonic seizures (decrease) [95, 96], in loss of consciousness (increased) [91], no change in mesial temporal lobe epilepsy [33]. It is possible that behavior and cognition occurs as a fusion emergence which can be thought as an ontological emergence which shows that generative atomism cannot be a universal method for representing behavior, cognition and disease or disorder (for a discussion on generative atomism see Santos [128]). It is possible that there exists a link between behavior and network metrics that is channeled through the intermediate level between ground anatomical/functional networks and the final emerging outcome at the higher level. It is possible that not every behavioral pattern corresponds to a unique microstructural level in a causal or statistical way and brain network dynamics may be representative of more information than behavioral and cognitive patterns. Higher level property or behavior may be built indeed from physical components, but how it is built determines if it is emergent or not and may display chaotic dynamics.

In an effort to relate graph theoretical measures to the cognitive and behavioral neurobiology in human subjects, an examination of neuroplasticity (adaptation of the brain synaptic and non-synaptic structures to new circumstances that arise out of behavioral and environmental changes), the dopaminergic reward system (mesolimbic pathway, including the ventral tegmental area, the nucleus accumbens, the cingulate cortex, and other structures), the orbitofrontal cortex (as the decision making center) and ways of bypassing it, is necessary. Sustained behavioral activities influence neuroanatomical structures and an effort to either identify the features of a behavioral/cognitive pattern or to propose a method to overcome health situations, such as addictions, will most likely require the elucidation of rewiring processes in relations to structural and functional networks. Luis Joshua Salés[1] has indicated that a deliberative will of the human subject contains implicitly in its definition the possibility of making poor choices (either by mistake or by habitual disposition) and it attaches certain negative valuations to state of affairs, not seeing things for their non-axiological features (non-valued characteristics), but through certain feeling and emotion that attaches to them according to the person's desire. Such valuations augment the pathological state of deliberately weighing different options before acting, by distorting the range of possibilities of action through coloring the perception of reality. This may lead to the development of problematic behavioral habits.

[1]Luis Joshua Salés, "Maximos and Neurobiology: A Neurotheological Investigation of Asceticism as Erosion of the Passions and the Gnomic Will,"Holodny Prize Recipient for most outstanding pre-professorial presentation, Sophia Institute Conference: New York, NY (December, 2013).

6 Conclusion

Network science offers a theoretical framework for the study of the nervous system. The connectome is a comprehensive description of the brain connectivity that can inform the design of global computational models of the human brain. Clinical applications of connectomics are expected to relate variations in connectivity to functional disturbance. Brain networks can be constructed in two ways: one to denote structural connectivity (pattern of structural connections between neurons, neuronal populations or brain regions), and one to denote functional connectivity (pattern of statistical dependencies, e.g. temporal correlations, between distinct neuronal elements). Regional parcellation schemes and/or voxel based parcellation schemes can be used and projects such as the Human Connectome Project use combinations of different imaging modalities and more than one parcellation. Several network metrics exist that can reveal brain network patterns intrinsically represented in the brain, and identify neural substrate relationships between intranetwork and internetwork activity during rest or/and task. Graph theoretical analysis of human brain networks requires the use of imaging modalities, such as structural MRI, diffusion MRI, functional MRI, and EEG/MEG. Complex brain networks have been shown to be sensitive to behavioral variability, cognitive ability, genetic information, experimental task, age, gender, drugs, blindness and diseases such as Alzheimer's disease, ADHD, etc. Quantitative analysis of patterns through the use of graph theoretical measures, such as measures of local connectivity (for processing), measures of global connectivity (global processing), measures of centrality (importance to processing), measures of community structure (for functional modules), to mention a few. Graph structures are extracted either by comparison to random networks, or by comparison among other real brain networks (e.g., other groups of subjects, experimental conditions, etc.). Segregation, integration and influence are first fundamental aspects of local and global connectivity that may model how connectivity goes beyond information channeling, to generate complex system-wide dynamics that enable local regions to participate to cognitive and behavioral tasks. Graph metrics are expected to vary across subjects and are likely to reflect behavioral and cognitive performances. However, graph analysis in neuroscience faces many methodological issues that remaining unresolved, limiting its interpretation and applicability [129]. Some examples are brain graph filtering, statistical variability of brain graphs due to noise, spatio-temporality of brain graphs (in long term scales related to plasticity and in short-time scales related cognitive/motor learning [130]). However, network modelling and graph theoretical methods can aid the detection of behavioral patterns that relate to brain function in health and disease and may help answering the question if behavior and cognition processing is inherited from the brain network substrate or instead arise from independent converging processes [131].

References

1. Stanley, M.L., M.N. Moussa, B.M. Paolini, R.G. Lyday, J.H. Burdette, and P.J. Laurienti. 2013. Defining Nodes in Complex Brain Networks. *Frontiers in Computational Neuroscience* 7: 169.
2. Boccaletti, S., V. Latora, Y. Moreno, M. Chavez, and D.U. Hwang. 2006. Complex Networks: Structure and Dynamics. *Physics Reports Review Section of Physics Letters* 424 (4–5): 175–308.
3. He, Y., and A. Evans. 2010. Graph Theoretical Modeling of Brain Connectivity. *Current Opinion in Neurology* 23 (4): 341–350.
4. Kelly, A.M., L.Q. Uddin, B.B. Biswal, F.X. Castellanos, and M.P. Milham. 2008. Competition Between Functional Brain Networks Mediates Behavioral Variability. *NeuroImage* 39 (1): 527–537.
5. Castellanos, F.X., and R. Tannock. 2002. Neuroscience of Attention-Deficit/Hyperactivity Disorder: The Search for Endophenotypes. *Nature Reviews Neuroscience* 3 (8): 617–628.
6. Stuss, D.T., K.J. Murphy, M.A. Binns, and M.P. Alexander. 2003. Staying on the Job: The Frontal Lobes Control Individual Performance Variability. *Brain* 126 (Pt 11): 2363–2380.
7. Girvan, M., and M.E. Newman. 2002. Community Structure in Social and Biological Networks. *Proceedings of the National Academy of Sciences of the United States of America* 99 (12): 7821–7826.
8. Clauset, A., M.E. Newman, and C. Moore. 2004. Finding Community Structure in Very Large Networks. *Physical Review. E, Statistical, Nonlinear, and Soft Matter Physics* 70 (6 Pt 2): 066111.
9. Guimera, R., M. Sales-Pardo, and L.A. Amaral. 2004. Modularity from Fluctuations in Random Graphs and Complex Networks. *Physical Review. E, Statistical, Nonlinear, and Soft Matter Physics* 70 (2 Pt 2): 025101.
10. Blondel, V.D., J.L. Guillaume, R. Lambiotte, and E. Lefebvre. 2008. Fast Unfolding of Communities in Large Networks. *Journal of Statistical Mechanics Theory and Experiment* 2008: P10008.
11. Fortunato, S., and M. Barthelemy. 2007. Resolution Limit in Community Detection. *Proceedings of the National Academy of Sciences of the United States of America* 104 (1): 36–41.
12. Shen, X., F. Tokoglu, X. Papademetris, and R.T. Constable. 2013. Groupwise Whole-Brain Parcellation from Resting-State fMRI Data for Network Node Identification. *NeuroImage* 82: 403–415.
13. Scheinost, D., J. Benjamin, C.M. Lacadie, B. Vohr, K.C. Schneider, L.R. Ment, X. Papademetris, and R.T. Constable. 2012. The Intrinsic Connectivity Distribution: A Novel Contrast Measure Reflecting Voxel Level Functional Connectivity. *NeuroImage* 62 (3): 1510–1519.
14. Smith, S.M., C.F. Beckmann, J. Andersson, E.J. Auerbach, J. Bijsterbosch, G. Douaud, E. Duff, D.A. Feinberg, L. Griffanti, M.P. Harms, M. Kelly, T. Laumann, K.L. Miller, S. Moeller, S. Petersen, J. Power, G. Salimi-Khorshidi, A.Z. Snyder, A.T. Vu, M.W. Woolrich, J. Xu, E. Yacoub, K. Ugurbil, D.C. Van Essen, M.F. Glasser, and W.U.-M.H. Consortium. 2013. Resting-State fMRI in the Human Connectome Project. *NeuroImage* 80: 144–168.
15. Lerch, J.P., K. Worsley, W.P. Shaw, D.K. Greenstein, R.K. Lenroot, J. Giedd, and A.C. Evans. 2006. Mapping Anatomical Correlations Across Cerebral Cortex (MACACC) Using Cortical Thickness from MRI. *NeuroImage* 31 (3): 993–1003.
16. Mechelli, A., K.J. Friston, R.S. Frackowiak, and C.J. Price. 2005. Structural Covariance in the Human Cortex. *Journal of Neuroscience* 25 (36): 8303–8310.
17. He, Y., Z. Chen, and A. Evans. 2008. Structural Insights into Aberrant Topological Patterns of Large-Scale Cortical Networks in Alzheimer's Disease. *Journal of Neuroscience* 28 (18): 4756–4766.

18. Bassett, D.S., E.T. Bullmore, B.A. Verchinski, V.S. Mattay, D.R. Weinberger, and A. Meyer-Lindenberg. 2008. Hierarchical Organization of Human Cortical Networks in Health and Schizophrenia. *Journal of Neuroscience* 28 (37): 9239–9248.
19. He, Y., A. Dagher, Z. Chen, A. Charil, A. Zijdenbos, K. Worsley, and A. Evans. 2009. Impaired Small-World Efficiency in Structural Cortical Networks in Multiple Sclerosis Associated with White Matter Lesion Load. *Brain* 132: 3366–3379.
20. Mori, S., and P.C.M. van Zijl. 2002. Fiber Tracking: Principles and Strategies – A Technical Review. *NMR in Biomedicine* 15 (7–8): 468–480.
21. Behrens, T.E.J., M.W. Woolrich, M. Jenkinson, H. Johansen-Berg, R.G. Nunes, S. Clare, P.M. Matthews, J.M. Brady, and S.M. Smith. 2003. Characterization and Propagation of Uncertainty in Diffusion-Weighted MR Imaging. *Magnetic Resonance in Medicine* 50 (5): 1077–1088.
22. Li, Y.H., Y. Liu, J. Li, W. Qin, K.C. Li, C.S. Yu, and T.Z. Jiang. 2009. Brain Anatomical Network and Intelligence. *PLoS Computational Biology* 5 (5): e1000395.
23. Shu, N., Y. Liu, J. Li, Y.H. Li, C.S. Yu, and T.Z. Jiang. 2009. Altered Anatomical Network in Early Blindness Revealed by Diffusion Tensor Tractography. *PLoS One* 4 (9): e1000395.
24. Gong, G.L., P. Rosa, F. Carbonell, Z.J. Chen, Y. He, and A.C. Evans. 2009. Age- and Gender-Related Differences in the Cortical Anatomical Network. *Journal of Neuroscience* 29 (50): 15684–15693.
25. van den Heuvel, M.P., C.J. Stam, R.S. Kahn, and H.E. Hulshoff Pol. 2009. Efficiency of Functional Brain Networks and Intellectual Performance. *The Journal of Neuroscience* 29 (23): 7619–7624.
26. Wang, L., Y.F. Li, P. Metzak, Y. He, and T.S. Woodward. 2010. Age-Related Changes in Topological Patterns of Large-Scale Brain Functional Networks During Memory Encoding and Recognition. *NeuroImage* 50 (3): 862–872.
27. Fair, D.A., A.L. Cohen, J.D. Power, N.U.F. Dosenbach, J.A. Church, F.M. Miezin, B.L. Schlaggar, and S.E. Petersen. 2009. Functional Brain Networks Develop from a "Local to Distributed" Organization. *PLoS Computational Biology* 5 (5): e1000381.
28. Supekar, K., M. Musen, and V. Menon. 2009. Development of Large-Scale Functional Brain Networks in Children. *PLoS Biology* 7 (7): e1000157.
29. Achard, S., and E.T. Bullmore. 2007. Efficiency and Cost of Economical Brain Functional Networks. *PLoS Computational Biology* 3 (2): 174–183.
30. Supekar, K., V. Menon, D. Rubin, M. Musen, and M.D. Greicius. 2008. Network Analysis of Intrinsic Functional Brain Connectivity in Alzheimer's Disease. *PLoS Computational Biology* 4 (6): e1000100.
31. Liu, Y., M. Liang, Y. Zhou, Y. He, Y.H. Hao, M. Song, C.S. Yu, H.H. Liu, Z.N. Liu, and T.Z. Jiang. 2008. Disrupted Small-World Networks in Schizophrenia. *Brain* 131: 945–961.
32. Wang, L., C.Z. Zhu, Y. He, Y.F. Zang, Q.J. Cao, H. Zhang, Q.H. Zhong, and Y.F. Wang. 2009. Altered Small-World Brain Functional Networks in Children With Attention-Deficit/Hyperactivity Disorder. *Human Brain Mapping* 30 (2): 638–649.
33. Liao, W., Z. Zhang, Z. Pan, D. Mantini, J. Ding, X. Duan, C. Luo, G. Lu, and H. Chen. 2010. Altered Functional Connectivity and Small-World in Mesial Temporal Lobe Epilepsy. *PLoS One* 5 (1): e8525.
34. Nakamura, T., F.G. Hillary, and B.B. Biswal. 2009. Resting Network Plasticity Following Brain Injury. *PLoS One* 4 (12): e8220.
35. Bassett, D.S., E.T. Bullmore, A. Meyer-Lindenberg, J.A. Apud, D.R. Weinberger, and R. Coppola. 2009. Cognitive Fitness of Cost-Efficient Brain Functional Networks. *Proceedings of the National Academy of Sciences of the United States of America* 106 (28): 11747–11752.
36. Micheloyannis, S., M. Vourkas, V. Tsirka, E. Karakonstantaki, K. Kanatsouli, and C.J. Stam. 2009. The Influence of Ageing on Complex Brain Networks: A Graph Theoretical Analysis. *Human Brain Mapping* 30 (1): 200–208.

37. Stam, C.J., W. de Haan, A. Daffertshofer, B.F. Jones, I. Manshanden, A.M.V. van Walsum, T. Montez, J.P.A. Verbunt, J.C. de Munck, B.W. van Dijk, H.W. Berendse, and P. Scheltens. 2009. Graph Theoretical Analysis of Magnetoencephalographic Functional Connectivity in Alzheimers Disease. *Brain* 132: 213–224.
38. de Haan, W., Y.A.L. Pijnenburg, R.L.M. Strijers, Y. van der Made, W.M. van der Flier, P. Scheltens, and C.J. Stam. 2009. Functional Neural Network Analysis in Frontotemporal Dementia and Alzheimer's Disease Using EEG and Graph Theory. *BMC Neuroscience* 10: 101.
39. Rubinov, M., S.A. Knock, C.J. Stam, S. Micheloyannis, A.W.F. Harris, L.M. Williams, and M. Breakspear. 2009. Small-World Properties of Nonlinear Brain Activity in Schizophrenia. *Human Brain Mapping* 30 (2): 403–416.
40. van Dellen, E., L. Douw, J.C. Baayen, J.J. Heimans, S.C. Ponten, W.P. Vandertop, D.N. Velis, C.J. Stam, and J.C. Reijneveld. 2009. Long-Term Effects of Temporal Lobe Epilepsy on Local Neural Networks: A Graph Theoretical Analysis of Corticography Recordings. *PLoS One* 4 (11): e8081.
41. Leistedt, S.J.J., N. Coumans, M. Dumont, J.P. Lanquart, C.J. Stam, and P. Linkowski. 2009. Altered Sleep Brain Functional Connectivity in Acutely Depressed Patients. *Human Brain Mapping* 30 (7): 2207–2219.
42. Damoiseaux, J.S., and M.D. Greicius. 2009. Greater Than the Sum of Its Parts: A Review of Studies Combining Structural Connectivity and Resting-State Functional Connectivity. *Brain Structure & Function* 213 (6): 525–533.
43. Hayasaka, S., and P.J. Laurienti. 2010. Comparison of characteristics between region-and voxel-based network analyses in resting-state fMRI data. *NeuroImage* 50 (2): 499–508.
44. Sanabria-Diaz, G., L. Melie-Garcia, Y. Iturria-Medina, Y. Aleman-Gomez, G. Hernandez-Gonzalez, L. Valdes-Urrutia, L. Galan, and P. Valdes-Sosa. 2010. Surface Area and Cortical Thickness Descriptors Reveal Different Attributes of the Structural Human Brain Networks. *NeuroImage* 50 (4): 1497–1510.
45. Wang, J.H., L. Wang, Y.F. Zang, H. Yang, H.H. Tang, Q.Y. Gong, Z. Chen, C.Z. Zhu, and Y. He. 2009. Parcellation-Dependent Small-World Brain Functional Networks: A Resting-State fMRI Study. *Human Brain Mapping* 30 (5): 1511–1523.
46. Zalesky, A., A. Fornito, I.H. Harding, L. Cocchi, M. Yucel, C. Pantelis, and E.T. Bullmore. 2010. Whole-Brain Anatomical Networks: Does the Choice of Nodes Matter? *NeuroImage* 50 (3): 970–983.
47. Vaessen, M.J., P.A.M. Hofman, H.N. Tijssen, A.P. Aldenkamp, J.F.A. Jansen, and W.H. Backes. 2010. The Effect and Reproducibility of Different Clinical DTI Gradient Sets on Small World Brain Connectivity Measures. *NeuroImage* 51 (3): 1106–1116.
48. Buckner, R.L., J. Sepulcre, T. Talukdar, F.M. Krienen, H.S. Liu, T. Hedden, J.R. Andrews-Hanna, R.A. Sperling, and K.A. Johnson. 2009. Cortical Hubs Revealed by Intrinsic Functional Connectivity: Mapping, Assessment of Stability, and Relation to Alzheimer's Disease. *Journal of Neuroscience* 29 (6): 1860–1873.
49. Deuker, L., E.T. Bullmore, M. Smith, S. Christensen, P.J. Nathan, B. Rockstroh, and D.S. Bassett. 2009. Reproducibility of Graph Metrics of Human Brain Functional Networks. *NeuroImage* 47 (4): 1460–1468.
50. He, Y., J.H. Wang, L. Wang, Z.J. Chen, C.G. Yan, H. Yang, H.H. Tang, C.Z. Zhu, Q.Y. Gong, Y.F. Zang, and A.C. Evans. 2009. Uncovering Intrinsic Modular Organization of Spontaneous Brain Activity in Humans. *PLoS One* 4 (4): e5226.
51. Honey, C.J., O. Sporns, L. Cammoun, X. Gigandet, J.P. Thiran, R. Meuli, and P. Hagmann. 2009. Predicting Human Resting-State Functional Connectivity from Structural Connectivity. *Proceedings of the National Academy of Sciences of the United States of America* 106 (6): 2035–2040.
52. Buckner, R.L., A.Z. Snyder, B.J. Shannon, G. LaRossa, R. Sachs, A.F. Fotenos, Y.I. Sheline, W.E. Klunk, C.A. Mathis, J.C. Morris, and M.A. Mintun. 2005. Molecular, Structural, and Functional Characterization of Alzheimer's Disease: Evidence for a Relationship Between Default Activity, Amyloid, and Memory. *Journal of Neuroscience* 25 (34): 7709–7717.

53. Filippini, N., B.J. MacIntosh, M.G. Hough, G.M. Goodwin, G.B. Frisoni, S.M. Smith, P.M. Matthews, C.F. Beckmann, and C.E. Mackay. 2009. Distinct Patterns of Brain Activity in Young Carriers of the APOE-Epsilon 4 Allele. *Proceedings of the National Academy of Sciences of the United States of America* 106 (17): 7209–7214.
54. Greicius, M.D., and V. Menon. 2004. Default-Mode Activity During a Passive Sensory Task: Uncoupled from Deactivation but Impacting Activation. *Journal of Cognitive Neuroscience* 16 (9): 1484–1492.
55. Zhang, Y., N. Schuff, G.H. Jahng, W. Bayne, S. Mori, L. Schad, S. Mueller, A.T. Du, J.H. Kramer, K. Yaffe, H. Chui, W.J. Jagust, B.L. Miller, and M.W. Weiner. 2007. Diffusion Tensor Imaging of Cingulum Fibers in Mild Cognitive Impairment and Alzheimer Disease. *Neurology* 68 (1): 13–19.
56. Byrge, L., O. Sporns, and L.B. Smith. 2014. Developmental Process Emerges from Extended Brain-Body-Behavior Networks. *Trends in Cognitive Sciences* 18 (8): 395–403.
57. Sugita, Y., and J. Tani. 2005. Learning Semantic Combinatoriality from the Interaction Between Linguistic and Behavioral Processes. *Adaptive Behavior* 13 (1): 33–52.
58. Almassy, N., G.M. Edelman, and O. Sporns. 1998. Behavioral Constraints in the Development of Neuronal Properties: A Cortical Model Embedded in a Real-World Device. *Cerebral Cortex* 8 (4): 346–361.
59. Mesulam, M.M. 1998. From Sensation to Cognition. *Brain* 121: 1013–1052.
60. Gong, D., H. He, W. Ma, D. Liu, M. Huang, L. Dong, J. Gong, J. Li, C. Luo, and D. Yao. 2016. Functional Integration between Salience and Central Executive Networks: A Role for Action Video Game Experience. *Neural Plasticity* 2016: 9.
61. Stanley, M.L., S.L. Simpson, D. Dagenbach, R.G. Lyday, J.H. Burdette, and P.J. Laurienti. 2015. Changes in Brain Network Efficiency and Working Memory Performance in Aging. *PLoS One* 10 (4): e0123950.
62. Tang, C.Y., E.L. Eaves, J.C. Ng, D.M. Carpenter, X. Mai, D.H. Schroeder, C.A. Condon, R. Colom, and R.J. Haier. 2010. Brain Networks for Working Memory and Factors of Intelligence Assessed in Males and Females with fMRI and DTI. *Intelligence* 38 (3): 293–303.
63. Rosenberg, M.D., E.S. Finn, D. Scheinost, X. Papademetris, X. Shen, R.T. Constable, and M.M. Chun. 2016. A neuromarker of Sustained Attention from Whole-Brain Functional Connectivity. *Nature Neuroscience* 19 (1): 165–171.
64. Whitfield-Gabrieli, S., S.S. Ghosh, A. Nieto-Castanon, Z. Saygin, O. Doehrmann, X.J. Chai, G.O. Reynolds, S.G. Hofmann, M.H. Pollack, and J.D. Gabrieli. 2016. Brain Connectomics Predict Response to Treatment in Social Anxiety Disorder. *Molecular Psychiatry* 21 (5): 680–685.
65. Davis, F.C., A.R. Knodt, O. Sporns, B.B. Lahey, D.H. Zald, B.D. Brigidi, and A.R. Hariri. 2013. Impulsivity and the Modular Organization of Resting-State Neural Networks. *Cerebral Cortex* 23 (6): 1444–1452.
66. Cao, M., J.H. Wang, Z.J. Dai, X.Y. Cao, L.L. Jiang, F.M. Fan, X.W. Song, M.R. Xia, N. Shu, Q. Dong, M.P. Milham, F.X. Castellanos, X.N. Zuo, and Y. He. 2014. Topological Organization of the Human Brain Functional Connectome Across the Lifespan. *Developmental Cognitive Neuroscience* 7: 76–93.
67. Dosenbach, N.U., D.A. Fair, F.M. Miezin, A.L. Cohen, K.K. Wenger, R.A. Dosenbach, M.D. Fox, A.Z. Snyder, J.L. Vincent, M.E. Raichle, B.L. Schlaggar, and S.E. Petersen. 2007. Distinct Brain Networks for Adaptive and Stable Task Control in Humans. *Proceedings of the National Academy of Sciences of the United States of America* 104 (26): 11073–11078.
68. Miraglia, F., F. Vecchio, P. Bramanti, and P.M. Rossini. 2015. Small-Worldness Characteristics and Its Gender Relation in Specific Hemispheric Networks. *Neuroscience* 310: 1–11.
69. DeSalvo, M.N., L. Douw, S. Takaya, H. Liu, and S.M. Stufflebeam. 2014. Task-Dependent Reorganization of Functional Connectivity Networks During Visual Semantic Decision Making. *Brain and Behavior: A Cognitive Neuroscience Perspective* 4 (6): 877–885.
70. Muller, V., and U. Lindenberger. 2014. Hyper-Brain Networks Support Romantic Kissing in Humans. *PLoS One* 9 (11): e112080.

71. Liu, K., G. Sun, B. Li, Q. Jiang, X. Yang, M. Li, L. Li, S. Qian, L. Zhao, Z. Zhou, K.M. von Deneen, and Y. Liu. 2013. The Impact of Passive Hyperthermia on Human Attention Networks: An fMRI study. *Behavioural Brain Research* 243: 220–230.
72. Qian, S., G. Sun, Q. Jiang, K. Liu, B. Li, M. Li, X. Yang, Z. Yang, and L. Zhao. 2013. Altered Topological Patterns of Large-Scale Brain Functional Networks During Passive Hyperthermia. *Brain and Cognition* 83 (1): 121–131.
73. Manelis, A., J.R. Almeida, R. Stiffler, J.C. Lockovich, H.A. Aslam, and M.L. Phillips. 2016. Anticipation-Related Brain Connectivity in Bipolar and Unipolar Depression: A Graph Theory Approach. *Brain* 139 (Pt 9): 2554–2566.
74. Luft, C.D., E. Pereda, M.J. Banissy, and J. Bhattacharya. 2014. Best of Both Worlds: Promise of Combining Brain Stimulation and Brain Connectome. *Frontiers in Systems Neuroscience* 8: 132.
75. Ash, J.A., and P.R. Rapp. 2014. A Quantitative Neural Network Approach to Understanding Aging Phenotypes. *Ageing Research Reviews* 15: 44–50.
76. Sadaghiani, S., J.B. Poline, A. Kleinschmidt, and M. D'Esposito. 2015. Ongoing Dynamics in Large-Scale Functional Connectivity Predict Perception. *Proceedings of the National Academy of Sciences of the United States of America* 112 (27): 8463–8468.
77. Tang, J., S. Zhong, Y. Chen, K. Chen, J. Zhang, G. Gong, A.S. Fleisher, Y. He, and Z. Zhang. 2015. Aberrant White Matter Networks Mediate Cognitive Impairment in Patients with Silent Lacunar Infarcts in Basal Ganglia Territory. *Journal of Cerebral Blood Flow and Metabolism* 35 (9): 1426–1434.
78. Lin, P., Y. Yang, J. Jovicich, N. De Pisapia, X. Wang, C.S. Zuo, and J.J. Levitt. 2016. Static and Dynamic Posterior Cingulate Cortex Nodal Topology of Default Mode Network Predicts Attention Task Performance. *Brain Imaging and Behavior* 10 (1): 212–225.
79. Godwin, D., R.L. Barry, and R. Marois. 2015. Breakdown of the Brain's Functional Network Modularity with Awareness. *Proceedings of the National Academy of Sciences of the United States of America* 112 (12): 3799–3804.
80. Paolini, B.M., P.J. Laurienti, S.L. Simpson, J.H. Burdette, R.G. Lyday, and W.J. Rejeski. 2015. Global Integration of the Hot-State Brain Network of Appetite Predicts Short Term Weight Loss in Older Adult. *Frontiers in Aging Neuroscience* 7: 70.
81. Khazaee, A., A. Ebrahimzadeh, A. Babajani-Feremi, and I. Alzheimer's Disease Neuroimaging. 2016. Classification of Patients with MCI and AD from Healthy Controls Using Directed Graph Measures of Resting-State fMRI. *Behavioural Brain Research* 322 (Pt B): 339–350.
82. Khazaee, A., A. Ebrahimzadeh, and A. Babajani-Feremi. 2015. Identifying Patients with Alzheimer's Disease Using Resting-State fMRI and Graph Theory. *Clinical Neurophysiology* 126 (11): 2132–2141.
83. Taya, F., Y. Sun, F. Babiloni, N. Thakor, and A. Bezerianos. 2016. Topological Changes in the Brain Network Induced by the Training on a Piloting Task: An EEG-Based Functional Connectome Approach. *IEEE Transactions on Neural Systems and Rehabilitation Engineering*. doi:10.1109/TNSRE.2016.2581809.
84. Barttfeld, P., A. Petroni, S. Baez, H. Urquina, M. Sigman, M. Cetkovich, T. Torralva, F. Torrente, A. Lischinsky, X. Castellanos, F. Manes, and A. Ibanez. 2014. Functional Connectivity and Temporal Variability of Brain Connections in Adults with Attention Deficit/Hyperactivity Disorder and Bipolar Disorder. *Neuropsychobiology* 69 (2): 65–75.
85. Cao, M., N. Shu, Q. Cao, Y. Wang, and Y. He. 2014. Imaging Functional and Structural Brain Connectomics in Attention-Deficit/Hyperactivity Disorder. *Molecular Neurobiology* 50 (3): 1111–1123.
86. Cao, Q., N. Shu, L. An, P. Wang, L. Sun, M.R. Xia, J.H. Wang, G.L. Gong, Y.F. Zang, Y.F. Wang, and Y. He. 2013. Probabilistic Diffusion Tractography and Graph Theory Analysis Reveal Abnormal White Matter Structural Connectivity Networks in Drug-Naive Boys with Attention Deficit/Hyperactivity Disorder. *The Journal of Neuroscience* 33 (26): 10676–10687.

87. Elton, A., S.P. Tripathi, T. Mletzko, J. Young, J.M. Cisler, G.A. James, and C.D. Kilts. 2014. Childhood Maltreatment Is Associated with a Sex-Dependent Functional Reorganization of a Brain Inhibitory Control Network. *Human Brain Mapping* 35 (4): 1654–1667.
88. Brown, J.A., K.H. Terashima, A.C. Burggren, L.M. Ercoli, K.J. Miller, G.W. Small, and S.Y. Bookheimer. 2011. Brain Network Local Interconnectivity Loss in Aging APOE-4 Allele Carriers. *Proceedings of the National Academy of Sciences of the United States of America* 108 (51): 20760–20765.
89. Jin, S.H., P. Lin, and M. Hallett. 2011. Abnormal Reorganization of Functional Cortical Small-World Networks in Focal hand Dystonia. *PLoS One* 6 (12): e28682.
90. Smit, D.J., M. Boersma, C.E. van Beijsterveldt, D. Posthuma, D.I. Boomsma, C.J. Stam, and E.J. de Geus. 2010. Endophenotypes in a Dynamically Connected Brain. *Behavior Genetics* 40 (2): 167–177.
91. Schroter, M.S., V.I. Spoormaker, A. Schorer, A. Wohlschlager, M. Czisch, E.F. Kochs, C. Zimmer, B. Hemmer, G. Schneider, D. Jordan, and R. Ilg. 2012. Spatiotemporal Reconfiguration of Large-Scale Brain Functional Networks During Propofol-Induced Loss of Consciousness. *The Journal of Neuroscience* 32 (37): 12832–12840.
92. Boveroux, P., A. Vanhaudenhuyse, M.A. Bruno, Q. Noirhomme, S. Lauwick, A. Luxen, C. Degueldre, A. Plenevaux, C. Schnakers, C. Phillips, J.F. Brichant, V. Bonhomme, P. Maquet, M.D. Greicius, S. Laureys, and M. Boly. 2010. Breakdown of Within- and Between-Network Resting State Functional Magnetic Resonance Imaging Connectivity During Propofol-Induced Loss of Consciousness. *Anesthesiology* 113 (5): 1038–1053.
93. Stamatakis, E.A., R.M. Adapa, A.R. Absalom, and D.K. Menon. 2010. Changes in Resting Neural Connectivity During Propofol Sedation. *PLoS One* 5 (12): e14224.
94. Vlooswijk, M.C., M.J. Vaessen, J.F. Jansen, M.C. de Krom, H.J. Majoie, P.A. Hofman, A.P. Aldenkamp, and W.H. Backes. 2011. Loss of Network Efficiency Associated with Cognitive Decline in Chronic Epilepsy. *Neurology* 77 (10): 938–944.
95. Zhang, Z., W. Liao, H. Chen, D. Mantini, J.R. Ding, Q. Xu, Z. Wang, C. Yuan, G. Chen, Q. Jiao, and G. Lu. 2011. Altered Functional-Structural Coupling of Large-Scale Brain Networks in Idiopathic Generalized Epilepsy. *Brain* 134 (Pt 10): 2912–2928.
96. Onias, H., A. Viol, F. Palhano-Fontes, K.C. Andrade, M. Sturzbecher, G. Viswanathan, and D.B. de Araujo. 2014. Brain Complex Network Analysis by Means of Resting State fMRI and Graph Analysis: Will It Be Helpful in Clinical Epilepsy? *Epilepsy & Behavior* 38: 71–80.
97. Askren, M.K., M. Jung, M.G. Berman, M. Zhang, B. Therrien, S. Peltier, L. Ossher, D.F. Hayes, P.A. Reuter-Lorenz, and B. Cimprich. 2014. Neuromarkers of Fatigue and Cognitive Complaints Following Chemotherapy for Breast Cancer: A Prospective fMRI Investigation. *Breast Cancer Research and Treatment* 147 (2): 445–455.
98. Jung, M.S., M. Zhang, M.K. Askren, M.G. Berman, S. Peltier, D.F. Hayes, B. Therrien, P.A. Reuter-Lorenz, and B. Cimprich. 2016. Cognitive Dysfunction and Symptom Burden in Women Treated for Breast Cancer: A Prospective Behavioral and fMRI Analysis. *Brain Imaging and Behavior*. doi:10.1007/s11682-016-9507-8.
99. Wixted, J.T., and L. Mickes. 2010. A Continuous Dual-Process Model of Remember/Know Judgments. *Psychological Review* 117 (4): 1025–1054.
100. Rugg, M.D., and K.L. Vilberg. 2013. Brain Networks Underlying Episodic Memory Retrieval. *Current Opinion in Neurobiology* 23 (2): 255–260.
101. Qiu, X., Y. Zhang, H. Feng, and D. Jiang. 2016. Positron Emission Tomography Reveals Abnormal Topological Organization in Functional Brain Network in Diabetic Patients. *Frontiers in Neuroscience* 10: 235.
102. Park, B.Y., J. Seo, and H. Park. 2016. Functional Brain Networks Associated with Eating Behaviors in Obesity. *Scientific Reports* 6: 23891.
103. Laird, A.R., P.M. Fox, S.B. Eickhoff, J.A. Turner, K.L. Ray, D.R. McKay, D.C. Glahn, C.F. Beckmann, S.M. Smith, and P.T. Fox. 2011. Behavioral Interpretations of Intrinsic Connectivity Networks. *Journal of Cognitive Neuroscience* 23 (12): 4022–4037.
104. Tewarie, P., E. van Dellen, A. Hillebrand, and C.J. Stam. 2015. The Minimum Spanning Tree: An Unbiased Method for Brain Network Analysis. *NeuroImage* 104: 177–188.

105. De Vico Fallani, F., V. Nicosia, R. Sinatra, L. Astolfi, F. Cincotti, D. Mattia, C. Wilke, A. Doud, V. Latora, B. He, and F. Babiloni. 2010. Defecting or Not Defecting: How to "Read" Human Behavior During Cooperative Games by EEG Measurements. *PLoS One* 5 (12): e14187.
106. Lv, J., D. Liu, J. Ma, X. Wang, and J. Zhang. 2015. Graph Theoretical Analysis of BOLD Functional Connectivity During Human Sleep Without EEG Monitoring. *PLoS One* 10 (9): e0137297.
107. Park, B., B. Roy, M.A. Woo, J.A. Palomares, G.C. Fonarow, R.M. Harper, and R. Kumar. 2016. Lateralized Resting-State Functional Brain Network Organization Changes in Heart Failure. *PLoS One* 11 (5): e0155894.
108. Wu, K., Y. Taki, K. Sato, H. Hashizume, Y. Sassa, H. Takeuchi, B. Thyreau, Y. He, A.C. Evans, X.B. Li, R. Kawashima, and H. Fukuda. 2013. Topological Organization of Functional Brain Networks in Healthy Children: Differences in Relation to Age, Sex, and Intelligence. *PLoS One* 8 (2): e55347.
109. Langer, N., A. Pedroni, L.R. Gianotti, J. Hanggi, D. Knoch, and L. Jancke. 2012. Functional Brain Network Efficiency Predicts Intelligence. *Human Brain Mapping* 33 (6): 1393–1406.
110. Crossley, N.A., A. Mechelli, P.E. Vertes, T.T. Winton-Brown, A.X. Patel, C.E. Ginestet, P. McGuire, and E.T. Bullmore. 2013. Cognitive Relevance of the Community Structure of the Human Brain Functional Coactivation Network. *Proceedings of the National Academy of Sciences of the United States of America* 110 (28): 11583–11588.
111. Stam, C.J. 2014. Modern Network Science of Neurological Disorders. *Nature Reviews Neuroscience* 15 (10): 683–695.
112. Doucet, G.E., R. Rider, N. Taylor, C. Skidmore, A. Sharan, M. Sperling, and J.I. Tracy. 2015. Presurgery Resting-State Local Graph-Theory Measures Predict Neurocognitive Outcomes After Brain Surgery in Temporal Lobe Epilepsy. *Epilepsia* 56 (4): 517–526.
113. Pandit, A.S., P. Expert, R. Lambiotte, V. Bonnelle, R. Leech, F.E. Turkheimer, and D.J. Sharp. 2013. Traumatic Brain Injury Impairs Small-World Topology. *Neurology* 80 (20): 1826–1833.
114. Pessoa, L. 2008. On the Relationship Between Emotion and Cognition. *Nature Reviews Neuroscience* 9 (2): 148–158.
115. Sporns, O. 2011. The Human Connectome: A Complex Network. *Annals of the New York Academy of Sciences* 1224: 109–125.
116. Sporns, O., G. Tononi, and R. Kotter. 2005. The Human Connectome: A Structural Description of the Human Brain. *PLoS Computational Biology* 1 (4): e42.
117. ——— 2005. The Human Connectome: A Structural Description of the Human Brain. PLoS Computational Biology 1 (4): 245–251.
118. Craddock, R.C., R.L. Tungaraza, and M.P. Milham. 2015. Connectomics and New Approaches for Analyzing Human Brain Functional Connectivity. *GigaScience* 4 (1): 1–12.
119. Felleman, D.J., and D.C. Van Essen. 1991. Distributed Hierarchical Processing in the Primate Cerebral Cortex. *Cerebral Cortex* 1 (1): 1–47.
120. Power, J.D., K.A. Barnes, A.Z. Snyder, B.L. Schlaggar, and S.E. Petersen. 2012. Spurious but Systematic Correlations in Functional Connectivity MRI Networks Arise from Subject Motion. *NeuroImage* 59 (3): 2142–2154.
121. Fox, M.D., D. Zhang, A.Z. Snyder, and M.E. Raichle. 2009. The Global Signal And Observed Anticorrelated Resting State Brain Networks. *Journal of Neurophysiology* 101 (6): 3270–3283.
122. Ugurbil, K., J. Xu, E.J. Auerbach, S. Moeller, A.T. Vu, J.M. Duarte-Carvajalino, C. Lenglet, X. Wu, S. Schmitter, P.F. Van de Moortele, J. Strupp, G. Sapiro, F. De Martino, D. Wang, N. Harel, M. Garwood, L. Chen, D.A. Feinberg, S.M. Smith, K.L. Miller, S.N. Sotiropoulos, S. Jbabdi, J.L. Andersson, T.E. Behrens, M.F. Glasser, D.C. Van Essen, E. Yacoub, and W.U.-M.H. Consortium. 2013. Pushing Spatial and Temporal Resolution for Functional and Diffusion MRI in the Human Connectome Project. *NeuroImage* 80: 80–104.

123. Vaughan, J.T., M. Garwood, C.M. Collins, W. Liu, L. DelaBarre, G. Adriany, P. Andersen, H. Merkle, R. Goebel, M.B. Smith, and K. Ugurbil. 2001. 7T vs. 4T: RF Power, Homogeneity, and Signal-to-Noise Comparison in Head Images. *Magnetic Resonance in Medicine* 46 (1): 24–30.
124. Yacoub, E., A. Shmuel, J. Pfeuffer, P.F. Van De Moortele, G. Adriany, P. Andersen, J.T. Vaughan, H. Merkle, K. Ugurbil, and X. Hu. 2001. Imaging Brain Function in Humans at 7 Tesla. *Magnetic Resonance in Medicine* 45 (4): 588–594.
125. Van Essen, D.C., S.M. Smith, D.M. Barch, T.E.J. Behrens, E. Yacoub, K. Ugurbil, and W.-M.H. Consortium. 2013. The WU-Minn Human Connectome Project: An Overview. *NeuroImage* 80: 62–79.
126. Barch, D.M., G.C. Burgess, M.P. Harms, S.E. Petersen, B.L. Schlaggar, M. Corbetta, M.F. Glasser, S. Curtiss, S. Dixit, C. Feldt, D. Nolan, E. Bryant, T. Hartley, O. Footer, J.M. Bjork, R. Poldrack, S. Smith, H. Johansen-Berg, A.Z. Snyder, D.C. Van Essen, and W.-M.H. Consortium. 2013. Function in the Human Connectome: Task-fMRI and Individual Differences in Behavior. *NeuroImage* 80: 169–189.
127. Jia, H., X. Hu, and G. Deshpande. 2014. Behavioral Relevance of the Dynamics of the Functional Brain Connectome. *Brain Connectivity* 4 (9): 741–759.
128. Santos, G.C. 2015. Ontological Emergence: How Is That Possible? Towards a New Relational Ontology. *Foundations of Science* 20 (4): 429–446.
129. De Vico Fallani, F., J. Richiardi, M. Chavez, and S. Achard. 2014. Graph Analysis of Functional Brain Networks: Practical Issues in Translational Neuroscience. *Philosophical Transactions of the Royal Society of London. Series B, Biological Sciences* 369 (1653): pii:20130521.
130. Bassett, D.S., N.F. Wymbs, M.A. Porter, P.J. Mucha, J.M. Carlson, and S.T. Grafton. 2011. Dynamic Reconfiguration of Human Brain Networks During Learning. *Proceedings of the National Academy of Sciences of the United States of America* 108 (18): 7641–7646.
131. Baronchelli, A., R. Ferrer-i-Cancho, R. Pastor-Satorras, N. Chater, and M.H. Christiansen. 2013. Networks in Cognitive Science. *Trends in Cognitive Sciences* 17 (7): 348–360.

Design of a Compensator Network to Stabilize Chaotic Tumor Growth

Michael Harney and Julie Seal

Abstract Tumorigenesis can be modeled as a system of chaotic, nonlinear differential equations. From the analysis of these equations in state space, a feedback compensator is designed to stabilize the system based on a desired response. The feedback array constants represent four transducer molecules which could be used for any tumor type that obeys the same dynamics as the model, reducing drug investment requirements for a wider range of cancer treatment.

Keywords Chaotic • Tumor • Stabilization • State-space

1 Introduction

Tumorigenesis has been shown in many cases to be modeled as a system of chaotic, nonlinear differential equations [1–5]. Ivancevic et al. [1], in particular, has shown a universal model based on a reaction-diffusion cancer growth model, expressed by:

$$\frac{\partial n}{\partial t} = d_{\mathrm{n}}\nabla^2 - \rho\nabla \bullet (\eta\nabla f) \tag{1}$$

$$\frac{\partial f}{\partial t} = \alpha\eta\,(m - f) \tag{2}$$

$$\frac{\partial m}{\partial t} = d_{\mathrm{m}}\nabla^2 m + \kappa n - \sigma m \tag{3}$$

$$\frac{\partial c}{\partial t} = d_{\mathrm{c}}\nabla^2 c + \nu f - \omega n - \phi c \tag{4}$$

M. Harney (✉)
Intermountain Healthcare Genomics, Salt Lake City, UT, USA
e-mail: mharney1268@yahoo.com

J. Seal
Brigham Young University, Provo, UT, USA
e-mail: coachjulieseal@yahoo.com

P. Vlamos (ed.), *GeNeDis 2016*, Advances in Experimental Medicine and Biology 988, DOI 10.1007/978-3-319-56246-9_2

Where n denotes the tumor cell density, f is the MM–concentration, m is the MDE–concentration, and c is the oxygen concentration. This model has been investigated previously by the author using simulations and Wolfram CDF to determine that low oxygen environments and high glucose conditions will slow down tumorigenesis [6]. Similar results have been confirmed by studying cancer chromosomal instability (CIN) [7]. The above model by Ivancevic will be simplified so as to allow useful parameters to be applied for analysis in state space. Using standard models of state-space control, the conversion of the equations above to the Laplace domain will be used to design a compensation network to modify pole placement for optimum stability.

Current approaches to cancer therapy involve the development of molecular inhibitors that target specific proteins that are considered essential to the operation of tumor growth. As this approach changes with the biochemical process associated with each tumor and tissue type, the number of inhibitors that are designed is proportional to the number of tumor and tissue types, which is inefficient considering the current expense of drug development. The approach outlined in this paper utilizes years of theory in state space control that has been shown to work across many dynamical systems with a common outcome—a simplified set of feedback constants to stabilize any system that is determined to be controllable. This process is scalable across many biochemical systems and is therefore more efficient than designing a unique inhibitor to block specific biochemical processes for each type of tumor. The end result is a reduced set of molecular transducers that will stabilize tumorigenesis in any tissue, which hypothetically covers the dynamics of the molecular process of all cancerous growth.

2 State Space Analysis

From the reaction-diffusion cancer growth model above, a simplified non-dimensional non-spatial derivative model was found. This model was further modified by adding four additional parameters (α, β, γ, δ) which represents tumor cell volume, glucose level, number of tumor cells, and diffusion saturation level from the surface, respectively:

$$\dot{\eta} = 0 \tag{5}$$

$$\dot{f} = \alpha\eta\,(m - f) \tag{6}$$

$$\dot{m} = \beta\kappa n + f\,(\gamma - c) - m \tag{7}$$

$$\dot{c} = \nu fm - \omega n - \delta\phi c \tag{8}$$

A state-space analysis of a system of interdependent equations can be shown to take the form [8]:

$$\dot{x} = Ax + Bu \tag{9}$$

Where x is the state vector $[n\,, f, m, c]$, $\dot{x}$ is the time derivative of the state vector, A is the state-transition matrix, B is the input matrix and u is the input which in this case we establish as a step-input for an example. This will not change the characteristic equation or resulting pole placement for the un-driven system. The equations in (5–8) produce the state-transition matrix A as follows:

$$A = \begin{bmatrix} 0 & 0 & 0 & 0 \\ \alpha m & -\alpha\eta & \alpha\eta & 0 \\ \beta\kappa & \gamma - c & -1 & -f \\ -\omega & \nu m & \nu f & -\delta\phi \end{bmatrix} \tag{10}$$

The characteristic equation of the system is found from the determinant of $(sI - A)$, where I is the identity matrix and s is the Laplace variable:

$$\begin{aligned} |sI - A| = {} & s^4 + s^3\,(1 + \delta\phi + \alpha\eta) \\ & + s^2\left(\delta\phi + \upsilon f^2 + \alpha\eta + \alpha\eta\delta\phi + \alpha\eta c - \alpha\eta\gamma\right) \\ & + s\left(\alpha\eta\delta\phi + \alpha\eta\upsilon f^2 + \alpha\eta c\delta\phi - \alpha\eta\gamma\delta\phi + \alpha\eta\upsilon m f\right) \end{aligned} \tag{11}$$

This characteristic equation has roots that shows the placement of poles in the s-domain which corresponds to the stability of the system. The following typical values (also used for the paper in [6]) are used to calculate the coefficients of (11): $\eta = 50$, $\nu = 0.5$, $\phi = 0.025$, $\alpha = 0.06\,, \gamma = 26.5\,, \delta = 40$. From this, the characteristic equation in (11) is:

$$s^4 + 5s^3 + 130s^2 + 1251s$$

which has the roots of:

$$s = -8.07803, s = 0, s = 1.53902 \pm 12.34892\mathbf{i}$$

With the last two roots of $s = 1.53902 \pm 12.34892\mathbf{i}$ lying in the right-hand side of the s-plane, making the system unstable and oscillatory (Fig. 4). The phase-space plot of this system shown in Fig. 1 diagrams the instability and associated chaotic attractor for the values used in this example, which corresponds to a cell with high oxygen and glucose conditions [6].

For poles in the right-hand side of the s-plane, the system is unstable and corresponds to metastatic growth. In order to control the system, we want to design a compensator that will place the poles in the left hand side of the plane at $s >= 0$ for stability.

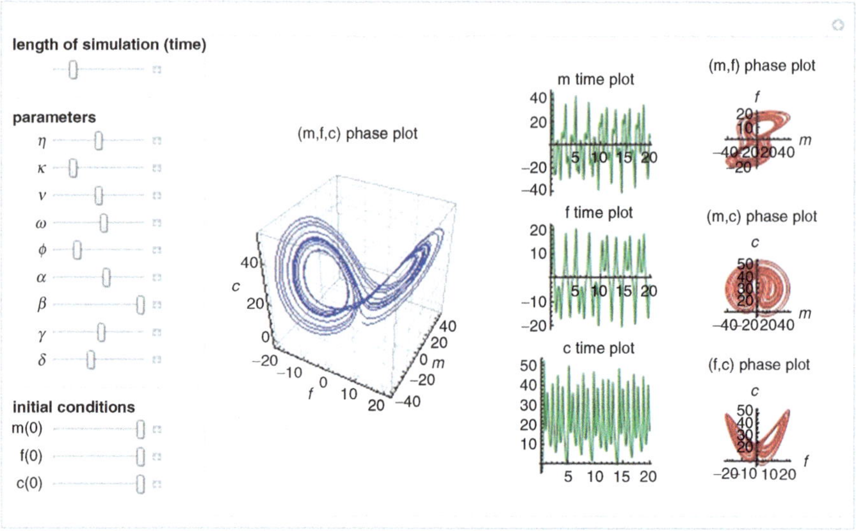

Fig. 1 State-space phase plot of unstable cellular growth, showing attractor

Fig. 2 State space control flow without compensator

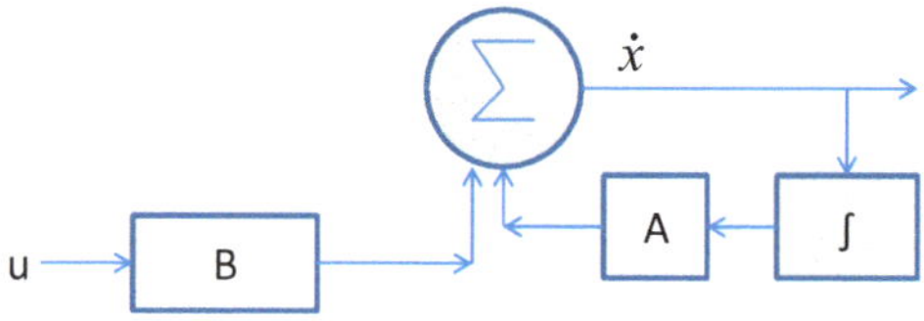

Fig. 3 State space control flow without compensator gain matrix G

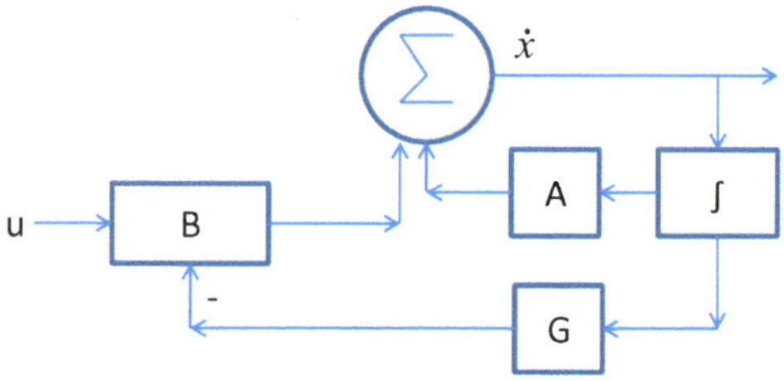

3 Design of a Compensator

The open loop system for the state-space equations in (9–11) is shown in Fig. 2.

By adding a feedback loop to the input block B with gain coefficients for the output (shown in Fig. 3), a compensator is established which effectively changes the internal gain of the system and alters pole placement and therefore stability.

The determination of the G matrix is found by first determining the desired response based on the characteristic equation in (11). The equation in (11) is of the form:

$$s^4 + a_1 s^3 + a_2 s^2 + a_3 s + a_4 \tag{12}$$

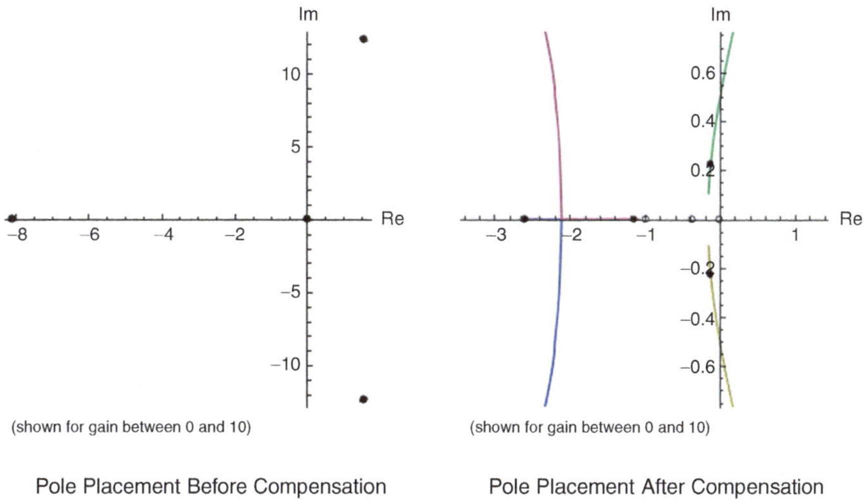

Fig. 4 Compensation moves poles to left-handed side of s-plane

Where a_1, a_2, a_3 and a_4 are the open-loop coefficients. Once the open-loop response is examined, the new coefficients are selected for the desired stability and these are represented by a_1', a_2', a_3' and a_4' for closed-loop response. For the example presented for (11), we had the values a_1, a_2, a_3 and a_4 given as (5, 130, 1251, 0) from the open-loop characteristic equation as $s^4 + 5s^3 + 130s^2 + 1251s$. The desired coefficients for stability (a_1', a_2', a_3' and a_4') may be set, for example, as (4, 4, 1, 0) which produces roots at s $= -0.38197$, $s = -2.6180$, $s = 0$ and $s = -1$. These roots are all on the left-hand side of the plane with no oscillatory response and are therefore stable (Fig. 4). The gain matrix is then found by equating the difference of the a vector with the a' vector:

$$G = \left[(QW)'\right]^{-1} (a' - a) \tag{13}$$

Where Q is the controllability test matrix given by the concatenation of the input matrix B and the product of the input matrix with the state transition matrix AB where A and B are given in (9 and 10):

$$Q = \left[B, AB, A^2B, A^3B\right] \tag{14}$$

The matrix W in (13) is given by:

$$W = \begin{bmatrix} 1 & a_1 & a_2 & a_3 \\ 0 & 1 & a_1 & a_2 \\ 0 & 0 & 1 & a_1 \\ 0 & 0 & 0 & 1 \end{bmatrix} \tag{15}$$

Where a_1, a_2, a_3 are given in (12) and the product $[(QW)']^{-1}$ in (13) is the inverse of the transpose of the product of Q and W. The resulting gain matrix in (13) stabilizes the system by moving the poles to the left-hand side of the plane in comparison to the uncompensated system, which has poles in the right hand side of the plane (Fig. 4).

4 Interpretation of Results

The gain matrix as determined in (13) is a set of coefficients that represent the negative feedback of the state variables x, through transformation and scaling through the G matrix, for input into the system through the B matrix. As such, the G matrix represents transducer molecules that would interact with the known state variables (n as tumor cell density, f as the MM–concentration, m as the MDE–concentration, and c as the oxygen concentration) and produce an output at the same unit level as the input to the system u, in (9). If the initial model in (1–8) is correct, the design of these four transducer molecules to act as a feedback mechanism is all that is required to close the loop of tumorigenesis in order to stabilize tumor growth and reduce metastasis. If the model requires modification, then the requirements for the transducer molecules represented by the gain matrix in (13) changes with the new model. Whatever model is used, the hope is that it is accurate enough to model tumor behavior across many tissue domains, allowing for the treatment of a wide variety of cancers with only the same four molecular transducers. In contrast, current approaches to cancer therapy usually involve the development of molecular inhibitors that target specific proteins that are considered essential to the operation of tumor growth. As this approach changes with the biochemical process associated with each tumor and tissue type, the use of a standard control methodology such as this one that has been successfully tested for many decades on a variety of dynamical systems, offers a substantial improvement in the investment of drug development.

References

1. Ivancevic, T. T., Bottema M. J., and Jain, L. C. A Theoretical Model of Chaotic Attractor in Tumor Growth and Metastasis. arXiv:0807.4272 in Cornell University Library's. arXiv.org.
2. Sole, R., and B. Goodwin. 2000. *Signs of Life: How Complexity Pervades Biology*. New York, NY: Basic Books.
3. Bar-Yam, Y. 2011. *Concepts: Chaos*. Cambridge, MA: New England Complex Systems Institute. http://www.necsi.edu/guide/concepts/chaos.html.
4. Guiot, C., P.G. Degiorgis, P.P. Delsanto, P. Gabriel, and T.S. Deisboeck. 2003. Does Tumor Growth Follow a Universal Law? *Journal of Theoretical Biology* 225 (2): 147–151. http://arxiv.org/ftp/physics/papers/0303/0303050.pdf.
5. Annibaldi, A., and C. Widmann. 2010. Glucose Metabolism in Cancer Cells, PubMed. http://www.ncbi.nlm.nih.gov/pubmed/20473153.

6. Harney, M., and W. Yim. 2014. Chaotic Attractors in Tumor Growth and Decay: A Differential Equation Model. *Advances in Experimental Medicine and Biology* 820: 193–206. http://link.springer.com/chapter/10.1007/978-3-319-09012-2_13.
7. Burrell, R.A., et al. 2013. Replication Stress Links Structural and Numerical Cancer Chromosomal Instability. *Nature* 494 (7438): 492–496. http://www.nature.com/nature/journal/v494/n7438/full/nature11935.html.
8. Friedland, B. 1986. *Control Systems Design – An Introduction to State Space Methods*. New York City, NY: McGraw-Hill.

Evaluation of Anti-Epileptic Effect of New Indole Derivatives by Estimation of Biogenic Amines Concentrations in Rat Brain

Konda Swathi and Manda Sarangapani

Abstract The new heterocyclic compounds are used to treat epilepsy. In the present work new indole derivatives i.e. 5-[2(3)-di alkyl amino alkoxy] Indole 2,3-di-one derivatives are synthesized and characterized and these compounds was subjected to acute toxicity and then screened for antiepileptic activity on Maximal Electroshock (MES) seizures model in albino wistar rats. In that study 5-[2-dimethyl amino ethoxy] Indole 2,3 dione and 5-[2-dimethyl amino ethoxy] Indole 2-one,3-semicarbazone(IVa) showed good antiepileptic activity and less neurotoxicity compared to phenytoin. The purpose of the present study is to investigate the effect of 5-[2-dimethyl amino ethoxy] Indole 2,3-di one and 5-[2-dimethyl amino ethoxy] Indole 2-one,3-semicarbazone(IVa) derivatives on biogenic amines concentrations in rat brain after induction of seizures by Maximal Electro Shock(MES) method. Our aim of study was relationship between seizure activities and altered the monoamines such as noradrenaline (NA), dopamine (DA), serotonin (5-HT) in forebrain of rats in MES seizure models. In MES model, study of 5-[2-dimethyl amino ethoxy] Indole 2,3 dione(IIIa) and 5-[2-dimethyl amino ethoxy] Indole 2-one,3-semicarbazone(IVa) (100 mg/kg) showed significantly restored the decreased levels of brain monoamines such as Noradrenaline, Dopamine & 5-Hydroxy Triptamine. Thus, this study suggests that study of 5-[2-Dimethyl amino ethoxy] Indole 2,3-dione(IIIa) and 5-[2-dimethyl amino ethoxy] Indole 2-one,3-semicarbazone(IVa) increased the monoamines on rat brain, which may be decreased the susceptibility to MES induced seizure in rats.

Keywords Synthesis • 5-[2-Dimethyl amino ethoxy] Indole 2,3-diones • 5-[2-Dimethyl amino ethoxy] Indole 2-one-3-semicarbazones • Antiepileptic activity • Biogenic amines • NA • DA • 5-HT

K. Swathi (✉) • M. Sarangapani
Medicinal Chemistry Laboratory, U.C.P.Sc., Kakatiya University, Wararagal 506009, Telangana, India
e-mail: kswathi84@yahoo.co.in

P. Vlamos (ed.), *GeNeDis 2016*, Advances in Experimental Medicine and Biology 988, DOI 10.1007/978-3-319-56246-9_3

1 Introduction

Surendranath pandya. reported the synthesis and anticonvulsant activity of some novel n-methyl/acetyl,5-(un)-substituted isatin-3-semicarbazones. In the last few years, Isatin derivatives have been discovered which show potential hypnotic [1], antibacterial [2–5], MAO inhibitory [6], antioxidant activity. We are reporting in the present communication the synthesis and characterization of some new compounds: 5-[2(3)-dialkyl amino alkoxy] Indole 2,3-diones,5-[2(3)-dialky amino alkoxy] Indole 2-one,3-semicarbazones.

In the present work some new 5-[2(3)-dialkylamino alkoxy] Indole 2, 3-dione derivatives were prepared from 5-hydroxy isatin and its semicarbazones. All the compounds were evaluated for anticonvulsant activity by Maximal electroshock induced convulsion method. These compounds were also evaluated for their neurotoxicity study by skeletal muscle relaxant activity method. 5-[2-dimethylamino ethoxy] Indole 2, 3-dione(IIIa) and 5-[2-dimethyl amino ethoxy] Indole 2-one, 3-semicarbazone(IVa) showed good anticonvulsant activity when compared with standard drug Phenytoin and all the compounds showed less neurotoxicity when compared with standard drug Diazepam. Since the antiepileptic effect of compound(IIIa) and compound(IVa) on brain has been experimentally not confirmed. Therefore, the aim of the present investigation was to evaluate the effect of compound(IIIa) and compound(IVa) in rat brain after induction of epilepsy by MES in albino wistar rats.

2 Materials and Methods

The compounds were mostly synthesized by conventional methods and described in experimental selection and also by the methods established in our laboratory.

2.1 Chemicals

Diazepam, Di alkylamino alkyl halides, Phenytoin, Semicarbazidehydrochloride purchased from Sigma- Aldrich Chemicals Private Limited, Hyderabad, India. p-Amino phenol, Hydroxylamine hydrochloride, Sodium sulfate were purchased from Merck Chemicals Private Limited, Hyderabad, India.

2.2 Preparation of 5-Hydroxy Indole 2-One,3-Semicarbazone(II)

As represented in Scheme-1,5-Hydroxyisatin(I) was heated under reflux in methanol containing two or three drops of acetic acid with semicarbazide hydrochloride for half an hour. The product 5-Hydroxy indole2-one,3-semicarbazone(II) thus separated was filtered and purified by recrystalization from suitable solvent.

2.3 Preparation of 5-[2(3)-Dialkyl Amino Alkoxy] Indole 2-One,3-Semicarbazone(III) and 5-[2(3)-Dialkyl Amino Alkoxy] Indole 2-One,3-Semicarbazone(IV)

As represented in Scheme-1 (Fig. 1), mixture of 5-Hydroxyindole2,3-dione(I)/5-Hydroxy indole2-one, 3-semicarbazone(II) (0.01 moles) and dialkylamino alkyl-halide (0.01 moles) placed in 10% alcoholic potassium hydroxide and this mixture was stirred at room temperature for 6 h. The alcohol was reduced to half of its volume and cooled. The product separated was filtered, washed with small portions of cold alcohol repeatedly and dried. It was purified by recrystalisation from hydro alcoholic mixtures to get a crystalline solid. The physical data of the title compounds were presented in Table 1. The compounds were characterized by spectral data.

2.4 Spectral Data

The compounds have been characterized by the spectral data IR, PMR and Mass.

Compound (IIIa) showed characteristic IR peaks at 3276(NH), 1651.96 (C=O), 1569.82(Ar,C=C), 1276(C-O-C), 807.93(Ar). Its PMR spectrum (DMSO,III) showed characteristic peaks at (d ppm) 300 MHz, DMSO) 10.36(s, 1H,-CONH), 7.01–7.29(m,3 H,Ar-H), 3.2 (t,2H,O-CH_2), 2.9 (t,2H,N-CH_2), 1.24(m,6H,CH_3-N-CH_3). Mass spectrum of compound IIIa showed molecular ion(M+) base peak at m/z 234 (100%). It also shows peak at m/z (71) may be due to the fragmentation of the alkyl chain from the molecule ion.

Compound (IVa) showed characteristic IR peaks at 3255.79 (NH), 1728.29 (C=O), 1509.02(C=C), 1133.15(C-O-Cether). Its PMR spectrum (DMSO,III) showed characteristic peaks at (d ppm) 300 MHz, DMSO) 10.36(s, 1H,-CONH), 6.65–7.29(m, 3H, Ar-H),3.29 (t,2H,O-CH_2), 2.93 (t,2H,N-CH_2), 7.41-7.46(d,2H, NH_2), 11.36(s,1H, NH_2), 1.24(m,6H,CH_3-N-CH_3). Mass spectrum of compound IIIa showed molecular ion (M+) base peak at m/z (291). The mass spectrum shows its base peak at m/z 77 (100%) may be due to the fragmentation of the semicarbazone from the molecule ion.

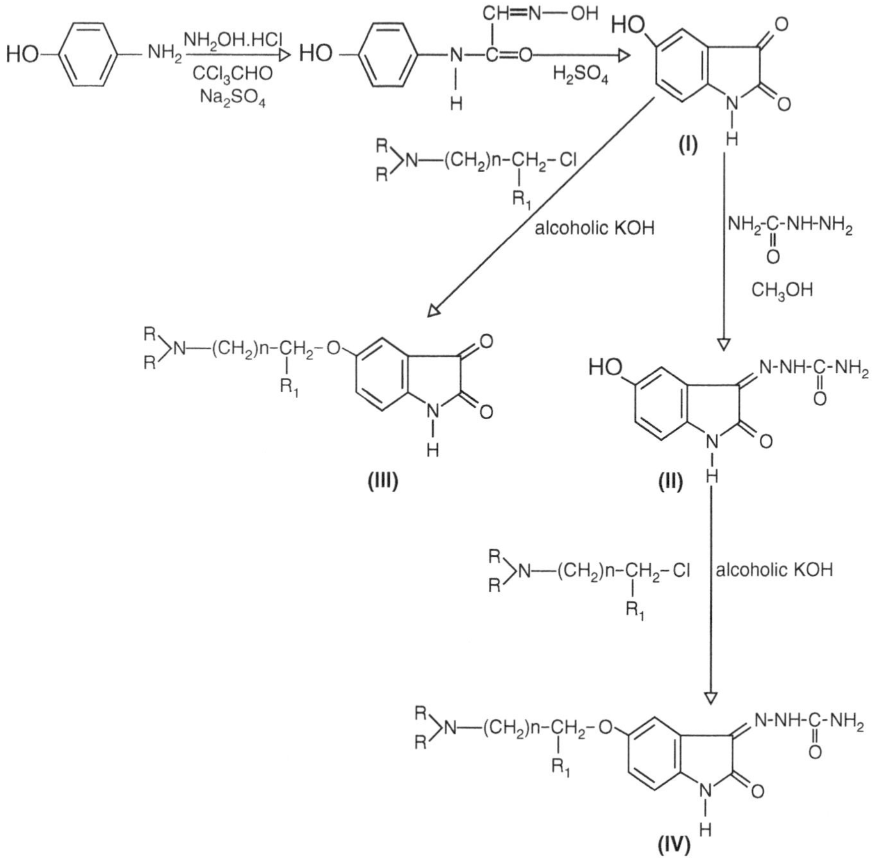

5-Hydroxyisatin derivatives (III)
IIIa R= CH3 ; R1= H; n=1
IIIb R= C2H5 ; R1= H; n=1
IIIc R= CH3 ; R1= H; n=2
IIId R= CH3 ; R1= CH3; n=1
IIIe R= CH3-CH- CH3; R1= H; n=1

5-Hydroxyisatin-3- semicarbazone derivatives (IV)
IVa R= CH3 ; R1= H; n=1
IVb R= C2H5 ; R1= H; n=1
IVc R= CH3 ; R1= H; n=2
IVd R= CH3 ; R1= CH3; n=1
IVe R= CH3- CH- CH3; R1= H; n=1

Fig. 1 Scheme1

3 Pharmacology

3.1 Anti-Epileptic Activity

Materials: 1% w/v SCMC, Test Compounds, Stop Watch, Cornealelectrodes, Phenytoin.

Table 1 Physical data of 5-[2(3)-dialkyl amino alkoxy] indole 2-one,3-semicarbazone(III) and 5-[2(3)-dialkyl amino alkoxy] indole 2-one,3-semicarbazone(IV)

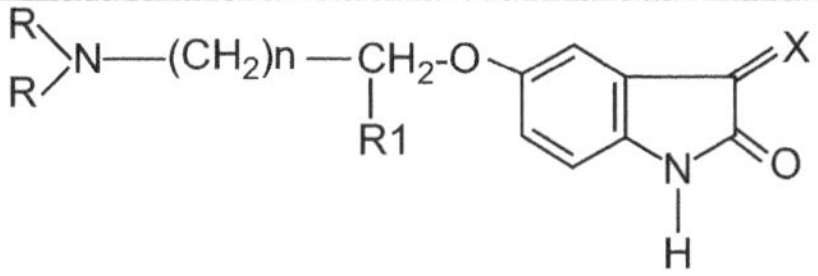

S.No	Compound	R	R_1	N	X	M.F	% Yeild	M.P	M.Wt
1	IIIa	CH_3	H	1	O	$C_{12}H_{14}N_2O_3$	91%	>320	234
2	IIIb	C_2H_5	H	1	O	$C_{14}H_{18}N_2O_3$	86%	>320	262
3	IIIc	CH_3	H	2	O	$C_{13}H_{16}N_2O_3$	93%	>320	248
4	IIId	CH_3	CH_3	1	O	$C_{13}H_{16}N_2O_3$	85%	>320	248
5	IIIe	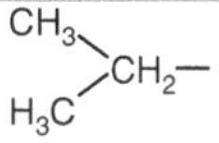	H	1	O	$C_{16}H_{24}N_2O_3$	81.8%	>320	292
6	IVa	CH_3	H	1	$NNHCONH_2$	$C_{13}H_{17}N_5O_3$	92%	>320	291
7	IVb	C_2H_5	H	1	$NNHCONH_2$	$C_{15}H_{21}N_5O_3$	83%	>320	319
8	IVc	CH_3	H	2	$NNHCONH_2$	$C_{14}H_{19}N_5O_3$	92%	>320	365
9	IVd	CH_3	CH_3	1	$NNHCONH_2$	$C_{14}H_{19}N_5O_3$	86%	>320	365
10	IVe	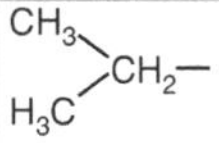	H	1	$NNHCONH_2$	$C_{17}H_{27}N_5O_3$	82%	>320	349

3.1.1 Maximal Electroshock Seizure (MES) Method & Neurotoxicity Study

Method: The antiepileptic activity was studied by Maximal Electroshock Induced Convulsion method [7] by using electro-convulsiometer. Healthy albino rats (150–220 g) were fasted for overnight and divided into groups of six animals each. The test compounds suspended in (1% w/v SCMC) were administered at a dose of 100 mg/kg body weight *i.p.* The control group animals received only vehicle (1% w/v SCMC). The test started 30 min after *i.p.* injection. Maximal seizures were induced by the application of electrical current to the brain via corneal electrodes. The stimulus parameter for mice was 50 mA in a pulse of 60 Hz for 200 ms. Abolition of the hind limb tonic extensor spasm was recorded as a measure of antiepileptic activity. The neurotoxicity [8] was studied by rotarod method by using diazepam as a standard, results are presented in Table 2.

5-[2-dimethyl amino ethoxy] Indole 2,3-dione(IIIa) and 5-[2-dimethyl amino ethoxy] Indole 2-one,3-semicarbazone(IVa) showed good anticonvulsant activity when compared with standard drug Phenytoin and other compounds. These two compounds showed less neurotoxicity when compared with standard drug Diazepam.

Since the antiepileptic effect of 5-[2-Dimethyl amino ethoxy] Indole 2, 3 di-one(IIIa), 5-[2-Dimethyl amino ethoxy] Indole 2-one,3-semicarbazone(IVa) on brain has been experimentally not confirmed. Therefore, the aim of the present

Table 2 Antiepileptic and neurotoxicity study of 5-[2(3)-dialkyl amino alkoxy] indole 2-one, 3-semicarbazone(IIIa-e), 5-[2(3)-dialkyl amino alkoxy] indole 2-one,3-semicarbazone(IVa-e)

S. No	Compound	MES induced convulsions	Neurotoxicity (%)
1.	IIIa	65.54 ± 0.341	7
2.	IIIb	58.61 ± 1.234	18
3.	IIIc	48.42 ± 1.134	6.7
4.	IIId	41.5 ± 1.345	3.4
5.	IIIe	37.46 ± 1.034	4.6
6.	IVa	67.18 ± 0.234	7
7.	IVb	66.68 ± 0.243	5
8.	IVc	54.76 ± 1.234	6.2
9.	IVd	38.18 ± 1.658	7.8
10.	IVe	37.44 ± 1.345	7
11.	Phenytoin	100	–
12.	Control	0	2
13.	Diazepam	–	88

Number of animals n = 6. $P < 0.05$. The compounds were tested at a dose of 100 mg/kg (b.w)

investigation was to evaluate the effect of 5-[2-Dimethylamino ethoxy] Indole 2, 3-dione and 5-[2-Dimethyl amino ethoxy] Indole 2-one, 3-semicarbazone(IVa) in rat brain after induction of epilepsy by MES in albino wistar rats.

3.2 Biogenicamines Estimation by Fluorimetric Micromethod

3.2.1 Experimental Design for Biogenicamine Estimation

Albino wistar rats were divided into four groups of six animals each. Group I received vehicle control (1% w/v SCMC, 1 ml/100 g) whereas Group-II received standard drug (Phenytoin, 25 mg/kg) *i.p*, Group-III and IV, received 5-[2-Dimethylamino ethoxy] Indole 2, 3-dione and 5-[2-Dimethyl amino ethoxy] Indole 2-one,3-semicarbazone(IVa) (100 mg/kg) *i.p* respectively for 14 days. On the 14th day, Seizures are induced to all the groups by using an Electro convulsiometer. The duration of various phases of epilepsy were observed.

3.2.2 A Fluorimetric Micromethod for the Simultaneous Determination of Serotonin, Noradrenaline and Dopamine

On the 14th day after observed the convulsions of all groups rats were sacrificed, whole brain was dissected out and separated the forebrain. Weighed quantity of tissue and was homogenized in 0.1 ml hydrochloric acid—butanol, (0.85 ml of 37% hydrochloric acid in *n*-butanol for spectroscopy) for 1 min in a cool environment.

The sample was then centrifuged for 10 min at 2000 rpm. 0.08 ml of supernatant phase was removed and added to an Eppendorf reagent tube containing 0.2 ml of heptane (for spectroscopy) and 0.025 ml 0.1 M hydrochloric acid. After 10 min of vigorous shaking, the tube was centrifuged under same conditions to separate two phases. Upper organic phase was discarded and the ethanol phase (0.02 ml) was used for estimation of Serotonin, Nor Adrenaline and Dopamine assay.

3.2.3 Nor-Adrenaline and Dopamine Assay

The assay represents a miniaturization of the trihydroxide method. To 0.02 ml of HCl phase, 0.05 ml 0.4 M and 0.01 ml EDTA/Sodium acetate buffer (pH 6.9) were added, followed by 0.01 ml iodine solution (0.1 M in ethanol) for oxidation. The reaction was stored after 2 min by addition of 0.01 ml Na_2SO_3 in 5 M NaOH. Acetic acid was added 1.5 min later. The solution was then heated to 100 for 6 min. When the sample again reached room temperature, excitation and emission spectra were read in the micro cuvette as with 5-HT: in some cases, the readings were limited to the excitation maxima. 395–485 nm for NA and 330–375 nm for DA uncorrected instrument values [9].

3.2.4 Serotonin Assay

As mentioned earlier some modifications in reagent concentration became necessary together with changes in the proportions of the solvent, in order to obtain in a good fluorescence yield with reduced volume for 5-HT determination, the O-pthaldialdehyde (OPT) method was employed. From the OPT reagent 0.025 ml were added to 0.02 ml of the HCl extract. The fluorophore was developed by heating at 100 °C for 10 min. After the samples reached equilibrium with the ambient temperature, excitation/estimation spectra or intensity reading at 360–470 nm were taken in the micro cuvette [10].

3.3 Statistical Analysis

The data were expressed as mean ± standard error mean (S.E.M). The Significance of differences among the group was assessed using one way and multiple way analysis of variance (ANOVA). The test followed by Dunnet's test p values less than 0.05 were considered as significance.

4 Results and Discussions

Synthesis and Charecterisation of 5-[2-dimethyl amino ethoxy] Indole 2,3-dione(IIIa) and 5-[2-dimethyl amino ethoxy] Indole 2-one,3-semicarbazone(IVa) was done by standard methods.

Effect 5-[2-dimethyl amino ethoxy] Indole 2,3-dione(IIIa) and 5-[2-dimethyl amino ethoxy] Indole 2-one,3-semicarbazone(IVa) showed good antiepileptic activity in seizure induced rats by MES. These compounds specifically evaluated for levels of biogenicamines in rat brain.

4.1 Noradrenaline

In MES model, Noradrenaline levels significantly ($p < 0.01$) decreased in forebrain of epileptic control animals. 5-[2-Dimethyl amino ethoxy] Indole 2,3 dione(IIIa) and 5-[2-Dimethyl amino ethoxy] Indole 2-one,3-semicarbazone(IVa) at the doses of 100 mg/kg, standard drugs phenytoin and diazepam treated animals showed a significantly ($p < 0.05$ & $p < 0.01$) increased in Noradrenaline levels in forebrain of rats (Table 3).

4.2 Dopamine

In MES model, Dopamine levels significantly ($p < 0.01$) decreased in forebrain of epileptic control animals. Test compounds(IIIa, IVa), phenytoin and diazepam treated animals showed a significantly ($p < 0.05$ & $p < 0.01$) increased in Dopamine levels in forebrain of rats (Table 3).

4.3 Serotonin

In MES model, Serotonin levels significantly ($p < 0.01$) decreased in forebrain of epileptic control animals were observed. 5-[2-Dimethyl amino ethoxy] Indole 2,3 dione(IIIa) and 5-[2-Dimethyl amino ethoxy] Indole 2-one,3-semicarbazone(IVa) at the doses of 100 mg/kg, standard drugs phenytoin and diazepam treated animals showed a significantly ($p < 0.05$ & $p < 0.01$) increased in Serotonin levels in forebrain of rats (Table 3).

Table 3 Effect 5-[2(3)-dialkyl amino alkoxy] Indole 2-one,3-semicarbazone(III), and 5-[2(3)-dialkyl amino alkoxy] indole 2-one,3-semicarbazone(IV) on neurotransmitters levels in rat brain after MES induced epilepsy

Group	Design of treatment	Noradrenaline	Dopamine	Serotonin
I	Vehicle control (SCMC 1 ml/100 gm)	664.25 ± 1.29	540.46 ± 3.17	184.39 ± 3.14
II	MES (SCMC 1 ml/100 gm)	428.17 ± 1.31 a**	449.17 ± 1.72a**	65.68 ± 21.25 a**
III	Phenytoin 25 mg/kg, *i.p*	546.15 ± 2.71 b**	625.27 ± 2.64 b**	25.27 ± 2.16 b**
IV	5-[2-Dimethylamino ethoxy] indole 2, 3-dione 100 mg/kg, *i.p*	522.41 ± 2.16b**	625.28 ± 4.12b**	95.14 ± 2.25b**
V	5-[2-Dimethylamino ethoxy] indole 2-one −3-semicarbazone 100 mg/kg, *i.p*	716.19 ± 2.13b*	524.33 ± 1.31b*	88.19 ± 1.28b*

Values are expressed as mean ± SEM of six observations. Comparison between: **a-** Group I Vs Group II, **b-** Group III Vs Group IV and Group V. Statistical significant test for comparison was done by ANOVA, followed by Dunnet's test $^{*}p < 0.05$; $^{**}p < 0.01$; Units = pg/mg of wet tissue

5 Conclusion

A new series of indole derivatives 5-[2(3)-dialkyl amino alkoxy] Indole 2-one,3-semicarbazone(III), 5-[2(3)-dialkyl amino alkoxy] Indole 2-one,3-semicarbazone(IV) were synthesized by reacting 5-hydroxyindole 2,3 dione/5-hydroxyindole 2,3 dione schiff bases with 2-N,N di alkylamino alkyl halides. Evaluation of these compounds as antiepileptic and skeletal muscle relaxant activity revealed that compounds **IVa, IIIa,** exhibited more promising activity. 5-[2-dimethyl amino ethoxy] Indole 2,3 dione(IIIa), 5-[2-Dimethyl amino ethoxy] Indole 2-one,3-semicarbazone(IVa) showed good anticonvulsant activity when compared with standard drug Phenytoin and all the compounds showed less neurotoxicity when compared with standard drug Diazepam.

The role of biogenic amines in epileptogenesis and in recurrent seizure activity is well-documented. Spontaneous and experimentally induced deficiencies in noradrenaline (NA), dopamine (DA) and/or serotonin (5-hydroxy- tryptamine or 5-HT). It has been implicated in the onset and perpetuation of many seizure disorders many experimental procedures designed to increase monoaminergic activity have proven antiepileptic properties [10–14]. In present study, the established antiepileptic drugs such as phenytoin restored the monoamine levels on brain [15]. Similarly compound(IIIa) and compound(IVa) significantly ($p < 0.05$ & $p < 0.01$) increased monoamines levels in forebrain of rats. Many drugs that increase the brain contents of GABA have exhibited anticonvulsant activity against seizures induced by MES [11]. MES is probably the best validated method for assessment of antiepileptic drugs in generalized tonic-clonic seizures [16].

In conclusion biogenic amines participate in the control of Maximal electroshock induced seizure in rat model. Our findings support the hypothesis that decreased the monoamines levels in rat brain after induction of seizure. Compound(IIIa) and compound(IVa) treated rats, monoamines such as NA, DA & 5-HT levels significantly restored on forebrain. Thus compound(IIIa) and compound(IVa) increases the seizure threshold and decreased the susceptibility to MES induced seizure in rats. Hence we suggest that new indole derivatives of 5-[2-Dimethylamino ethoxy] Indole 2, 3-dione(IIIa) and 5-[2-Dimethyl amino ethoxy] Indole 2-one, 3-semicarbazone(IVa) possess antiepileptic properties that may be due to restored the biogenic amines in rat brain.

Acknowledgments The First author would like to thank the UGC, New Delhi for providing financial support. Authors are thankful to Principal University College of Pharmaceutical Sciences, Kakatiya University, Warangal for providing facilities.

References

1. Pandeya, S.N., and A. Senthil Raja. 2002. *Journal of Pharmaceutical Sciences* 5 (3): 275.
2. Padhy, A.K., S.K. Sahu, P.K. Panda, D.M. Kar, and P.K. Misro. 2004 . *Indian Journal of Chemistry*43B : 971 .
3. Raviraj, A., Kusanur, et al. 2004 . *Journal of Chemical Sciences*116 (5): 265.
4. Gupta, S., et al. 2004 . *Asian Journal of Chemistry*16 (2): 779 – 783.
5. Ajitha, M., et al. 2002. *Pharmazie* 57 (12): 796.
6. Krall, R.L., et al. 1978. *Epilepsia* 19: 409.
7. H. Gerhard Vogel ed. 2002. Drug Discovery and Evaluation of Pharmacological Assays, 2nd Edition: 398.
8. Schlumpf, M., et al. 1974. A Fluorimetric Micromethod for the Simultaneous Determination of Serotonin, Noradrenaline and Dopamine in Milligram Amounts of Brain Tissue. *Biochemical Pharmacology* 23: 2337–2346.
9. Applegate, C.D., et al. 1986. Kindling Antagonism: Effects of Norepinephrine Depletion on Kindled Seizure Suppression After Concurrent, Alternate Stimulation in Rats. *Experimental Neurology* 94: 379–390.
10. Corcoran, M.E. 1988. Characteristics of Accelerated Kindling After Depletion of Noradrenaline in Adult Rats. *Neuropharmacology* 27: 1081–1084.
11. McIntyre, D.C., and N. Edson. 1989. Kindling-Based Status Epilepticus: Effects of Norepinephrine Depletion with 6-Hydroxydopamine. *Experimental Neurology* 104: 10–14.
12. Pelletier, M.R., and M.E. Corcoran. 1993. Infusions of a2 Noradrenergic Agonists and Antagonists into the Amygdala: Effects on Kindling. *Brain Research* 632: 29–35.
13. Yan, Q.S., et al. 1995. Further Evidence of Anticonvulsant Role for 5- Hydroxy Tryptamine in Genetically Epilepsy-Prone Rats. *British Journal of Pharmacology* 115: 1314–1318.
14. Zis, A.P., et al. 1992. Neurochemical Effects of Electrically and Chemically Induced Seizures: An In Vivo Microdialysis Study in Hippocampus. *Neuropsychopharmacology* 7: 189–195.
15. Applegate, C.D., et al. 1992. Kindling Antagonism: Effects of Norepinephrine Depletion on Kindled Seizure Suppression After Concurrent, Alternate Stimulation in Rats. *Experimental Neurology* 94: 379–390.
16. Pandeya, S.N., P. Yogeeswari, and J.P. Stables. 2000. *European Journal of Medicinal Chemistry* 35: 879–886.

Does Health Perception, Dietary Habits and Lifestyle Effect Optimism? A Quantitative and Qualitative Study

Aikaterini Kargakou, Athanasios Sachlas, Georgios Lyrakos, Sofia Zyga, Maria Tsironi, and Andrea Paola Rojas Gil

Abstract The aim of the study was to investigate the relationship between optimism, general health perception, nutritional habits and lifestyle. A quantitative study was conducted on 500 Greek adults. A standardized questionnaire was used which consisted of the General Health Self-Assessment Questionnaire (GHSAQ), the Life Orientation Test-Revised (GrLOT-R), the dietary habits and lifestyle questionnaire. A qualitative study was conducted through interviews and focus groups. Participants' average score of GrLot-R was 20.47 (±4.017) units. The highest GrLot-R score was statistically correlated to more frequent consumption of fruits, salads, dairy products, olive oil, high-fibre cereals and water and to the lower consumption of canned products. Multivariate analysis showed that optimism is significantly positively predicted by the factors "Vitamin K & A", "Vitamin C", and negatively by "Preservatives." There was also a positive correlation between GrLot-R and GHSAQ. The qualitative study analysis showed that physical exercise, duration/quality of sleeping, family and stress affected participants' optimism. It seems that intrinsic factors as vitamins, preservatives, a healthy lifestyle and extrinsic factors as the family background and the environment that someone is raised, affect the level of optimism.

A. Kargakou (✉) • A.P. Rojas Gil (✉)
Faculty of Human Movement and Quality of Life Sciences, Department of Nursing, Efstathiou & Stamatikis Valioti and Plateon, University of Peloponnese, Sparti, 23100, Greece
e-mail: kargakou@gmail.com; apaola71@yahoo.com.mx; arojas@uop.gr

A. Sachlas
Faculty of Finance and Statistics, Department of Statistics and Insurance Science, University of Piraeus, 80, M. Karaoli & A. Dimitriou Street, Piraeus, 18534, Greece
e-mail: asachlas@unipi.gr

G. Lyrakos
General Hospital of Nikaia "Agios Panteleimon", 3 D. Mandouvalou St, Nikaia, 18454, Piraeus, Greece
e-mail: g.lyrakos@cityu.gr

S. Zyga • M. Tsironi
Department of Nursing, Faculty of Human Movement and Quality of Life Sciences, University of Peloponnese, Sparta, Greece
e-mail: zygas@uop.gr; tsironi@uop.gr

P. Vlamos (ed.), *GeNeDis 2016*, Advances in Experimental Medicine and Biology 988, DOI 10.1007/978-3-319-56246-9_4

Keywords Optimism • Health • Health self-assessment • Dietary habits • Lifestyle

1 Introduction

Diet can be a determinant factor regarding the incidence of chronic diseases and it severely affects the development of diseases such as cardiovascular diseases, obesity, diabetes, cancer etc. [1].

The association between dietary habits and physical health is quite familiar compared to the association between dietary habits, mental health and mood. Cross-sectional and prospective studies have been carried out with the aim of investigating the relationship of diet and non-communicable diseases, part of which are mental disorders [2]. Individuals who have experienced mental disorders like depression, bipolar disorder, schizophrenia and obsessive compulsive disorder, seem to present a deficiency of many nutrients and vitamins, minerals, omega-3 fatty acids etc. [3].

Over the last years, the interest of the scientific community has increased regarding the study of positive emotions and their effect on people's health. One emotion considered as "positive" is optimism. The most widely used definition of optimism is that of Scheier and Carver's [4], who conceptualize it as "the belief that good, as opposed to bad, things will generally occur in one's life."

The level of optimism, regardless of the theoretical approach used, has been associated with lower risk of mortality in general medical patients [5], in coronary heart disease patients [6], with slower progress of atherosclerotic carotid disease [7] and AIDS [8], with lower pain sensitivity [9], the incidence of lower levels of inflammation indicators and haemostasis [10] and anxiety [11].

Studies with the object of the association of optimism, dietary habits and lifestyle are not enough in order to lead to firm conclusions. A study of Boehm et al. [12] showed that the highest score in optimism correlated negatively with smoking and alcohol consumption and positively with dietary habits, healthier lipids profile and physical exercise. Kelloniemi et al. [13] demonstrated that optimist people consumed more fresh vegetables and salads, fruits, low-fat cheese than the pessimist ones.

The purpose of the present study was to investigate the association between optimism, dietary habits, health self-perception and lifestyle on a sample of the general Greek population.

2 Study Design

The present study was conducted by applying the triangulation method [14]. The methodological triangulation was chosen in order to obtain more reliable and valid results, to assure each method's positive data, to reduce weaknesses and to avoid methodological limitations.

Taking into consideration the fact that it is the first study in Greece which explores the relationship between optimism, dietary habits, health and lifestyle, a qualitative study was applied in order to interpret quantitative data but mainly to investigate issues that could not be assessed through a structured questionnaire.

An anonymous reliable and valid questionnaire combined of three standardized questionnaires was used [15] for the quantitative study. The questionnaire was consisted of the General Health Self-Assessment Questionnaire (GHSAQ), the Life Orientation Test-Revised (GrLOT-R) [16], the dietary habits and lifestyle questionnaire and demographic data. There were 565 questionnaires distributed to adults of the general population in Lakonia, Greece, of which 500 were returned completed (response rate 88.5%).

The qualitative study consisted of semi-structured interviews and focus group discussions (FGDs). Before conducting face-to-face interviews, two personal interviews were piloted in order to check the semi-structured questions guide and any difficulties. The process followed for interviews included: (a) thematize the study and design the interview guide following the interviews, (b) transcription, (c) analyzing process and finally and (d) verification of findings. Six (6) women and four (4) men were participated. Regarding Focus Group Discussions (FGDs), three groups were conducted, each of four individuals. Time and location of FGDs were decided jointly. Participants were placed in a circle in order to have eye contact and better interaction. A 10-point questions guide was used based on the study's hypotheses. Sample was selected randomly from the quantitative study's participants and all interviews were recorded both in notes and tape with the consent and verification of the interviewees.

2.1 Data Analysis

Absolute and relevant frequencies for qualitative variables, mean and standard deviation for quantitative variables are presented. Pearson's correlation coefficient was analyzed, t-test was used on two independent samples and one-way analysis of variance was applied. Welch t-test was applied in the cases of deviation from the equality of variances hypothesis. In cases where normality was not possible, Mann-Whitney test and Kruskal-Wallis test were applied. Exploratory factor analysis through Varimax rotation and multivariate analysis regression was performed in order to aggregate dietary habits and physical exercise. Multivariate regression analysis was used to determine the factors affecting the GrLot-R score. The statistical analysis was conducted through IBM SPSS Statistics 22.0 (SPSS, Chicago, IL, USA) and all results were considered statistically significant at $p < 0.05$.

Qualitative data were analyzed using a content analysis method. The information was coded following themes and the analysis was carried out by the researchers. More specifically, the researchers read the notes, arranged the content following the topic, coded, analyzed into themes, and presented integrated with quantitative data. Finally, all participants reviewed and verified their responses, both recorded and written ones.

3 Results

3.1 Quantitative Findings

The socio-demographic characteristics of the 500 individuals who participated in the study are presented in Table 1.

Life Orientation Test-Revised and General Health Self-Assessment Questionnaire Study

The total score of GrLot-R was equal to 20.47 (±4.017) and it was affected by marital status ($p = 0.005$), working status ($p = 0.004$) and education ($p < 0.001$). According to the mean values, married people, freelancers and post-graduate degree holders scored higher on the optimism scale. Moreover, GrLot-R score correlated negatively statistically significant with participant's age ($r = -0.155$; $p = 0.001$).

The level of optimism correlated statistically significant, although weak, with water consumption ($r = 0.092$; $p = 0.039$) as well as with the weekly frequency of various food consumption (Table 2). According to the mean values, individuals who consumed at least four times a week the above foods, achieved higher score on the optimism scale. The same is observed at individuals who consumed two to

Table 1 Socio-demographic characteristics

		N (%)
Gender	Male	226 (45.2)
	Female	274 (54.8)
Age		35.27 (11.850)[a]
Marital status	Single	261 (52.2)
	Married	197 (39.4)
	Cohabitation	18 (3.6)
	Separated	3 (0.6)
	Divorced	16 (3.2)
	Widowed	5 (1.0)
Education	Primary education	13 (2.6)
	Lower secondary education	33 (6.6)
	Higher secondary education	107 (21.4)
	Undergraduate student	64 (12.8)
	University graduate	200 (40.4)
	Postgraduate studies	83 (16.6)
Work status	Unemployed	112 (22.4)
	State employee	174 (34.8)
	Private employee	87 (17.4)
	Freelancer	46 (9.2)
	Farmer	29 (5.8)
	Retiree	9 (1.8)
	Other	43 (8.6)

[a]Mean (±SD)

Table 2 Statistically significant associations between weekly dietary habits and GrLOT-R score

How many times a week do you consume	Consumption frequency At most once			2–4 times			>4 times			
	N	Mean	SD	N	Mean	SD	N	Mean	SD	p-value
Fruits	98	19.19	4.268	190	20.18	3.832	212	21.33	3.880	<0.001[a]
Salads	45	18.62	4.716	159	19.58	3.747	296	21.23	3.864	<0.001[a]
High-fibre cereals	278	20.08	4.066	109	20.72	4.105	113	21.20	3.708	0.033[a]
Canned products	464	20.59	3.973	–	–	–	–	–	–	0.041[a]
Lettuce	143	19.66	4.065	262	20.60	3.981	95	21.34	3.855	0.005[a]
Pepper	256	19.87	4.042	190	21.11	3.916	54	21.09	3.877	0.003[a]
Tomato	84	19.67	3.551	239	20.03	4.235	177	21.46	3.740	<0.001[a]
Apple	155	20.04	4.391	230	20.37	3.648	115	21.27	4.113	0.038[a]
Watermelon	172	19.71	3.991	214	20.75	4.041	114	21.10	3.865	0.007[a]
Carrot	187	19.92	3.887	231	20.60	4.235	82	21.38	3.491	0.019[a]
Apricot	230	19.77	4.145	198	20.98	3.705	72	21.33	4.097	0.001[a]
Orange	77	19.55	4.194	197	19.74	3.906	226	21.42	3.857	<0.001[a]
Mandarin	163	19.58	4.207	196	20.65	3.956	141	21.25	3.694	0.001[a]
Yellow—Orange pepper	374	20.18	3.964	99	21.70	4.009	27	19.96	4.052	0.046[b]
Pineapple	464	20.54	3.947	24	21.08	4.149	12	16.75	4.901	0.046[b]
Dairy products	16	18.75	4.074	129	20.24	3.844	355	20.63	4.064	0.001[b]
Raw olive	61	18.75	3.837	134	20.13	3.687	305	20.96	4.092	<0.001[a]

[a] ANOVA
[b] Kruskal-Wallis

four times a week green, red, and yellow pepper, pineapple and once a week at most canned food.

The analyses showed a moderate positive correlation between GHSAQ and GrLot-R score ($r = 0.388$, $p < 0.001$).

Exploratory Factor Analysis of Dietary Habits

By applying factor analysis with Varimax rotation, 12 statistically significant factors emerged which describe participants' food preference/consumption and interpret 60.28% of the total dispersion. The foods comprising each factor and the percent of variance that each factor interprets are presented in Table 3. The factors are consisted of food groups that contain a common nutrient [17].

Multivariate Analysis

Multivariate analysis showed that GrLot-R score is significantly predicted by the factors "Vitamin K & A" ($\beta = 0.885$, $t(48) = 5.125$, $p < 0.001$) "Vitamin C" ($\beta = 0.538$, $t(48) = 3.117$, $p = 0.002$) and "Preservatives," ($\beta = -0.533$, $t(48) = -3.089$, $p = 0.002$). These factors interpret a significant variance rate of GrLot-R score ($R^2 = 0.079$, $F(3.496) = 15.173$, $p < 0.001$).

Table 3 The factors describing participants' food preference/consumption

Factors	Variable	% of variance interpreted
Flavonoids	Pepper, red cabbage, carrot, yellow-orange peppers, corn	8.210
Vitamin C	Apple, watermelon, pomegranate, cherries, apricots, oranges, mandarins	7.845
Vitamin K & A	Fruits, salads, lettuce, carrot, raw olive oil	6.920
Iron	Spinach, vegetables cooked with olive oil, legumes, fish	5.639
Polyphenols	Green tea, dried fruits, nuts, whole grains	5.250
Saturated fat	Junk food, red meat, fried food	4.894
Preservatives	Canned food, pineapple, dairy	4.042
Folate and plant lignans	Spinach, broccoli, brown bread	3.939
Pectin	Apple, brown rice	3.754
Carbohydrates and cholesterol	Sweets, egg, pasta	3.404
Alcohol	Alcohol	2.949

3.2 *Qualitative Findings*

3.2.1 Interview Results

Participants ranged in age from 26 to 50, with a mean age of 39.8 (±7.177). A majority of the participants were female (60%).

Optimism and Development Factors

Regarding the factors that affect the level of optimism, participants' opinions varied. The main factor mentioned by the majority of participants (80%) was the effect (positive/negative) of the family environment in which a person is raised. Half of the participants (50%) mentioned stress as a deterrent whereas their opinions did not coincide on the issue of the negative effect of financial status.

> Difficulties in financial issues, work and livelihood make me feel pessimistic.

Four participants (40%) reported that the feeling of companionship, love and communication as well as having children are considered substantial factors of positive effect on optimism whereas loneliness is a negative one. Of total interviewed, 50% reported weather, culture, faith in God and mass media were mentioned as important factors for optimism/pessimism's development.

> When I feel I have people by my side who support, help and love me I am very optimistic.
> My marriage, my wife and my children strengthen my optimism.

Only one (10%) participant noticed that when his/her diet included specific food such as fruits and vegetables his/her level of optimism was affected positively. A significant finding is the ignorance and lack of "self-observation" of the effect that dietary and smoking habits (by the 90% and 100% of participants respectively) have on mental health and specifically on the level of optimism, while it was observed a positive correlation between the level of optimism with physical exercise and duration/quality of sleeping by 40% of interviewed.

> My mood is better and I feel optimistic when I eat healthy ... also I feel good when I consume food that I consider tasty.

Association Between Health Self-Assessment and Optimism

The association and interaction between health and optimism and generally mood was observable and familiar to all interviewees (100%). The effect of optimistic and positive feelings regarding the significance and expression of symptoms, the development of chronic diseases and the incidence of psychosomatic symptoms was estimated very important.

> Poor health affects negatively my optimism and being pessimistic affects my body. I suppose these are interconnected.

3.2.2 Focus Groups Discussions

Participants ranged in age from 25 to 54, with a mean age of 39.91 ($\pm$10.448) and 50% were male.

Optimism and Development Factors

Regarding the level of optimism, it emerged that the majority of the participants (83.3%) considered themselves as optimistic. Through discussions participants supported that factors which affect their level of optimism are: health condition, the adequate duration and quality of sleeping, family background in which someone is raised, feeling satisfaction and love in their marriage, having communication in marriage, the possible genetic origin of optimism, mass media, culture, stress and weather.

> Family environment affects children's optimism. For instance, I notice that I have adopted behaviors and mentality from my parents.

Moreover, there was a correlation between the lack of observing the effect of water consumption and smoking had on the level of optimism whereas the opposite happened for physical exercise.

> When I exercise my optimism is increased and my mood is better, it cheers me up. Exercise and, as far as I'm concerned, hiking is optimism, life, liveliness.

Finally, it was not observed a realization regarding the effect of diet both on optimism and mood whereas they agreed that consumption of unhealthy food made them feel regrets.

> The anticipation of a good meal, a meal that you like, makes you happy and cheers you up. Even when it is junk food. Of course you regret it later.

Optimism and Health

There was unanimity that positive mood and optimism affect positively health and vice versa.

> I am very optimistic and it affects my health. At times that I might be less optimistic, I've noticed I have headaches and my back hurts. Moreover, I think that any physical problem, I might face, gets worse if I don't feel optimistic.

4 Discussion

The present study was the first attempt to investigate the association between optimism, general health self-assessment, dietary habits and lifestyle in Greece.

According to the findings, the mean score of GrLot-R was up to 20.47 (±4.017) which categorizes the participants as very optimists. This result is consistent with Roy's et al. [10] multinational study on general population, where the mean was 19.9, but is in opposition to the results of other Greek studies like Lyrakos et al. [16] (on a sample of nursing staff) and Tsakogia et al. [18] (on a sample of patients) where the mean was 14.4 and 15.5, respectively. The above differentiation occurs probably due to sample differences e.g. differences regarding the job, the geographical location the study was conducted, whether it concerns patients or general population etc.

Association Between Optimism and Demographic Data

The study showed that the higher the age the less optimism the participants felt. This result is consistent with Lachman's et al. [19] study and it possibly associates with the gradual realization of reaching the end of life cycle and the negative feelings of old age [20]. On the contrary, Isaacowitz's [21] study in USA shows that older people interpreted their lives' events from a more optimistic perspective compared to young people. The fact that the results regarding the age-optimism correlation do not coincide, points out that this particular correlation is still quite debatable.

As far as the working status is concerned, the results showed that freelancers scored higher at GrLot-R, which is in accordance with Blanchflower and Oswald's [22] results where "self-employed might be inherently more optimistic and happy than others." In addition, the above result might be associated with the fact that freelancers feel more satisfied in their lives than employees [23].

With regard to marital status, it was found that married people scored higher at GrLot-R which is consistent with Coll and Draves' [24] study and probably is associated with Carr's [25] view "Couple formation and marriage leads to the development of a series of important relationships: the marital relationship, the kinship relationships, and later, the parent-child relationships, all contribute to happiness and well-being" (p. 287). The qualitative study revealed that it was not marriage itself that affected positively the participants' optimism but the existence of satisfaction, love and communication within it. Furthermore, it showed that people with kids felt they had a highly important purpose in their lives which contributed positively to their optimism.

Correlation Between Optimism and Diet

According to the results of this study the highest score at GrLot-R correlated positively with the consumption of tomato, lettuce, apricot, apple, watermelon, pineapple, oranges, tangerines and raw olive oil. A possible interpretation of these results could be the positive effect of antioxidants –especially those of carotenoids-on the level of optimism as it was observed at Boehm's et al. [26] study, while at Kelloniemi's et al. [13] study in Finland showed that a diet rich in fruits and vegetables is more frequent among optimists than pessimists. Moreover, the above foods are rich in vitamin C which reduces stress, "mental" and "psychological" fatigue and improves mood [27], while raw olive oil has an antioxidant effect due to biophenol, which affects positively the mood [28].

In addition, individuals who consumed cereals and dairy products more than four times a week scored higher at GrLot-R. This might be associated with the fact that deficiency of magnesium, where the above foods are significant source of it, is associated with depressive disorders and other mental health problems like irritability, sadness and stress [29]. The same association was observed with the consumption of water, where people who consumed more water on daily basis were more optimists. This can happen due to the negative effect of mild dehydration on psychosomatic functionality [30].

The multivariate analysis showed that the GrLot-R score is predicted statistically significant by the factors "Vitamin K & A" and "Vitamin C," which is in accordance with the aforementioned studies [13, 26, 27, 31]. Regarding the factor "Preservatives," which as it came out from the analysis affected negatively the above score, could be linked with bisphenol A (BPA), which is widely spread in plastic containers and canned foods. BPA exposure has been adversely associated with some sociosexual behaviors and anxiety at relatively low doses [32].

The qualitative study showed that individuals had not observed any effect of specific foods consumption on their optimism but it was found that the adoption of healthy diet contributes generally in the creation of positive emotions. In addition, as the participants reported, the consumption of junk food made them feel a temporary satisfaction at the beginning but afterwards felt guilty and regrets. The relation of junk food and satisfaction could be associated with the fact that consumption of food rich in carbohydrates and sugar leads to insulin release which in turn causes serotonin production that contributes to a temporary satisfaction, pleasant mood and relaxation [33].

Correlation Between General Health Self-Assessment and Well-Being and Between Physical Exercise and Optimism

In regard to the positive correlation of the total score on GrLot-R and total score on GHSAQ, from the present study it was shown that optimists have a more positive perspective regarding their health, they feel that their organism is more resistant to infections and viruses, they use natural ways that strengthen their organism and they don't feel often fatigue. This result is consistent with Roy's et al. [10] study, conducted on individuals from 45 to 84 years old, where optimists appeared to have lower concentrations of inflammatory markers than pessimists. At Maruta's et al. [5] study on 723 patients, was found that pessimism is correlated significantly with mortality on a rate of 19% which highlights the determinant effect of optimism.

From the qualitative study emerged that the level of optimism was affected by physical exercise and especially the type of physical exercise. This finding could be interpreted by applying the Self-Determination Theory and the intrinsic-extrinsic motivation in physical exercise, according to which every person chooses to exercise based on his/her personal needs (social, psychological etc.) that wants to fill [34].

Other Factors Affecting Optimism

One of the most important findings of the qualitative study was that the majority of the participants considered that the family background and the environment that someone is raised and developed, is one of the most significant factors that develops optimism and that optimist parents raise optimist children. This finding is in line with the theories and studies regarding the association of parents' and children's optimism. It has been suggested that children's behavior is harmonized with the environment they grow up, they copy and adopt their parents' attitude towards life as well as the way they interpret it [35].

A factor that was correlated negatively with participants' optimism in the present qualitative study was stress. This result coincides with Dewberry and Richardson's [11] study in which stressed individuals were less optimistic than the non-stressed ones. The results of Zenger's et al. [36] study, performed on 427 patients, were similar, according to which the highest score on Life Orientation Test correlated reversely with stress.

Finally, some additional factors were reported from participants during the qualitative study, which affected positively their level of optimism such as the adequate duration and quality of sleeping, "nice" weather and faith in God, whereas media mass correlated negatively, results that have been observed at previous studies [35, 37–39].

5 Conclusion

Obviously, the issue of health and optimism is multidimensional and as it seemed from the study's results is affected from both intrinsic factors like diet and extrinsic like family. It seems that a diet rich in vitamin C, K and A, and with less preservatives affects positively optimism. In addition, from the qualitative study it emerged that participants had not realized any effect of their dietary habits on optimism although in the quantitative study there was a statistically significant correlation. Also, the fact that the participants of the qualitative study confused healthy dietary habits with diet and consciously consumed unhealthy food for temporary pleasure, leads to the conclusion that there is a lack of awareness and education in these specific areas.

It makes necessary and imperative the need of applying specially designed programs, which will be addressed to all ages and will run preventive, educationally and supportively for the purpose of reducing risk factors regarding physical and mental health.

References

1. Fardet, A., and Y. Boirie. 2014. Associations Between Food and Beverage Groups and Major Diet-Related Chronic Diseases: An Exhaustive Review of Pooled/Meta-Analyses and Systematic Reviews. *Nutrition Reviews* 72 (12): 741–762.
2. Jacka, F.N., A. Mykletun, and M. Berk. 2012. Moving Towards a Population Health Approach to the Primary Prevention of Common Mental Disorders. *BMC Medicine* 14: 149.
3. Lakhan, S.E., and K.F. Vieira. 2008. Nutritional Therapies for Mental Disorders. *Nutrition Journal* 7: 2.
4. Scheier, M.F., and C.S. Carver. 1993. On the Power of Positive Thinking: The Benefits of Being Optimistic. *Current Directions in Psychological Science* 2: 26.
5. Maruta, T., R. Colligan, M. Malinchoc, and K. Offord. 2000. Optimists vs Pessimist: Survival Rate Among Medical Patients Over a 30-Year Period. *Mayo Clinic Proceedings* 75 (2): 140–143.
6. Giltay, E.J., M.H. Kamphuis, S. Kalmijn, F.G. Zitman, and D. Kromhout. 2006. Dispositional Optimism and the Risk of Cardiovascular Death: The Zutphen Elderly Study. *Archives of Internal Medicine* 166: 431–436.
7. Matthews, K.A., K. Räikkönen, K. Sutton-Tyrrell, and L.H. Kuller. 2004. Optimistic Attitudes Protect Against Progression of Carotid Atherosclerosis in Healthy Middle-Aged Women. *Psychosomatic Medicine* 66: 640–644.
8. Ironson, G., E. Balbin, R. Stuetzle, M.A. Fletcher, C. O'Cleirigh, et al. 2005. Dispositional Optimism and the Mechanisms by Which It Predicts Slower Disease Progression in HIV: Proactive Behavior, Avoidant Coping, and Depression. *International Journal of Behavioral Medicine* 12 (2): 86–97.
9. Geers, A.L., J.A. Wellman, S.G. Helfer, S.L. Fowler, and C.R. France. 2008. Dispositional Optimism and Thoughts of Well-Being Determine Sensitivity to an Experimental Pain Task. *Annals of Behavioral Medicine* 36: 304–313.
10. Roy, B., A.V. Diez-Roux, T. Seeman, N. Ranjit, S. Shea, et al. 2010. Association of Optimism and Pessimism with Inflammation and Hemostasis in the Multi-Ethnic Study of Atherosclerosis (MESA). *Psychosomatic Medicine* 72 (2): 134–140.
11. Dewberry, C., and Richardson. 1990. Effect of Anxiety on Optimism. *The Journal of Social Psychology* 130 (6): 731–738.
12. Boehm, J.K., D.R. Williams, E.B. Rimm, C. Ryff, and L.D. Kubzansky. 2013b. Relation Between Optimism and Lipids in Midlife. *The American Journal of Cardiology* 111 (10): 1425–1431.
13. Kelloniemi, H., E. Ek, and J. Laitinen. 2005. Optimism, Dietary Habits, Body Mass Index and Smoking Among Young Finnish Adults. *Apetite* 45 (2): 169–176.
14. Creswell, J.W., and V.L. Plano Clark. 2010. *Designing and Conducting Mixed Methods Research*. 2nd ed. Thousand Oaks: SAGE Publications.
15. Kargakou, A., A. Sachlas, N.G. Lyrakos, S. Zyga, M. Tsironi, et al. 2015. Association Between Optimism, Dietary Habits, Lifestyle and General Health Self-Assessment: A Pilot Study. *International Journal of Health and Psychology Research* 1 (3): 13–28.

16. Lyrakos, G.N., D. Damigos, V. Mavreas, G. Kostopanagiotou, and I. Dimoliatis. 2010. A Translation and Validation Study of the Life Orientation Test Revised in the Greek Speaking Population of Nurses Among Three Hospitals in Athens and Ioannina. *Social Indicators Research* 95 (1): 129–142.
17. USDA. 2015. National Nutrient Database for Standard Reference Release. Available at: http://ndb.nal.usda.gov/ndb/search. Accessed 08 Jan 2016).
18. Tsakogia, Z., G.N. Lyrakos, D. Damigos, V. Mayreas, and I.D.K. Dimoliatis. 2010. The Effect of Dispositional Optimism in HRQOL in Patients with Chronic Musculoskeletal Pain Conditions in Greece. *ISQOLS* 6 (1): 53–70.
19. Lachman, M.E., C. Rocke, C. Rosnick, and C.D. Ryff. 2008. Realism and Illusion in Americans' Temporal Views of Their Life Satisfaction. *Psychological Science* 19: 889–897.
20. Warnick, J. 1995. *Listening with Different Ears: Counseling People Over Sixty*. Ft. Bragg CA: QED Press.
21. Isaacowitz, D. 2005. Correlates of Well-Being in Adulthood and Old Age: A Tale of Two Optimisms. *Journal of Research in Personality* 39: 224–244.
22. Blanchflower, D.G., and A.J. Oswald. 1998. What Makes an Entrepreneur? *Journal of Labor Economics* 16 (1): 26–60.
23. Binder, M., and A. Coad. 2013. Life Satisfaction and Self-Employment: A Matching Approach. *Small Business Economics* 40 (4): 1009–1033.
24. Coll, J.E., and P.R. Draves. 2008. An Examination of the Relationship Between Optimism and Worldview Among University Students. *College Student Journal* 42 (2): 395–401.
25. Carr, A. 2004. *Positive Psychology: The Science of Happiness and Human Strengths*. Hove, East Sussex: Brunner-Routledge.
26. Boehm, J.K., D.R. Williams, E.B. Rimm, C. Ryff, and L.D. Kubzansky. 2013a. Association Between Optimism and Serum Antioxidants in the Midlife in the United States Study. *Psychosomatic Medicine* 75 (1): 2–10.
27. Kennedy, D., R. Veasey, A. Watson, F. Dodd, E. Jones, et al. 2010. Effects of High – Dose B Vitamin Complex with Vitamin C and Minerals on Subjective Mood and Performance in Healthy Males. *Psychopharmacology* 211 (1): 55–68.
28. Saija, A., and N. Uccella. 2000. Olive Biophenols: Functional Effects on Human Wellbeing. *Trends in Food Science & Technology* 11 (9–10): 357–363.
29. Eby, G.A., and K.L. Eby. 2006. Rapid Recovery from Major Depression Using Magnesium Treatment. *Medical Hypotheses* 67 (2): 362–370.
30. Armstrong, L.E., M.S. Ganio, D.J. Casa, E.C. Lee, B.P. McDermott, et al. 2012. Mild Dehydration Affects Mood in Healthy Young Women. *The Journal of Nutrition* 142 (2): 382–388.
31. Rebec, V., and C. Pierce. 1994. A Vitamin as Neuromodulator: Ascorbate Release Into the Extracellular Fluid of the Brain Regulates Dopaminergic and Glutamatergic Transmission. *Progress in Neurobiology* 43: 537–565.
32. Watson, C.S., R.A. Alyea, A. Kathryn, K.A. Cunningham, and Y.J. Jeng. 2010. Estrogens of Multiple Classes and Their Role in Mental Health Disease Mechanisms. *International Journal of Women's Health* 2: 153–166.
33. Wurtman, J., and N.T. Frusztajer. 2006. *The Serotonin Power Diet: Use Your Brain's Natural Chemistry to Cut Cravings Curb Emotional Overeating, and Lose Weight*. New York: Rodale Inc.
34. Hagger, M., and N. Chatzisarantis. 2007. *Intrinsic Motivation and Self-Determination in Exercise and Sport*. Champaign, IL: Human Kinetics.
35. Peterson, C., and T.A. Steen. 2002. Optimistic Explanatory Style. In *Handbook of Positive Psychology*, ed. R.C. Snyder and J.S. Lopez, 248–250. New York: Oxford University Press.
36. Zenger, M., C. Brix, J. Borowski, J.U. Stolzenburg, and A. Hinz. 2010. The Impact of Optimism on Anxiety, Depression and Quality of Life in Urogenital Cancer Patients. *Psycho-Oncology* 19 (8): 879–886.

37. Lee, D.E., E.W. Neblett Jr., and V. Jackson. 2014. The Role of Optimism and Religious Involvement in the Association Between Race-Related Stress and Anxiety Symptomatology. *Journal of Black Psychology* 41 (3): 221–246.
38. Lemola, S., K. Räikkönen, V. Gomez, and M. Allemand. 2013. Optimism and Self-Esteem Are Related to Sleep. Results from a Large Community-Based Sample. *International Journal of Behavioral Medicine* 20 (4): 567–571.
39. Prodan, I. 2013. *The Effect of Weather on Stock Returns: A Comparison Between Emerging and Developed Markets*. Hamburg: Anchor Academic Publishing.

Moderating Nutritious Habits in Psychiatric Patients Using Transtheoretical Model of Change and Counseling

Konstantina Anastopoulou, Evangelos C. Fradelos, Evdokia Misouridou, Michael Kourakos, Aristea Berk, Ioanna V. Papathanasiou, Christos Kleisiaris, and Sofia Zyga

Abstract Motivational Interviewing provides the opportunity to health professionals to have an effective strategy to increase the level of readiness to change health behaviors. Along with the Transtheoretical Model (Stages of Change Model) compose the theoretical base of intervention in psychiatry settings. Objective: This study was aimed to change nutritious behavior of psychiatric patients using a specific Model of Change and Counseling implementing a health education program. Methodology: A quasi-experimental design was adopted on a random sample of 60 psychiatric patients at Military Hospital of Athens. Patients were

K. Anastopoulou
Naval Hospital of Athens, Athens, Greece

E.C. Fradelos (✉)
2nd Psychiatric Department, State Mental Hospital of Attica "Daphne", Athens, Greece

Nursing Department, Faculty of Human Movement and Quality of Life, University of Peloponnese, Tripoli, Greece
e-mail: evagelosfradelos@hotmail.com

E. Misouridou
Nursing Department, Technological Educational Institute of Athens, Athens, Greece

M. Kourakos
General Hospital, Asklipieion Voula, Athens, Greece

A. Berk
Daycare Facility, "Eginition" Hospital, First Psychiatric Clinic, University of Athens, Athens, Greece
e-mail: ariaberk@gmail.com

I.V. Papathanasiou
Nursing Department, Technological Educational Institute of Thessaly, Thessaly, Greece

C. Kleisiaris
Nursing Department, Technological Educational Institute of Crete, Heraklion, Greece

S. Zyga
Department of Nursing, Faculty of Human Movement and Quality of Life Sciences, University of Peloponnese, Sparta, Greece
e-mail: zygas@uop.gr

P. Vlamos (ed.), *GeNeDis 2016*, Advances in Experimental Medicine and Biology 988, DOI 10.1007/978-3-319-56246-9_5

divided into two groups as follows; (a) Intervention Group (four sessions of counseling and encouraging motivation for modification of their nutritious habits), and (b) Control Group (simple information sessions about the principles of healthy alimentation). Results: The mean age of Intervention Group (IG) was 43.9 ± 9.5 and Control Group (CG) 46.1 ± 9.1, ranging from 40 to 55 years old. Also, 26.7% of the participants were female, 23.3% were married and, 10% divorced. Our analyses showed that IG patients were significantly loss weight post-intervention compared to CG patients. Specifically, IG patients were significantly moderated the intake of starchy foods in every meal ($p < 0.001$) and the intake of fruits and vegetables ($p < 0.001$). Similarly, IG patients were moderated the intake of low fat dairy foods while they changed the full fat dairy foods with low fat ($p < 0.001$). Also important, IG patients showed significant enhance (80%) regarding drugs compliance, suggesting that 34% of the CG patients often forgot to take their medication. Finally, IG patients reported a positive attitude towards moderating unhealthy nutritious behaviors ($p = 0.032$). Conclusions: Our results confirms that health educational and promotional Interventions may change behavior of psychiatric patients and thus may positively influence their nutritious habits.

Keywords Health education program • Nutritious habits • Psychiatric patients • Ttranstheoretical model of change • Motivational interviewing • Moderating

1 Introduction

Motivational Interviewing is a therapeutic approach which incorporates the principles of therapeutic relationship in Carl Rogers's humanistic therapy (1951) and the more active cognitive-behavior techniques adjusted at individual's stage of change [1]. It is settled as human centered method that raises individual's motive for change, by expressing and resolving the ambivalence for adaptation and maintenance healthy behaviors [2]. Motivational Interviewing are short interventions (5–60 min and a frequency of 1–5 sessions), and mainly used to raise patients' knowledge about health matters to think and adapt changes improving their health [3]. On the other hand, Transtheoretical Model is an integrative model of behavior change [4]. Behavior change was often construed as an event, such as quitting smoking, drinking, or over-eating. The Transtheoretical Model construes change as a process involving progress through a series of five stages. In particular, Precontemplation is the stage in which people are not intending to take action in the foreseeable future usually measured at the next 6 months [5]. Contemplation constitutes the stage in which people are intending to change in the next six months and, Preparation is the stage in which people are intending to take action in the immediate future, usually measured as the next month [6].

It is widely recognized that comorbid diseases such as cardiovascular diseases and metabolic syndrome are more prevalent among patients with chronic mental health disorders especially in ageing. It has also been suggested that obesity is the common and major health problem in several psychiatric disorders such as major depression, bipolar disorder and schizophrenia [7] and, has been associated with the use of psychotropic, antipsychotic and antidepressant medicine that raises body weight and thus metabolic disorder [8, 9]. However, excluding obesity, there are many other non-modified risk factors such as hypertension, smoking, diabetes, hyperlipidemia, lack of exercise and unhealthy eating habits and that can be modified improving psychiatric patients' quality of life. Therefore, health education programs regarding Nutritious Behavior adopting the Transtheoretical Model of Change and Counseling to these patients are considered appropriate as well as helps individuals to make decisions for adapting behaviors and acting in accordance with their health needs [10].

Consequently, this study was aimed to change behavior of psychiatric patients moderating their nutritious habits. Our secondary aims were the evaluation of patients' classification at stage (Stages of Change) and the evaluation of patients' satisfaction using motivational interviewing, concerning the effectiveness of the health education program.

2 Material and Methods

2.1 Study Design

In this comparative study a quasi-experimental design was adopted on a random sample of 60 psychiatric hospitalized patients at Military (Naval) Hospital of Athens. Particularly, we studied already diagnosed (DSM criteria) patients with bipolar disorders, anxiety, panic attack disorders and schizophrenia. All included patients were asked to participate in health education program and were wiling participated after their informed consent. Participants were accidentally divided into two groups as follows; (a) Intervention Group (four sessions of counseling and encouraging motivation for modification of their nutritious habits), and (b) Control Group (simple information sessions about the principles of healthy alimentation). Excluded only patients that already were participated in similar health educational program at the same time-period. Demographic characteristics such as age, educational level and socioeconomic characteristics were recorded.

2.2 Study Instruments

Health Questionnaire It is self-administrative questionnaire that includes questions concerning personal and family status, medical conditions and, health behaviors or habits such as smoking, alcohol consumption, and alimentation, taken medicines.

Questionnaire of Stages of Change It is constituted by 6 sections and classifies patients in stages of change as follows; precontemplation, contemplation, preparation, action and maintenance of healthy nutritious habits [11].

The Change Questionnaire It is consisted of 12 questions in six categories related to motivation for change: desire, ability, important reasons, need, commitment and taking action for behavior modification [12].

Short Form-12 It is include 12 questions concerning physical and mental Health-Related Quality of Life (HRQoL).

Questionnaire of Process Evaluation Evaluates the post-intervention process and the effectiveness of the health education program [13].

2.3 Study Implementation

The study has been held in three phases:

Preliminary Phase Evaluation of the health parameters and health behaviors using the questionnaires that filled-out at their visit and follow-up at Liaison Psychiatry department. The CG patients were received a short telephone communication (5–10 min) that they were informed about the results and the evaluation of the questionnaires and also about the choices of therapeutic interventions as well as the benefits from the modification of their nutritious behavior.

Intervention Phase At this phase IG patients' behavior profile was evaluated (what they consume, how many meals per day, how much salt, etc.) including patients' stage of change into four personal sessions of 45 duration. These sessions have been achieved (a) in first contact after the evaluation, (b) the 2nd week (c) the 4th week (d) 6th week. The consultant was communicating with the participants once per week between the sessions and with patients' voluntarily consent.

Evaluation Phase IG and CG patients were asked to complete both; the health and the behavior change questionnaire once again.

3 Results

Demographic characteristic are presented in Table 1. Specifically, the mean age of IG patients and CG patients was 43.9 ± 9.5 and 46.1 ± 9, respectively. The majority of IG patients were male (73.3%), 23% pensioners, 50% have University degree and, 63% were married. On the other hand, 40% of the CG patients were male, 16% pensioners, 60% have University degree and, 73% were married.

The distribution of the patients according to the stages of change and the transtheoretical model before and after the intervention is presented in Table 2.

Table 1 Demographic characteristic of the study participants (Intervention group—Control group)

		Intervention group		Control group	
		Mean ± Standard deviation	*Minimum–Maximum*	Mean ± Standard deviation	Minimum–Maximum
Age		*43.9 ± 9.5*	*20–58*	46.1 ± 9.1	22–58
Weight		*81.2 ± 14.1*	*59–115*	83.6 ± 8.3	66–100
Gender		*Frequency*	*Percentage (%)*	Frequency	Percentage (%)
	Male	22	73.3	12	40.0
	Female	8	26.7	18	60.0
Profession	Pensioner	7	23.3	5	16.6
	Public sector	16	53.3	17	56.6
	Private sectors	2	06.7	4	13.3
	Free lancer	3	10.0	4	13.3
	House hold	2	06.7	0	00.0
Educational status	Primary school	1	03.3	0	00.0
	High school	14	46.7	12	40.0
	University	15	50.0	18	60.0
Marital status	Single	7	23.3	0	00.0
	Married	19	63.3	22	73.3
	Divorced	3	10.0	8	26.6
	Widowed	1	03.3	0	00.0
	Total	30	100.0	30	100.0

Table 2 Distribution of the patients according to the stages of change(Intervention group—Control group)

	Pre intervention Number of patients	n	Post intervention Number of patients	n
	Pre-contemplation	13	Contemplation	5
Intervention Group	Contemplation	17	Preparation	3
			Action	22
	Total	30	Total	30
Control Group	Pre-contemplation	14	Pre-contemplation	7
	Contemplation	16	Contemplation	10
			Preparation	3
			Action	10
	Total	30	Total	30

We found that Intervention Group had a positive attitude in moderating unhealthy nutritious behaviors ($p = 0.032$) compared to Control Group, suggesting that IG patients were positively increased their health behaviors. Also, IG patients moderated the intake of starchy foods in every meal ($p = 0.001$) and the intake of fruits and vegetables ($p = 0.001$). Similarly they moderated the intake of low fat dairy foods while they changed the full fat dairy foods with low fat ($p < 0.001$). Moreover, IG patients were significantly decreased alcohol consumption ($p = 0.017$) and thus, improved their "bad" habits that also positively were related to their HRQoL ($p = 0.019$) compared to CG patients. Also important, 34% of the control group patients forgot to take their medicines in comparison to the intervention-group patients that improved their compliance (80%).

Furthermore, 30% of the participants were reported that modifying their behavior as helpful and 67.7% find this intervention program real helpful. In addition, 20% of both patients groups; helped them to improve quite their daily living, 53.3% adequate and 26.7% more adequate. The majority of the patients (96.7%) would recommend this program to a friend and, 80% of them evaluated their decision to participate to this program as positive. Moreover, 66.7% of them stated that this program fulfilled their expectations quite, 16.7% adequate and 16.7% more adequate.

4 Discussion

In this study we modified the nutritional habits of psychiatric patients and evaluated the effectiveness of health interventions conducting a health education program. It is widely accepted that psychiatric patients tend to have an unhealthy diet, rich in saturated fats [14]. Also, unhealthy nutritional habits, overweight and obesity are prevalent in psychiatric patients and constitute risk factors for multiple chronic and life threatening conditions [15].

The main finding of the present study was that the Motivational interviewing model was resulted as an effective intervention on the modification of nutritional habits. This finding is in line with the results of similar studies as well as this model-intervention was adequately effective in various patient populations [16–18]. Moreover, the effectiveness of this intervention was also confirmed by numerous studies' outcomes concerning psychiatric populations that had participated in similar interventions program using motivational interviewing [19–21]. Similarly, the effectiveness of motivational interviewing for behavior modification has been demonstrated in the international literature [22, 23]. However, the figures among studies are conflicted due to a different methodological approaching and thus, the effectiveness of motivational interviewing for nutritious behavior modification is considerably biased [24].

We also found change on stage, suggesting that classification in stages was differ according to expectations and suggestions of the Transtheoretical Model—patients moved from one stage to next, after the completion of the intervention. This is in agreement with the results of the Rollnick and Cohen's [25] study that patients were classified in stage readiness closer to the achievement of modification 6 months after the intervention. In contrary, Colby, Mont and Barnett [26], found no significant differences between intervention and control group in relation to patients' movement from one stage to another at a three month follow-up. However, this finding may be attributed to the fact that this intervention included only one session. In the view of that, it may be hypothesized that more sessions are necessary in order to motivate patient's movement from one stage to another. Importantly, patients that were classified in a stage more close to behavior modification (stage of preparation), the modification of their nutritious habits was much easier. This awareness is reinforced by Steinberg, Ziedonis, Krejci study [27] in which motivational interviewing was found to be more effective in patients that mentioned small or no intention to moderate their behavior.

The lengthening of the duration of the intervention and the possibility of giving a larger number of sessions has been considered from the majority of the patients (90%) as a way of supporting them in achieving their goals in relation to moderate unhealthy food consumption. Additionally, it is positive that patients had moved on into the stages of change and managed to reduce fat and starchy food consumption after the completion of the intervention. Remarkably, this finding supports the use of motivational interviewing as an appropriate model in behavior modification as well as our control group have received only a short telephone communication. Also important, intervention-group has managed to lose weight (81.2 ± 14.1 kg: Pre intervention and 78.8 ± 13.3 kg post-intervention).

Also important, intervention group patients were significantly reported positive attitudes in relation to health education program integrating in health care settings. It is essential that the relationship between patients and health professionals is depended on the collaboration of both; and not only to professional's authority. Although, there is no evidence about consulting technique that have an important role to change of patients' behavior, it is however obvious that motivational interviewing has succeeded patients' empowerment for adaptation healthy behaviors.

5 Conclusion

This study confirms that motivational interviewing in health educational programs may change behavior of psychiatric patients and thus may positively influence their nutritious habits increasing their HRQoL. Moreover, the Transtheoretical Model of Change and Counseling can be strongly used by health professionals as an instrument for nutritious behavioral programs.

Acknowledgements Authors would like to thanks all patients that participated in the study.

Conflict of interest All authors declare that there is no financial or personal conflict of interest related to this paper.

Funding There was no funding for the current work.

References

1. Prochaska, J.O., C.C. Diclemente, and J.C. Norcross. 1992. In Search of How People Change: Applications to Addictive Behaviours. *American Psychologist* 47: 1102–1114.
2. Miller, W.R., and S. Rollnick. 1991. *Motivational Interviewing: Preparing People to Change Addictive Behaviors*. New York: Guilford Press.
3. Higgins, J., D. Babor, and R. Biddle. 1996. A Randomized Clinical Trial of Brief Interventions in Primary Health Care. *Disease Control Priorities Project* 47: 13–20.
4. DiClememte, C.C., and J.O. Prohaska. 1998. Toward a Comprehensive Transtheoritical Model of Change: Stages of Change and Addictive Behaviors. In *Motivational Interviewing: Enhancing Motivation for Change – A Learners Manual for the American/Indian/Alaska Native Counselor*, ed. K. Tomlin, R. Walker, J. Grover, W. Arquette, and P. Stewart. Portland, OR: One Sky Center.
5. Peppas, E. 2006. In *Alcohol Abuse and Dependency*, ed. I. Liappas, E. Mellos, and B. Pomini. Athens: Itaca Publications Greek Department.
6. Tomlin, K., R. Walker, J. Grover, et al. 2014. *Motivational Interviewing: Enhancing Motivation for Change – A Learner's Manual for the American Indian/Alaska Native Counselor*. Portland, OR: One Sky Center.
7. MacLean, L., N. Edwards, M. Garrard, N. Sims-Jones, K. Clinton, and L. Ashley. 2009. Obesity, Stigma and Public Health Planning. *Health Promotion International* 24: 88–93.
8. Hasler, G., D.S. Pine, A. Gamma, et al. 2004. The Associations Between Psychopathology and Being Overweight: A 20-Year Prospective Study. *Psychiatric Medicine* 34: 1047–1057.
9. John, U., C. Meyer, H. Rumpf, et al. 2005. Relationships of Psychiatric Disorders with Overweight and Obesity in an Adult General Population. *Obesity Research* 13: 101–109.
10. Tountas, Y. 1994. In *Preventive medicine and health promotion: Health education, Athens University, Epidimiological and Health Laboratory*, ed. E. Kaklamanaki and Y. Fragoulis-Koumantakis. Athens: Paschalides Medical Publications.
11. Laforge, R.G., J. Maddock, and J. Rossi. 1998. Using the Transtheoretical Model for Population Based Approaches to Health Promotion and Disease Prevention. *Homeostasis in Health and Disease* 40 (5): 174–195.
12. Miller, W.R., T.B. Moyers, and P. Amrhein. 2005. The Change Questionnaire. Department of Psychology, University of New Mexico, unpublished.
13. Lochoro, P. 2004. Measuring Patients' Satisfaction in U.C.M.B. Health Institutions. *Health Policy and Development* 2 (3): 243–248.

14. Dipasquale, S., C.M. Pariante, P. Dazzan, et al. 2013. The Dietary Pattern of Patients with Schizophrenia: A Systematic Review. *Journal of Psychiatric Research* 47 (2): 197–207.
15. Chouinard, V.A., S.M. Pingali, G. Chouinard, et al. 2016. Factors Associated with Overweight and Obesity in Schizophrenia, Schizoaffective and Bipolar Disorders. *Psychiatry Research* 237: 304–310.
16. Hardcastle, S., A. Taylor, M. Bailey, et al. 2008. A Randomised Controlled Trial on the Effectiveness of a Primary Health Care Based Counselling Intervention on Physical Activity, Diet and CHD Risk Factors. *Patient Education and Counseling* 70 (1): 31–39. doi:10.1016/j.pec.2007.09.014.
17. Karatay, G., Y. Akkus, N. Demirci, and B. Ozturk. 2016. Short-Term Effects of the Stage-Matched Multicomponent Lifestyle Intervention on Weight Control. *Middle East Journal of Rehabilitation and Health.*, (in press). doi:10.17795/mejrh-36533.
18. Lee, W.W.M., K.C. Choi, R.W.Y. Yum, et al. 2016. Effectiveness of Motivational Interviewing on Lifestyle Modification and Health Outcomes of Clients at Risk or Diagnosed with Cardiovascular Diseases: A Systematic Review. *International Journal of Nursing Studies* 53: 331–341.
19. Resnicow, K., R. Davis, and S. Rollnick. 2006. Motivational Interviewing for Pediatric Obecity: Conceptual Issues and Evidence Review. *Journal of the American Dental Association* 106: 2024–2033.
20. Steptoe, A., E. Rink, and S. Kerry. 2000. Psycosocial Predictors of Changes in Physical Activity in Overweight Sedentary Adults Following Counselling in Primary Care. *Preventive Medicine* 31: 183–194.
21. Wilson, G.T., and T.R. Schlam. 2004. The Transtheoretical Model and Motivational Interviewing in the Treatment of Eating and Weight Disorders. *Clinical Psychology Review* 24: 361–378.
22. Dunn, C., L. Deroo, and F.P. Rivara. 2001. The Use of Brief Interventions Adapted from Motivational Interviewing Across Behavioral Domains: A Systematic Review. *Addiction* 96: 1725–1742.
23. Woolard, J., L. Beilin, T. Lord, I. Puddey, D. MacAdam, and I. Rouse. 1995. A Controlled Trial of Nurse Counseling on Lifestyle Change for Hypertensives Treated in General Practice: Preliminary Results. *Clinical and Experimental Pharmacology and Physiology* 22: 466–468.
24. Hettema, J., J. Steele, and W. Miller. 2005. Motivational Interviewing. *Annual Review Clinical Psychology* 1: 91–111.
25. Rollnick, S.R., N. Heather, and A. Bell. 1992. Negotiating Behavioral Change in Medical Settings: The Development of Brief Motivational Interviewing. *Journal of Mental Health* 1: 25–37.
26. Colby, S.M., P.M. Monti, N.P. Barnett, et al. 1998. Brief Motivational Interviewing in a Hospital Setting for Adolescent Smoking: A Preliminary Study. *Journal of Consulting and Clinical Psycholog* 66: 574–578.
27. Steinberg, M.L., D.M. Ziedonis, J.A. Krejci, et al. 2004. Motivational Interviewing with Personalized Feedback: A Brief Intervention for Motivating Smokers with Schizophrenia to Seek Treatment for Tobacco Dependence. *Journal of Consulting and Clinical Psychology* 72 (4): 723–728.

Approaches on Generating Optimized Sequences of Items Used in Assessment

Doru Anastasiu Popescu, Daniel Nijloveanu, and Nicolae Bold

Abstract Assessment is a key element in education. Given its multiple methods, assessment contains the aspect of correctness and honesty from both the assessor and the assessed. These methods can be classified and their types are numerous, starting from single questions to tests and practical work. Concerning tests, their vast usage in practice raises problems such as progressing difficulty, concept understanding regardless the test morphology or test sort depending on subject. Generally, we will refer at questions within a test or tests within a group of tests as items. In this paper, we present several approaches for generating items, defining or not some restrictions, in order to obtain optimized sequences of items (OSI). These restrictions consist in connections between items regarding degrees of difficulty or sorting depending on item subject. These cases will be solved using nature-inspired algorithms (genetic algorithms), arborescent structures and lexical usage (keywords).

Keywords Tree • Genetic • Test • Question • Keyword

1 Introduction

Assessment is an important issue in the educational environment, due to its function of checking the gained knowledge. Assessment can be formal or informal, depending on its purpose. Whatever its type, assessment depends on four basic elements: the assessor (the person which makes the evaluation), the assessed (the person which is evaluated), the method of evaluation and the context.

Usually, the assessor can identify with the assessed, in the case of auto-evaluation. However, in most cases, the assessor is the teacher and the assessed

D.A. Popescu (✉) • N. Bold
Faculty of Mathematics and Computer Sciences, University of Pitesti, Pitești, Romania
e-mail: dopopan@yahoo.com; bold_nicolae@yahoo.com

D. Nijloveanu
Faculty of Management, Economic Engineering in Agriculture and Rural Development, University of Agronomic Sciences and Veterinary Medicine Bucharest, Slatina Branch, Romania
e-mail: nijloveanu_daniel@yahoo.com

P. Vlamos (ed.), *GeNeDis 2016*, Advances in Experimental Medicine and Biology 988, DOI 10.1007/978-3-319-56246-9_6

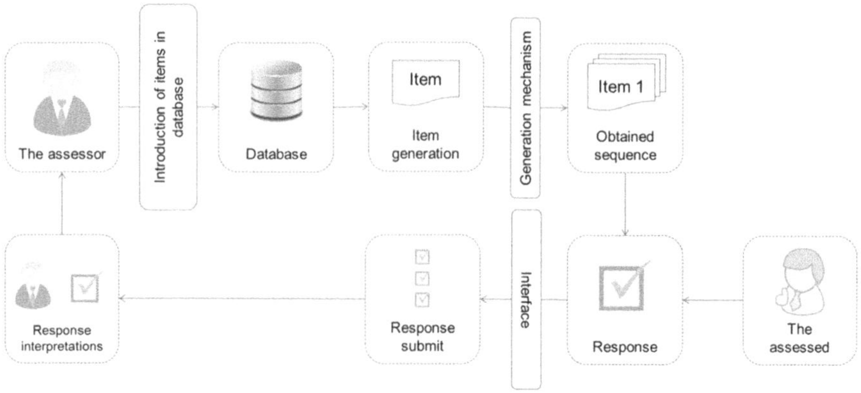

Fig. 1 The scheme of assessment for our method

persons are the students. These two categories have different perceptions on assessment, as in the study presented in paper [1]. This difference leads to the trial of new methods, including the technology-based one. This raises implications about both the positive and negative effects on assessment, as shown in [2]. Examples of good effects are the one given by the methods described in this paper, which consist in time savings, better management of assessment and an increased efficiency. Other examples and a comparative study between traditional and technology-based methods have been made in [3].

Including technology in assessment led to the creation of two directions of technology usage. On one hand, this led to new improved methods of assessment, whose description is the object of the paper and which assist the traditional methods of assessment. On the other hand, it led to the creation of web environments, which replace the traditional assessment. Examples of these environments are found in [4–6]. All these types of methods must be optimized on a long term, using principles such as the ones presented in [7, 8].

We will focus in this paper on the improved methods of assessment. Thus, the assessor can use a variety of methods to test the assessed. The most used types of methods are tests formed of a set number of questions. Our method of assessment can be summarized in Fig. 1.

Usually, in practice, we use for assessment a large number of tests and questions, grouped in battery of tests or of questions. What happens in the case in which the items are related through degrees of difficulty or the assessor wants to select items described by certain subjects from a battery of items? What if the assessor just wants to generate random sequences of tests which respect certain restrictions?

The main objective of this paper is to present some modalities of generating optimized sequences of items. These items are grouped differently, based on certain criteria. Optimization consists in respecting the given restrictions and giving the best results in terms of respecting the restrictions in the same time. In addition, optimization means that a fitness function must exist.

For solving the problem, we used notions and concepts from different areas. Depending on the structure of the items, we used arborescent structures (trees), genetic algorithms, random notions and word processing. We used in the paper algorithms defined by network and nature-inspired concepts. We could have also used backtracking algorithms, but the complexity restrictions limit the usage for large amount of items.

The solutions described in this paper have a wide range of applications, besides education. They can be used in domains that need the presence of at least two operations: selection of specific items from a large pool of items and optimization of the result. Generally, these domains are the ones which use evaluation.

2 Issue Approaches

Before starting to present the approaches, we must familiarize with the notions used in this paper. The definitions that follow will clarify these notions. We must mention that the items can be sequentially connected or isolated, meaning that two cases can be possible:

- the items are connected when they are related through a type of connection, one depending on the previous question (sequential character), the connection depending on characteristics such as subject, domain, degree of difficulty etc.
- the items are isolated, meaning that they do not depend one from another, the relatedness of items having a random character.

Remark 1 We will state by optimized sequence of items and denote by OSI an optimized sequence of questions (OSQ) or an optimized sequence of tests (OST). This will be made for an easier and clearer language, as we will also refer at items in general.

Definition 1 Let $q_1, q_2, \ldots, q_n$ be n questions. A test is a sequence of n questions which are sequentially connected or isolated. A test is denoted by T.

Obs.: The questions will be codified with numbers from 1 to n (e.g., $q_1 \rightarrow 1$) and referred as numbers in the paper.

Definition 2 Let $T_1, T_2, \ldots T_n$ be n tests. A battery of tests is a sequence of n tests which are sequentially connected or isolated. A battery of tests is denoted by BT.

Observation 1 The tests will be codified with numbers from 1 to n (e.g., $T_1 \rightarrow 1$) and referred as numbers in the paper.

Remark 2 We will state by item and denote by I_i a question q_i within a test or a test T_i within a BT. This will be made for an easier and clearer language, as we will also refer at items in general.

Definition 3 Let T be a test according to Definition 1 which contains n questions $\{q_1, q_2, \ldots, q_n\}$ and a fitness function $f:\{1,2,\ldots,n\} \rightarrow \mathbb{R}$. For k sequences, an optimized sequence of questions (denoted by OSI_i) is a sequence of m questions (m ≤ n) for which $f(OSQ_i) \geq f(OSQ_j), \forall j = \overline{1, K}$ or $f(OSQ_i) \leq f(OSQ_j)$, $\forall j = \overline{1, k}$.

Definition 4 Let BT be a battery of tests according to Definition 2 which contains n tests $\{T_1, T_2, \ldots T_n\}$ and a fitness function f:{1,2,...,n} → $\mathbb{R}$. An optimized sequence of tests (denoted by OST) is a sequence of m tests for which $f(OST_i) \geq f(OST_j), \forall j = \overline{1, K}$ or $f(OST_i) \leq f(OST_j), \forall j = \overline{1, k}$.

Remark 3 The fitness function, used as in the context of a genetic algorithm, is a mathematical function with a definite form whose values are calculated for a sequence of items and which characterize it. For example, the fitness function for a sequence of items can be the degree of difficulty of the sequence of items, given the separate degrees of difficulty of the component items of the sequence summed up to form the general difficulty of the sequence (for a sequence OSI = $\{I_1, I_2, \ldots, I_m\}$, $f(OSI) = \sum_{i=1}^{m} value\ of\ (I_i), \forall m \leq n$).

Observation 2 Depending on restrictions, f tends to maximum or minimum. Thus, f is a function of maximum/minimum.

Remark 4 Let I = $(I_1, I_2, \ldots I_n)$ be a test T or a battery of tests BT. For simplification, I is considered to be (1, 2, ..., n), where I_1 is codified with 1, I_2 with 2 and so on. Thus, essentially, the OSIs will be sequences of numbers.

As we presented in the introduction, the cases of generating tests have several instances. We identified four cases of generation which can be divided in sub-cases:

- depending on the items that are studied, we can generate questions or tests;
- depending on the connection between items, the items can be sequentially connected or isolated;
- depending on the existence of characterization, we can generate items that are characterized by keywords;
- depending on item timing, we can generate items characterized by solving time, depending on a given global time.

Depending on the items that can be studied, the problem can be referred in case of tests and questions. Questions within a test can be permuted, as well as the choices of a questions (if they exist), in order to form different variants of the same test. This permutation is made to verify if the information contained by a question was understood in a correct way. Another usage of the solution of this case is to rearrange the questions or variants for the teacher in order to prevent forgery or to avoid the memorization of the variants.

The second criterion of classification is the degree of connection between the items. Thus, the items can be connected through a degree of difficulty (the items are situated on levels of difficulty) or they can be isolated, forming a heterogeneous mix of items. The solutions made in these cases are useful when the assessment is made following a progressive increase of difficulty or the items have different degrees of difficulty.

Another criterion of the approach is the one in which the items are characterized or not by keywords. The solution in these cases is useful for situations when the assessor wants to select several items characterized by certain keywords or labels. These keywords mean the assessor wants to test only for certain topics from a certain domain or to avoid items characterized by certain keywords.

The latter criterion is the time given for a certain item. In this case, given a fixed amount of global time and times for each item, the assessor wants to find optimized sequences of items which fit in the global time.

This classification is purely for clarifying the approaches that we studied for this problem. In practice, the criteria can interchange and mix. For example, the assessor wishes to generate optimized sequences of tests which contain untimed isolated tests characterized by keywords.

For generating all kinds of items, we can use tree-based algorithms or genetic algorithms. The general algorithm used in all four cases will be a genetic algorithm, except for the case of connection between items. We will also use in all cases notions of random algorithms. Depending on the structures and algorithms used, the tests will be considered nodes within a tree or genes in a chromosome. Every item will be described by a list of keywords (if case) and a time (if case). The generation is made using random algorithms or genetic algorithm, depending on case.

Genetic algorithms are used in many domains: traffic [9], chemistry [10], web applications [11, 12], agriculture [13], design [14], optimization problems and even fashion [15]. We chose genetic algorithms for the fact of outputting many solutions using large input data and for their character of optimization.

Arborescent structures are used in many types of application. Examples of domains which use trees are optimization problems [16], sort or web applications [17]. We used them for the easiness of application of tree structures in our problem and the nature of the problem to be solved whose similarity with tree structures is noticed.

3 Applied Strategies and Models

Respecting the classification above, we used different models to mathematically represent the used structures, connections and strategies to solve the problems. Basically, the cases presented above can be solved using two major methods: using arborescent structures (trees) or genetic algorithms. Some variables and notations are used in the paper and they are shown in Table 1.

In addition, we will use for genetic algorithm some arrays (T and TN), whose significance is given in Table 3.

The mathematical model when a tree is used is shown in Table 2.

The mathematical model for the genetic-based method is shown in Table 3.

Table 1 Variables and structures used in the paper

Using trees	Using genetic algorithms
Common variables	
• N—the total number of items • M—the desired number of OSI to be output • nKW—number of keywords given by the user to be found • TG[nKW]—keywords given by the user to be found • keywords for each item (stored as binaries in T and TN).	
Uncommon variables	
• L – the number of levels in a tree • T[n] – the parents node array • sol[L] – the array that store the solution (the OSI)	• Gen—number of generations • NG—number of chromosomes in the population • nKWN—number of keywords given by the user to be avoided • TG[nKW]—keywords given by the user to be avoided • Tt—the given global solving time • Tt[n]—the solving time for each item

Table 2 Mathematical model for connected items (using trees and genetic algorithms /random generation)

Input data	Possible restrictions	Output data
• N	• no. of generated tests = L/M	• sol[L]
• M	• L of sol[i] $\neq$ L of sol[j], i,j $\leq$ L (nodes from OSI on different levels)	
• T[n]	• sol[i] =/$\neq$ T[sol[i + 1]], i $\leq$ L	
• L	• the nodes from the OSI are maximally linked (the number of missing edges between the nodes in the OSI is minimal)	

Table 3 Mathematical model for isolated items (using genetic algorithms)

Input data	Possible restrictions	Output data
• N, M, NG, keywords for each item (iKW_i), nKW, nKWN	*Obs*: T[i][j] is 1 if TG[j] is found amongst keywords of test i; analogous for TN[i][j].	– pop[NG][m]
• TG[nKW]	– no. of generated tests = m	
• TGN[nKWN]	– $\sum$T[pop[i][j]] maximal, i $\leq$ n, j $\leq$ m	
• T[N][nKW] – stores the keywords for the item	– $\sum$(T[pop[i][j]]-TN[pop[i][k]]) minimal, i $\leq$ N, j,k $\leq$ M	
• TN[n][nKWN] – stores the keywords for the item to be avoided	– pop[i][j] $\neq$ pop[i][k], i $\leq$ NG, j,k $\leq$ M	
• Tt	– $\sum$Tt[pop[i][j]] $\leq$ Tt and minimal, i $\leq$ NG	
• Tt[n]		

Table 4 contains the main models and strategies used for solving the problem in all cases, as well as other characteristics of them. As seen from the columns, the main cases are considered depending on the existence of connections between the items. The two cases can be combined (e.g., the items are represented as nodes in a tree and the OSI is obtained as a result of a genetic algorithm).

The restrictions can be limitless. In Table 4 we gave some studied cases of restrictions and combinations of restrictions. Considering an OSI sol = (sol[1], sol[2], ..., sol[L]), for the tree-based algorithms, other restrictions that we studied in previous papers are:

- L of sol[i] $\neq$ L of sol[j], i,j $\leq$ L (nodes from OSI on different levels) and sol[i] = T[sol[i + 1]] [1];
- L of sol[i] $\neq$ L of sol[j], i,j $\leq$ L, sol[i] =/$\neq$ T[sol[i + 1]] and the keywords of the items in the OSI must be found among the keywords given by the user (which paper is in course of publication).

Considering the sequence pop[i] = (pop[i][1], pop[i][2], ..., pop[i][M]), for the genetic algorithms, we studied the next restrictions:

- $\sum$T[pop[i][j]] maximal, i $\leq$ n, j $\leq$ m and pop[i][j] $\neq$ pop[i][k], i $\leq$ NG, j,k $\leq$ M [2];
- $\sum$(T[pop[i][j]]-TN[pop[i][k]]) minimal, i $\leq$ N, j,k $\leq$ M and pop[i][j] $\neq$ pop[i][k], i $\leq$ NG, j,k $\leq$ M [3];
- $\sum$Tt[pop[i][j]] $\leq$ Tt and minimal and pop[i][j] $\neq$ pop[i][k], i $\leq$ NG [4].

A special work is made in paper [18], where we present the generation of several different tests using the same questions using genetic algorithms. In this case, optimization means the finding of a higher number of distinct tests which resemble as least as possible, using the same questions.

The choice of one of the methods presented above is made depending on the case. Thus:

- the tree-based method can be applied mostly in cases when the items are connected. The main advantage of this method is based on the fact that items can be placed on levels, which is a requirement in cases of sequentially-connected tests or batteries of tests. Another advantage is the discovery of the most optimal solution for reasonable amount of input data.
- the genetic-based method is most useful in cases of items that are not connected. However, a great advantage is that this method can be also used for the opposite case, another advantage being the possibility of usage of this method for large input data.

Given the two reduced cases, we will show the steps of the algorithms for each one in the next section.

Table 4 Main characteristics of the strategies and models used for solving the generation problem

Case	Using trees	Using genetic algorithms
Strategy and model descriptions	The items are labeled with numbers from 1 to total number of items. The OSI are output as sequences of numbers. The two models can as well be combined	
	An item is a node and the connections are edges. Sequences of items are paths from root to a leaf. Using the mentioned methods, we start from a leaf to the root, verifying if the node corresponds to the restrictions	An item is a gene and a sequence of items is a chromosome. Using the mentioned methods, genes are generated, respecting the restrictions given by the user
Methods	Breadth-first search (BFS), depth-first search (DFS), random algorithms, other methods	Random generation, mutation, crossover with one point
Structures used	Trees, nodes	Genes, chromosomes
Optimization function	A function which numbers the items in an OSI which correspond to a restriction (e.g., the lowest number of edges which does not exist between the nodes)—can be of maximum and minimum	
Graphic representation	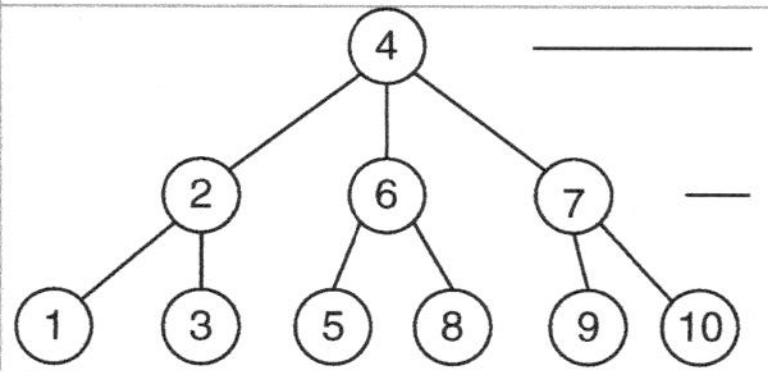	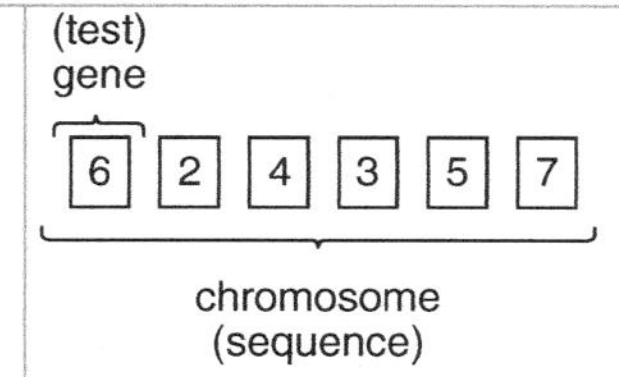

4 Steps of the Algorithms

The models presented above were materialized and structured in algorithms. Based on these algorithms, we will solve the problem depending on restrictions.

The general form of a tree-based algorithm is presented in Table 5.

We present the steps for the algorithm which uses arborescent structure:

Step 1. Input data is read.
Step 2. The tree is built from the array of parent nodes and the leaves are determined (optionally).
Step 3. Starting from the root or from a leaf, the sequence is built. The passage to the next level is made through breadth-first search or depth-first search or using any other methods. The restrictions for the current node are verified.
Step 4. The sequence is output.

The general form of the genetic-based algorithm is presented in Table 6.

Now, when using a genetic algorithm, the steps of this are:

Step 1. Input data is read.
Step 2 (depending on case). The keywords of tests which are found among the given keywords are stored in an array.
Step 3. The initial population of chromosomes is generated gene by gene, using random generation.
Step 4. The fitness function is calculated for each chromosome of the initial population. Each time we calculate the fitness, we verify the actual restrictions

Table 5 The tree-based algorithm

The tree-based algorithm
read(); //Reads the input data (N, array of parent nodes)

leafs(); //The leafs from the tree are determined
while *root is not reached* or *leaf is not reached* do
   generation(); //a node is generated using random or genetic-based generation
   if *restrictions are respected* then
   store_node(); //If the restrictions are respected, the node is stored output_solution(); //Found solutions are output

Table 6 The genetic-based algorithm

The genetic-based algorithm
read(); // Reads the input data (N, M, NG, arrays TG, TGN, Tt)

generation(); //Generates the initial population of chromosomes
for i = 1,Gen do
   mutation(); //Mutation operation is applied
   crossover();//Crossover operation is applied
output_solutions(); //Solutions are output

which are summarized by the fitness function. Then, the chromosomes are sort depending on this value.

The next two steps are iterated for a set number of times which represents the number of generations. Step 5 and 6 can be followed in an order.

Step 5. The mutation operation is applied (in various ways) on the current population. The chromosomes are sort.

Step 6. The crossover operation is applied (in various ways) on the current population. The chromosomes are sort.

Step 7. The first chromosomes are output.

As genetic methods, we used mutation and crossover with one point. The mutation has different forms. A form used in the paper consists in the replacement of one gene with a randomly-generated one (C) in a randomly-generated position (k).

$$pop\,[i] = (pop\,[i]\,[1]\,, pop\,[i]\,[2]\,, \ldots, pop\,[i]\,[k]\,, \ldots, pop\,[i]\,[M]) \rightarrow$$

$$pop\,[i] = (pop\,[i]\,[1]\,, pop\,[i]\,[2]\,, \ldots, C, \ldots, pop\,[i]\,[M]) \tag{1}$$

As well as the mutation, the crossover may have different forms. The form we used in the paper uses two chromosomes (k1 $\neq$ k2).

$$pop\,[i] = (pop\,[i]\,[1]\,, pop\,[i]\,[2]\,, \ldots, pop\,[i]\,[k1]\,, \ldots, pop\,[i]\,[M])$$

$$pop\,[j] = (pop\,[j]\,[1]\,, pop\,[j]\,[2]\,, \ldots, pop\,[j]\,[k2]\,, \ldots, pop\,[j]\,[M])$$

$$\downarrow$$

$$pop\,[i] = (pop\,[i]\,[1]\,, pop\,[i]\,[2]\,, \ldots, pop\,[j]\,[k1]\,, \ldots, pop\,[j]\,[M])$$

$$pop\,[j] = (pop\,[j]\,[1]\,, pop\,[j]\,[2]\,, \ldots, pop\,[j]\,[k2]\,, \ldots, pop\,[i]\,[M]) \tag{2}$$

These two methods help to increase the fitness function values for the future populations.

5 Discussion on Results

Using these algorithms (with variations of restrictions), we obtained some good results for input data. Although this is the primary objective of the problem, we would also like to obtain optimized results for large amounts of data. Thus, we will study also the runtime of the algorithms.

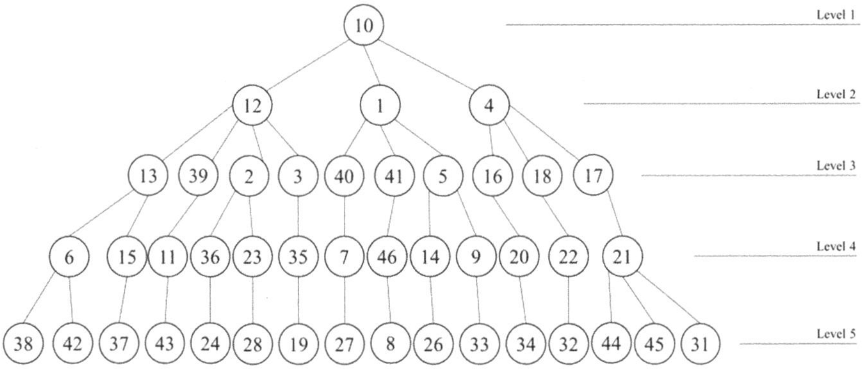

Fig. 2 Battery of 46 tests for our example

There exist approaches on this problem in the literature. However, they present the problem using different types and structures, based especially on specially-defined coefficients and using other methods. We will compare our results for genetic algorithms with the ones presented in these papers. As shown in paper [19], a divide-and-conquer memetic algorithm for a pretty similar issue has a similar trend (even if the runtime cannot be compared for obvious reasons related to development of technology), both of them appearing to increase proportionally with the number of items, ours according to the function in [20]. Other approaches based on genetic algorithms also showed similar behaviors [21, 22]. Still, restrictions and approaches are quite different with the ones presented in this paper and a strict comparison is hard and quite inappropriate to be done. Moreover, the approaches described in this paper are singular. This introduction was done in order to show the up-to-a-point similarity between approaches.

In case of tree-based algorithms, because of the short amount of space and for a purely functioning example, we will take a simple example of a tree with 46 nodes, shown in Fig. 2. The restrictions are:

- tests must be different in an OSI;
- each level must be represented by a test.

Some desired output can be (10, 12, 2, 23, 28) or (10, 1, 41, 46, 8). The runtime for this algorithm was approximately 0.015 s. Paper [23] shows the variation of runtime depending on the number of tests. We can observe that the value of runtime slightly increases as n grows.

As for the algorithm based on a genetic method, we will take an example of a battery of tests with 500 items and an example of generating variants of tests using the same questions.

For our first example, the variables have 500 items, 15 items desired to be output, 200 generations and the next values, presented in Table 7.

Table 7 The form of items for the first example

Item	Keyword number	Keywords
Item 1	2	Operating
Item 2	2	Java, taskbar
Item 3	5	Taskbar, operating, window, Android, partition
Item 4	2	partition, engine
...	...	...
Item 148	4	Window, folder, Android, file
...	...	...
Item 216	5	Folder, Linux, taskbar, terms, drive
...	...	...
Item 499	5	Website, boot, path, search, engine
Item 500	5	Engine, path, Internet, window, taskbar
Given by the user (TG)	10	Operating, Android, Windows, Linux, system, file, folder, path, start, taskbar

Table 8 Results for the first example

Optimized sequence of items (a number is an item—148 represents the item 148)	Number of keywords from the ones given by the user
148 216 284 124 70 30 344 285 434 480 398 53 483 301 312	35
53 70 398 301 264 344 30 312 434 148 284 483 285 124 480	34
264 285 53 124 148 284 344 434 30 483 398 312 70 301 480	34
148 398 53 124 264 344 30 312 434 483 70 285 301 284 480	34
284 70 148 124 264 312 30 285 434 483 53 344 398 301 480	34
148 70 284 124 434 344 30 285 264 483 398 301 53 312 480	34
148 70 434 124 264 344 30 53 285 483 398 312 284 480 301	34
483 301 264 124 284 344 398 285 434 70 30 148 312 53 480	34
148 53 312 124 264 344 284 285 434 483 30 398 70 301 480	34
312 70 284 285 124 483 148 264 434 30 398 53 344 480 301	34

The algorithm outputs the sequences which have the highest sum of the keywords found in items, keywords which coincide with those given by the user (TG). In the example, for the first sequence, 35 is the sum of the keywords of item 148 found in TG (which is 3) plus the keywords of item 216 found in TG (which is 3) and so on. Table 8 presents the first 10 OSI output by the algorithm.

The runtime for this example was 2.869 s. The number of keywords from the second column in Table 8 can grow. For example, for 700 generations, we obtained 38 keywords. Runtimes for genetic algorithms can be studied in papers [20, 24, 25].

As for the second example, the questions and their choices can be permuted in order to form a desired number of tests. For 10 questions, number of variants of each question equal to (4, 4, 4, 3, 3, 4, 4, 2, 5, 3) and 5 tests desired to be output, we obtained the next values presented in Table 9.

Table 9 Results for the second example

Test 1		Test 2		Test 3		Test 4		Test 5	
It.	Variants	It.	Variants	It.	Variants	It.	Variants	It.	Variants
1	1 3 4 2	1	1 3 4 2	1	1 4 3 2	1	1 4 2 3	1	1 4 3 2
3	1 4 3 2	10	1 2 3	3	1 3 2 4	6	1 4 3 2	6	1 2 4 3
2	1 3 4 2	2	1 4 3 2	2	1 3 2 4	10	1 3 2	5	1 2 3
4	1 3 2	3	1 4 2 3	9	1 4 2 5 3	3	1 2 3 4	8	1 2
9	1 3 2 5 4	5	1 2 3	10	1 2 3	9	1 4 3 2 5	10	1 3 2
5	1 3 2	9	1 4 3 5 2	5	1 2 3	7	1 4 2 3	9	1 4 5 2 3
10	1 3 2	7	1 3 2 4	7	1 4 2 3	2	1 4 2 3	4	1 3 2
8	1 2	6	1 2 4 3	4	1 3 2	5	1 3 2	3	1 4 2 3
6	1 4 3 2	8	1 2	8	1 2	8	1 2	2	1 2 4 3
7	1 2 3 4	4	1 2 3	6	1 3 4 2	4	1 2 3	7	1 4 3 2

The runtime of this algorithm is approximately 0.0077 s. For example, for 10 questions and 3200 output tests, the runtime was 29.1506 s, as shown in paper [18].

6 Conclusions

The variety of the restrictions can lead to different types of algorithms and values. The restrictions can be unlimited or structured on the needs of the assessor. As future work, we would like to create a graphic interface that would sum up all our work on this domain. This future work would concentrate all the work in this domain and unify the concepts and structures used until now in this domain.

The future web application would mainly consist in generating items to form sequences of items (i.e., a test or a series of tests). This would make the connection between the assessor and the assessed. From a database of items which is completed in time, the assessor generates items and form tests/series of tests with these items using the approaches presented in this paper. This newly-formed test/series of tests is then shown to the assessed and proposed to be solved. After solving, the assessed person submits the answers. The answers are then corrected by the assessor and the answer is given to the assessed person. These will be made using a web interface and login restrictions.

The technology is used in many aspects and domains. The key to balance with these approaches is the rational usage of technology-based methods in order to obtain the expected results and to take into account the human factor, which is the most important in the process of education, and the restrictions given by the users.

References

1. MacLellan, E. 2001. Assessment for Learning: The Differing Perceptions of Tutors and Students. *Assessment & Evaluation in Higher Education* 26 (4): 307–318.
2. Ştefănescu, V., C. Ştefănescu, and O. Roşu Stoican. 2015. The Influence of Using ICT on the Quality of Learning. *International Conference on Virtual Learning – ICVL*: 169–172.
3. Schulman, A.H., and R.L. Sims. 1999. Learning in an Online Format Versus an In-Class Format: An Experimental Study. *The Journal* 26 (11): 54–56.
4. Popescu, E. 2010. Adaptation Provisioning with Respect to Learning Styles in a Web-Based Educational System: An Experimental Study. *Journal of Computer Assisted Learning* 26 (4): 243–257.
5. Holotescu, C. 2015. A Conceptual Model for Open Learning Environments. International Conference on Virtual Learning – ICVL, pp. 54–61.
6. Baron, C., A. Şerb, N.M. Iacob, and C.L. Defta. 2014. IT Infrastructure Model Used for Implementing an E-learning Platform Based on Distributed Databases. *Quality-Access to Success Journal* 15 (140): 195–201.
7. Boud, D., and N. Falchikov. 2007. *Rethinking Assessment in Higher Education: Learning for the Longer Term*. London: Routledge Publishing.
8. Graff, M. 2003. Cognitive Style and Attitudes Towards Using Online Learning and Assessment Methods. *Electronic Journal of e-Learning* 1 (1): 21–28.
9. Rahmani, S., S.M. Mousavi, and M.J. Kamali. 2011. Modeling of Road-Traffic Noise with the Use of Genetic Algorithm. *Applied Soft Computing* 11 (1): 1008–1013.
10. Darby, S., M.J. Thomas, L. Johnston Roy, and C. Roberts. Theoretical Study of Cu–Au Nanoalloy Clusters Using a Genetic Algorithm. *The Journal of Chemical Physics* 116 (4): 1536, 2002.
11. Popescu, D.A., and D. Radulescu. 2015. Approximately Similarity Measurement of Web Sites, ICONIP, Neural Information Processing, Proceedings LNCS, Springer, 9–12 November, 2015.
12. Popescu, D.A., and I.A. Popescu. 2015. Model of Determination of Coverings with Web Pages for a Website. *International Conference on Virtual Learning*: 279–283.
13. Popescu, D.A., Radulescu, D.. 2015. Monitoring of Irrigation Systems Using Genetic Algorithms. ICMSAO IEEE Xplorer, pp. 1–4.
14. Gama, C.L., and K. Norford Leslie. 2002. A Design Optimization Tool Based on a Genetic Algorithm. *Automation in Construction ACADIA 99* 11 (2): 173–184.
15. Hee-Su, K., and C. Sung-Bae. December 2000. Application of Interactive Genetic Algorithm to Fashion Design. *Engineering Applications of Artificial Intelligence* 13 (6): 635–644.
16. Domşa, O., E. Ceuca, and A. Râşteiu. 2003. Algorithm to Find a Tree with Maximal Terminal Nodes. 1st Balkan conference in informatics, Thessaloniki, Greece, November 21–23, pp. 113–122.
17. Popescu, D.A. 2010. Reducing the Navigation Graph Associated to a Web Application. *Buletin Stiintific - Universitatea din Pitesti, Seria Matematica si Informatica* 16: 125–130.
18. Domşa, O., and N. Bold. 2016. Generator of Variants of Tests Using the Same Questions. The 12th international scientific conference eLearning and software for education, Bucharest, April 21–22.
19. Nguyen, M.L., S.C. Hui, and A.C.M. Fong. March 2012. Divide and Conquer Memetic Algorithm for Online Multi-Objective Test Paper Generation. *Regular Research Paper, Memetic Computing* 4 (1): 33–47.
20. Popescu, D.A., N. Bold, and D. Nijloveanu. 2016. A Method Based on Genetic Algorithms for Generating Assessment Tests Used for Learning. 17th international conference on intelligent text processing and computational linguistics, April 3–9.
21. Hwang, G.J., M.T. Bertrand, Hsien-Hao Lin, and Tsung-Liang Lin Tseng. 2005. On the Development of a Computer-Assisted Testing System with Genetic Test-Sheet Generating Approach. *IEEE Transactions on Systems Man and Cybernetics* 35 (4): 590–594.

22. Wang, F., W. Wang, H. Yang, and Q. Pan. 2009. A Novel Discrete Differential Evolution Algorithm for Computer-Aided Test-Sheet Composition Problems. Information Engineering and Computer Science, 2009. ICIECS 2009, Wuhan, 19–20 December, pp. 1–4.
23. Nijloveanu, D., N. Bold, and A.C. Bold. 2015. A Hierarchical Model of Test Generation within a Battery of Tests. International conference on virtual learning, pp. 147–153, Timişoara.
24. Popescu, D.A., N. Bold, and O. Domşa. 2016. Generating Assessment Tests with Restrictions Using Genetic Algorithms. 12th IEEE international conference on control & automation, Kathmandu, Nepal, June 1–3.
25. Nijloveanu, D., N. Bold, and I.A. Popescu. 2016. Model of Evaluation Using Questions with Specified Solving Time. The 12th international scientific conference eLearning and software for education, Bucharest, April 21–22.

Evaluation of Pulse Oximetry Knowledge of Greek Registered Nurses

John Stathoulis, Maria Tsironi, Nikolaos Konofaos, Sofia Zyga, Victoria Alikari, Evangelos C. Fradelos, Helen Bakola, and George Panoutsopoulos

Abstract Proper use and evaluation of the pulse oximeter readings in everyday clinical practice are related to patient safety and quality of provided patient healthcare. Purpose of this study was the evaluation of Greek registered nurses' knowledge in pulse oximetry before and after an educational intervention implemented in a 2-h educational intervention (workshop). Anonymous self-administered validated questionnaire consisted of two parts was used to collect the data, after the written consent of the author, in a sample consisted of 78 participants (12 men and 66 women) and the output data were analyzed with SPSS v. 19.0 (SPSS Inc., Chicago, IL). The quantitative variables are expressed as mean values (SD) or as median values (interquartile range = IQR) while the qualitative variables are expressed as absolute and relative frequencies. For the comparison of the proportions of the correct answers before and after the intervention, Mc Nemar tests were used. A knowledge score was computed for every participant from all correct answers and converted to a scale from 0 to 100 (where 0 = none correct answer and 100 = all answers were correct). Paired Student's t-tests were used for the comparison of the knowledge score before and after the intervention. All reported p values are two-tailed and the statistical significance was set at $p < 0.05$. The mean knowledge score increased significantly from 73.4 to 80.5% after the intervention. The study showed that the implementation of educational programs contributes positively to update registered nurses' Knowledge on clinical issues, which cannot be replenished only through undergraduate education and experience.

J. Stathoulis (✉) • M. Tsironi • S. Zyga • V. Alikari • H. Bakola • G. Panoutsopoulos
Department of Nursing, Faculty of Human Movement and Quality of Life Sciences, University of Peloponnese, Sparta, Greece
e-mail: johnstathoulis@yahoo.gr; jstath@hol.gr; tsironi@uop.gr; zygas@uop.gr

N. Konofaos
Department of Informatics at the Aristotle University of Thessaloniki, Thessaloniki, Greece

E.C. Fradelos
2nd Psychiatric Department, State Mental Hospital of Attica "Daphne", Athens, Greece

Nursing Department, Faculty of Human Movement and Quality of Life, University of Peloponnese, Tripoli, Greece
e-mail: evagelosfradelos@hotmail.com

P. Vlamos (ed.), *GeNeDis 2016*, Advances in Experimental Medicine and Biology 988, DOI 10.1007/978-3-319-56246-9_7

Keywords Pulse oximetry • Nurses • Safety procedures • Nursing education • Clinical engineering

1 Introduction

A bloodless and non invasive method which plays a crucial role for the assessment of peripheral oxygen saturation (SpO_2) is pulse oximetry. Its measurement is based on the pulse created by the circulation system of human body [1, 2].The photo detection method is used for measuring the level of SpO_2. Particularly, two differing wavelengths of light (red light at 660 nm and infrared light at 940 nm) radiate through a well-perfused body area thus measuring the changing absorbance for each separate wavelength. By comparing the amount of the absorbed infrared and red light, the oxyhaemoglobin/deoxyhaemoglobin ratio is calculated by the device's algorithm and thus the percentage of saturation of arterial haemoglobin displayed on its screen. Using the above mentioned physical principle, pulse oximeters, were first developed in Japan during mid-1970s. Only a decade later, when the multiple use probe (sensor) was introduced, pulse oximeters became commercially available, [3, 4]. Because of the early identification of hypoxemia, its simplicity and cost-effectiveness, pulse oximetry is most preferable by clinicians. It has been widely used in much different clinical settings, especially in intensive care units, emergency departments and home care also.

Therefore, the necessity of health care professionals' knowledge and especially nurses' knowledge on pulse oximetry is urgent in order to achieve patient safety and to avoid misinterpretation of SpO_2 values. The topics in which nurses should be aware of are the oxyhaemoglobin dissociation curve, method's basic principles, technical characteristics, as well as the conditions under which its readings can be affected [5, 6]. Several studies have found that there is a lack on respiratory physiology, basic principles and limitations of pulse oximetry [7–14]. Remarkable knowledge deficits are observed in subjects related to what exactly pulse oximetry counts and what the normal range of values is.

2 Materials and Methods

This study comes to evaluate the knowledge of Greek registered nurses in pulse oximetry before and after an educational intervention. Anonymous self-administered validated questionnaire consisted of two parts was used to collect the data, after the written consent of the author [15]. The questionnaire consisted of two parts: the first contained demographical questions (ex. sex, age, marital status, educational level, overall working experience, etc) and the second part consisted of 21 questions (True of False type) investigating registered nurses' knowledge on the principles of oximeter's operation and the factors that affect the accuracy and reliability of its measurements. Before conducting the specific study, permission from the Hellenic Data Protection Authority was requested and obtained (Approval Number 4038, 06/08/2015). In the application form, the names of researchers who took part in the survey, the purpose and form of the study along with how the output data was going to be used, were mentioned ensuring the anonymity of the participants and the confidentiality of the results. This study followed all the fundamental principles of research. Specifically, all the information about the participants was completely anonymous and confidential. Commitment given that the information and the extracted data will be used solely for the purposes of this study. The whole educational intervention was implemented in a 2-h educational workshop including lectures, case studies and interaction questions between lecturers and workshop participants.

Concerning analysis, quantitative variables are expressed as mean values (SD). Qualitative variables are expressed as absolute and relative frequencies. For the comparison of the proportions of the correct answers before and after the intervention, Mc Nemar tests were used. A knowledge score was computed for every participant from all correct answers and converted to a scale from 0 to 100 (0 = none correct answer and 100 = all answers were correct). Paired Students' t-tests were used for the comparison of the knowledge score before and after the intervention. All reported p values are two-tailed. Statistical significance was set at $p < 0.05$ and analyses were conducted using SPSS v. 19.0 (SPSS Inc., Chicago, IL).

3 Results

Our intervention sample consisted of 78 participants (12 men and 66 women). Our sample is a sample of convenience and as in all studies conducted in Nurses in Greece the majority of participants were women. Our sample was collected within two conducted Pan-Hellenic Nursing Conferences and consisted of Nurses of many different disciplines and workplaces derived from hospitals throughout the entire Greek territory. The sample characteristics are presented throughout in Table 1. Almost half of the participants (47.4%) aged less than 29 years, 32.1% aged from 30 to 39 years and 20.5% aged from 40 to 49 years. Most of the participants were Technological Education Registered Nurses (70.5%). Additionally, 80.8% of the sample was Registered Nurses, 11.5% were Shift Managers and 7.7% were Head Nurses. Also, overall Experience as Registered Nurse for most of the participants (62.8%) was more than 4 years.

The proportion of correct answers before and after the intervention, along with the knowledge score are presented in Table 2. A significant increase in the proportion of correct answers was found after the intervention for all questions except for those numbered with 4, 11, 15, 17 and 18. The mean knowledge score (Fig. 1) increased significantly ($p < 0.001$) from 73.4 to 80.5% after the intervention.

Table 1 Sample characteristics

	N (%)
Sex	
Men	12 (15.4)
Women	66 (84.6)
Age	
20–29	37 (47.4)
30–39	25 (32.1)
40–49	16 (20.5)
Family status	
Single	53 (67.9)
Married	22 (28.2)
Divorced	3 (3.8)
Number of children, median (IQR)	0 (0–1)
Educational status	
Technological education registered nurses	55 (70.5)
University education registered nurses	13 (16.7)
MSc	9 (11.5)
PhD	1 (1.3)
Nurse specialty	8 (10.3)
Second degree	3 (3.8)
Degree in English	76 (97.4)
Hospital location	
Athens	33 (42.3)
Thessaloniki	14 (17.9)
Other prefecture	31 (39.7)
Job position	
Head nurse	6 (7.7)
Shift manager	9 (11.5)
Registered nurse	63 (80.8)
Professional status	
State employee	63 (80.8)
Private law employee	9 (11.5)
Contract employee	6 (7.7)
Overall experience as registered nurse	
0–4 years	29 (37.2)
5–9 years	21 (26.9)
10–14 years	11 (14.1)
15–19 years	11 (14.1)
>20 years	6 (7.7)
Working experience in present nursing department	
0–4 years	44 (56.4)
5–9 years	24 (30.8)
10–14 years	8 (10.3)
>15 years	2 (2.6)

Table 2 Proportion of correct answers before and after the intervention

Question	Correct answers		P Mc Nemar test
	Before	After	
	N (%)	N (%)	
1	69 (88.5)	72 (92.3)	**0.375**
2	35 (44.9)	50 (64.1)	**0.025**
3	18 (23.1)	31 (39.7)	**0.011**
4	53 (67.9)	62 (79.5)	0.064
5	40 (51.3)	49 (62.8)	**0.212**
6	66 (84.6)	63 (80.8)	**0.607**
7	72 (92.3)	74 (94.9)	**0.754**
8	71 (91.0)	74 (94.9)	**0.508**
9	66 (84.6)	60 (76.9)	**0.180**
10	65 (83.3)	77 (98.7)	**0.002**
11	74 (94.9)	71 (91.0)	0.508
12	60 (76.9)	72 (92.3)	**0.017**
13	72 (92.3)	73 (93.6)	**1.000**
14	34 (43.6)	52 (66.7)	**0.001**
15	41 (52.6)	42 (53.8)	1.000
16	55 (70.5)	55 (70.5)	**1.000**
17	66 (84.6)	68 (87.2)	0.791
18	53 (67.9)	59 (75.6)	0.263
19	60 (76.9)	72 (92.3)	**0.002**
20	66 (84.6)	72 (92.3)	**0.070**
21	67 (85.9)	71 (91.0)	**0.388**
Knowledge score (%), mean (SD)	73.4 (13.5)	80.5 (13.7)	**<0.001**[a]

[a]Paired t-test

Fig. 1 Mean knowledge score before and after the intervention

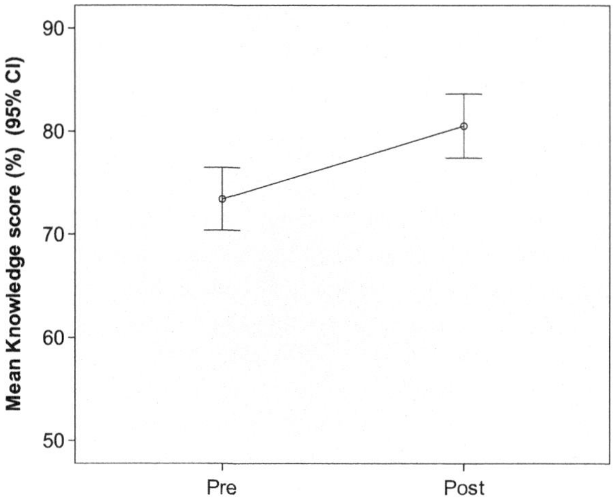

4 Discussion

This particular study, aimed to assess the level of knowledge of pulse oximetry among registered nurses, working in hospital wards and critical care units of all kind of hospitals, from medical centres through tertiary hospitals and who use pulse oximetry in everyday clinical nursing practice. The sample of 78 participants (12 men and 68 women), although it seems questionable for its size in order to make relevant conclusions, seems to be enough comparing with other similar size studies using as a tool the same questionnaire and which were targeted only at specific nursing departments like the one contacted by Kiekkas et al. [15]. However, in other international studies, there are also cases that had a much more sample size than ours [7, 16] and which also showed the same results. Today, there are several instruments for the assessment of knowledge on pulse oximetry [6, 7, 13] on the other hand, we opted to use a modified version of questionnaire previously utilized by Kiekkas et al. in his similar research contacted in 2012, for two major reasons: (a) because the specific questionnaire could be completed in a short period of time and (b) because it contained items in the form of closed questions "true/false". For the above reasons this modified questionnaire appeared to be very motivating for nurses to participate in this survey.

The findings of our study with regard to questions related to the physiology of the respiratory system and the questions related to the principles of operation of the pulse oximeter are in agreement with the results of relevant studies conducted by Harper in 2004 to a sample of 19 nurses working post-anesthesia care units (PACU), in Greece by Kiekkas et al. [15] held in a sample of 53 nurses working in anesthesiology departments [15], also by the another study of Kiekkas et al. [14] where it was held in a sample of 207 nurses working at Anesthesia Departments, Emergency Departments and ICU's and finally with the latest study held in to the same parts with the Kiekkas et al. [14] study [17]. With regard to questions related to the clinical presentation of the patient, the results are also similar.

Also, our study results are comparable to those of other relevant studies and from the literature review that conducted for the purposes of this research found the existence of relevant studies conducted in Paediatric Departments and NICU staff [10] and a group of studies in which there was a knowledge comparison between doctors and nurses on the same topics concerning pulse oximetry [7, 9, 18].

From all the above seen there is an intense research interest on knowledge of a seemingly easy-to-use medical device, but nevertheless it demonstrated that there is deficit of knowledge on this issue. The deficit related knowledge, as shown in all the relevant studies with undergraduate education of nurses and the lack of continuing education programs on the subject of pulse oximetry. Instead, we see at the same studies that the findings related to clinical experience and years of service in specialized areas (e.g., Intensive Care Unit, Emergency Department, Anesthesia departments, etc.) show no statistical significance.

5 Conclusion

Summing up, it seemed insufficient knowledge to the understanding of the physiology of the respiratory system, the principles and operation modes of the pulse oximeter and the factors that can affect the accuracy of its measurements. This leads to limited clinical value and reliability of this device. Although the use of pulse oximetry has been established internationally for many years now, it seems that in our country there is a need to develop educational programs and undergraduate education enrichment, so as to ensure safe patient care and improved in treatment quality.

References

1. Hernandez, L. 2001. Pulse Oximetry Monitoring: Clinical Implications. *Connect: The World of Critical Care Nursing* 1 (2): 59–60.
2. Howell, M.M. 2002. Pulse Oximetry: An Audit of Nursing and Medical Staff Understanding. *British Journal of Nursing (Mark Allen Publishing)* 11 (3): 191–197. doi:10.12968/bjon.2002.11.3.10066.
3. Kyes, K. 2007. Pulse Oximetry: Developmental Milestones: A Look at the Development of Pulse Oximetry. *RT Mag: Journal of Respiratory Care Practitioners* 6: 30–33.
4. Westhorpe, R.N., and C. Ball. 2008. The Pulse Oximeter. *Anaesthesia and Intensive Care* 36 (6): 767.
5. Elliott, M., and A. Coventry. 2012. Critical Care: The Eight Vital Signs of Patient Monitoring. *British Journal of Nursing* 21 (10): 621–625. doi:10.12968/bjon.2012.21.10.621.
6. Giuliano, K., and L. Liu. 2006. Knowledge of Pulse Oximetry Among Critical Care Nurses. *Dimensions of Critical Care Nursing* 25 (1): 44–49.
7. Attin, M., S. Cardin, V. Dee, L. Doering, D. Dunn, K. Ellstrom, V. Erickson, M. Etchepare, A. Gawlinski, T. Haley, E. Henneman, M. Keckeisen, M. Malmet, and L. Olson. 2002. An Educational Project to Improve Knowledge Related to Pulse Oximetry. *American Journal of Critical Care* 11 (6): 529–534.
8. Bilgin, H., O. Kutlay, D. Cevheroglu, and G. Korfali. 2000. Knowledge About Pulse Oximetry Among Residents and Nurses. *European Journal of Anaesthesiology* 17 (10): 650–651.
9. Elliott, M., R. Tate, and K. Page. 2006. Do Clinicians Know how to use Pulse Oximetry? A Literature Review and Clinical Implications. *Australian Critical Care* 19 (4): 139–144. doi:10.1016/S1036-7314(06)80027-5.
10. Fouzas, S., P. Politis, E. Skylogianni, T. Syriopoulou, K.N. Priftis, A. Chatzimichael, and M.B. Anthracopoulos. 2010. Knowledge on Pulse Oximetry Among Pediatric Health Care Professionals: A Multicenter Survey. *Pediatrics* 126 (3): e657–e662. doi:10.1542/peds.2010-0849.
11. Harper, J. 2004. Post-Anesthesia Care Unit Nurses' Knowledge of Pulse Oximetry. *Journal for Nurses in Staff Development* 20 (4): 177–180.
12. Huijgen, Q.C.A., T.W. Effing, K.L. Hancock, T.R. Schermer, and A.J. Crockett. 2011. Knowledge of Pulse Oximetry Among General Practitioners in South Australia. *Primary Care Respiratory Journal* 20 (4): 456–458. doi:10.4104/pcrj.2011.00088.
13. Johnson, C.L., M.A. Anderson, and P.D. Hill. 2012. Comparison of Pulse Oximetry Measures in a Healthy Population. *Medsurg Nursing* 21 (2): 70–75. quiz 76.
14. Kiekkas, P., A. Alimoutsi, F. Tseko, N. Bakalis, N. Stefanopoulos, T. Fotis, and E. Konstantinou. 2013. Knowledge of Pulse Oximetry: Comparison Among Intensive Care, Anesthesiology and Emergency Nurses. *Journal of Clinical Nursing* 22 (5-6): 828–837. doi:10.1111/j.1365-2702.2012.04151.x.

15. Kiekkas, P., F. Tseko, A. Alimuci, N. Stefanopoulos, M. Papadimitriou, M. Karga, and E. Konstantinou. 2012. Evaluation of Pulse Oximetry Knowledge of Nurses Employed in Anesthesiology Departments. *Perioperative Nursing-Quarterly Scientific, Online Official Journal of GORNA* 1 (3 September–December 2012): 94–102.
16. Walters, P.T. 2007. Pulse Oximetry Knowledge and Its Effects on Clinical Practice. *British Journal of Nursing* 16 (21): 1332–1340. doi:10.12968/bjon.2007.16. 21.27721.
17. Milutinovic, D., G. Repic, and B. Arandelovic. 2016. Clinical Nurses' Knowledge Level on Pulse Oximetry: A Descriptive Multi-Centre Study. *Intensive and Critical Care Nursing*. doi:10.1016/j.iccn.2016.05.006.
18. Koutsouki, S., and D. Kosmidis. 2007. Knowledge of Pulse Oximetry Among Medical and Nursing Staff. *To Vima tou Asklipiou* 6 (2): 1–7.

Psychodynamic Leadership Approach and Leader-Member Exchange (LMX): A Psychiatric Perspective on Two Leadership Theories and Implications for Training Future Psychiatrist Leaders

Christos Plakiotis

Abstract An increased emphasis in recent years on psychiatrists as healthcare leaders has not only drawn attention to the skills they can bring to this role but has also raised questions about how to best train and prepare them to assume leadership responsibilities. Such training should not be conducted in isolation from, and oblivious to, the wide-ranging expertise in human behaviour and relationships that psychiatrists can bring to the leadership arena. The aim of this theoretical paper is to draw attention to how psychiatrists can use their existing knowledge and skill set to inform their understanding of leadership theory and practice. In particular, the Psychodynamic Leadership Approach and Leader-Member Exchange theory are compared and contrasted to illustrate this point. The former represents a less well-known approach to leadership theory and practice whereas the latter is a widely familiar, conventional theory that is regularly taught in leadership courses. Both are underpinned by their emphasis on leader-follower relationships—and human relationships more broadly—and are intuitively appealing to psychiatrists endeavouring to understand aspects of organisational behaviour in the healthcare settings in which they work and lead. The application of these theories to assist reflection on and understanding of professional and personal leadership behaviours through leadership-oriented Balint-style groups and 360-degree appraisal is proposed. It is hoped that this paper will serve to stimulate thought and discussion about how leadership training for future psychiatrists can be tailored to better harness their existing competencies, thereby developing richer formative learning experiences and, ultimately, achieving superior leadership outcomes.

C. Plakiotis (✉)
Monash Ageing Research Centre (MONARC), Monash University, Melbourne, VIC, Australia

Department of Psychiatry, School of Clinical Sciences at Monash Health, Monash University, Melbourne, VIC, Australia

Aged Persons Mental Health Service, Monash Health, Melbourne, VIC, Australia

Aged Psychiatry Academic Unit (Monash University), c/o Medical Administration, Kingston Centre, Warrigal Road, Cheltenham, VIC, 3192, Australia
e-mail: Chris.Plakiotis@monash.edu

P. Vlamos (ed.), *GeNeDis 2016*, Advances in Experimental Medicine and Biology 988, DOI 10.1007/978-3-319-56246-9_8

Keywords Psychiatrist • Leadership • Management • Leadership training • Psychodynamic leadership approach • Leader-member exchange (LMX) • Balint groups • 360-degree appraisal

1 Introduction

Leadership of mental health services is a multidisciplinary endeavour involving professionals from a range of clinical and non-clinical backgrounds. As clinical experts across the spectrum of mental health and illness, psychiatrists are potentially important players in the leadership arena but, tasked with extensive clinical responsibilities, may find themselves all too easily deferring to non-medical colleagues to administer the health services in which they work. Not utilising psychiatrists' skills to their full potential, including in the leadership domain, may be to the detriment of mental health services and the patients whom they treat.

In recent years, there has been an increased interest in psychiatrists as leaders of mental health services [1–3]. The important role that psychiatrist leaders can play—in partnership with their clinical counterparts—in facilitating clinically sound organisational change and innovation has been identified [4]. In discussing the contribution that doctors in general can make to the leadership and management of healthcare services, the need to provide doctors with specialised training in this area, as distinct to their existing clinical skills, is frequently emphasised [5, 6]. Goodall [7] noted, however, that leaders' 'expert knowledge' (technical competence) has been shown to highly correlate with organisational performance in many settings, and for this reason proposed that placing psychiatrists in leadership positions might lead to better organisational performance in mental healthcare.

This increased emphasis on psychiatrists as healthcare leaders not only draws attention to the skills they can bring to this role but also raises questions about how to best train and prepare them to assume leadership responsibilities. Adopting a generic approach to leadership training for psychiatrists may lead to their unique understanding of diverse areas pertaining to human behaviour, psychological factors, communication, interpersonal relationships and systems theory being overlooked or underutilised in developing formative learning experiences in this field.

2 Objectives

The aim of this theoretical paper is to draw attention to how psychiatrists can use their existing knowledge and skill set to inform their understanding of leadership theory and practice. A psychiatric perspective on two leadership theories will be provided as an example of this approach. In particular, the key principles of the Psychodynamic Leadership Approach and Leader-Member Exchange (LMX) theory will be reviewed as a basis for comparing and contrasting the similarities and differences and advantages and limitations of these two theories. Consideration will also be given to how these theories can be applied to improving leadership practice. While the paper is focussed on psychiatrists as leaders, the ideas presented are equally relevant to clinicians from non-medical disciplines who are involved in, or aspire to, senior leadership roles in mental health services.

3 Psychodynamic Leadership Approach

3.1 Why Apply Psychodynamic Theory to Leadership?

An understanding of human behaviour in organisations and the skill to motivate others to willingly do things are central to effective leadership. Successful leaders are able to grasp group processes and calm collective anxieties while satisfying followers' needs and inspiring them to attain their hopes and aspirations. While most leadership theory and practice focusses on phenomena that are observable, measurable, rational and conscious, organisations are an extension of everyday life in which unconscious and potentially irrational factors drive human behaviour [8, 9]. The psychodynamic leadership approach, as espoused by proponents such as Zaleznik [10] and Maccoby [11], applies the principles of depth psychology and psychoanalytic theory to leadership and management research and practice [12, 13]. It offers a perspective on understanding and managing the complexities of organisational life by examining the underlying subtexts of human behaviour [9].

3.2 Basic and Applied Psychodynamic Concepts

An individual's first encounter with leadership—as both follower and leader—is through interaction with one's *family of origin* (parents) in early life and the socialisation experiences this provides [13]. Past interactions with significant others are repeatedly played out in a scripted manner in an *inner theatre* of the mind and may result in responses to the perceived (rather than real) intentions of people at work [9]. Parental values are carried into adulthood but may be altered through *maturation* and *individualisation* [13].

Exact behavioural outcomes are not easily foreseen. Either domineering or inclusive leadership styles may mirror or contrast with strict or permissive parenting experiences. In turn, employees may respond in a *dependent*, antagonistic or *independent* manner towards leaders depending on their own early life experiences. Dependency may see employees unconsciously seeking refuge in a 'parent-like' leader to guard against a sense of vulnerability [9]. An independent response entails challenging the limits of the leader-follower relationship and questioning and disregarding non-meaningful directions [13].

Suppression of socially unacceptable thoughts, feelings and behaviours (e.g. urge towards physical aggression) deeply within one's conscience and replacement with more socially acceptable alternatives (e.g. verbal debate) is encouraged through socialisation but may only partially succeed. Thus while a leader may perceive any authoritarian tendencies as successfully suppressed, followers may continue to recognise these in displayed behaviours (as a *shadow of the ego*) [13].

Despite training leaders in responding appropriately to followers, *regression* may occur under pressure towards familiar but maladaptive behavioural patterns originating in childhood. Taught leadership techniques can only overlap with, and not supplant, these deep-seated patterns [13]. Employees may also resort to regressive *social defence mechanisms* and problematic behaviours to cope with work stressors that are not effectively contained by the leader. Employees may divide colleagues and supervisors into allies or adversaries (*splitting*) and display hostility through absenteeism, work avoidance and resignation (*fight or flight* responses). Protection against perceived weakness may be sought by *pairing* up with others, further exacerbating conflict. In turn, senseless rules and regulations may be relied upon by leaders to help employees distance themselves from their anxieties [9].

Leaders may concentrate on boosting their approval among employees (*idealisation*) rather than advancing organisational objectives, in a mutually flattering mirroring process. *Identification with the aggressor* may emerge among employees as a defence against hostile leadership tactics and honesty and truth may be compromised in a *folie à deux* (joint delusional) process to avoid disputes with a leader who is out of touch with reality [9, 14, 15].

3.3 A Psychodynamic Perspective on Leadership Development and Styles

Leadership manifests when a group of people project their values onto an idealised individual via projective identification [16–18]. As long as the group's identification needs are met, the leader is overvalued and negative aspects overlooked. However, this unwavering support may be replaced with condemnation where the group's expectations are no longer satisfied [13, 19].

From a psychodynamic perspective, many organisations directly or indirectly treat followers as child-like, immature individuals whereas leaders are designated the adult role [20]. Complementary transactions (e.g. between a leader-parent and employee-child) are associated with stability, and mismatching transactions (e.g. between a leader-parent and employee-adult) with instability, in leader-follower relationships [13, 21, 22].

Narcissism, encompassing the spectrum of rational self-respect to unfettered selfishness, provides leaders with the confidence needed to rouse others' allegiance [9, 23]. Maccoby [24] asserted that narcissistic (proud, self-protective, aggressive, pursuing good impressions) and narcissistic-coercive (meticulous and self-assured) leadership styles are most consistent with the concept of capable leadership. Through Freud's processes of identification and idealisation, such leaders symbolise their followers' principles while rising above them [13].

There are limitations to the benefits of narcissism, however. *Constructive, or healthy, narcissists* have had positive childhood experiences and are able to inspire others towards high achievements in a reflective, empathic manner. Conversely, *reactive, or excessive, narcissists* are preoccupied with their own status, accomplishments and ability to dominate others as a means of redressing childhood trauma. Removed from reality, they can inflict chaos upon an organisation [9].

3.4 Implications for Leadership Training

Peer group coaching under the guidance of a skilled facilitator can be used advantageously in psychodynamic leadership training to create reflective leaders, by allowing collective consideration of relational and leadership styles, work habits, problem-solving and decision-making in the wider organisational environment. Appropriate defence mechanisms, emotional expression and reality orientation can be addressed in this context [8, 9, 25].

4 Leader-Member Exchange (LMX) Theory

4.1 Premises of Leader-Member Exchange (LMX) Theory

The central concept of the Leader-Member Exchange (LMX) theory is that an exchange relationship gradually forms between a leader and follower founded on the degree of like-mindedness and the follower's skill and reliability. It is a role-making process whereby the follower's role is jointly established [26–28]. The theory presupposes that leaders lead and followers follow because they derive gains from each other in a mutually beneficial exchange relationship [29], the quality of which is the focus of investigation [13, 30].

4.2 Quality of Leader-Member Relationships

LMX theory describes vertical dyadic linkages between a superior and his or her subordinates as 'low-quality' or 'high-quality' leader-member relations. The former occur within the constraints of formal contractual agreements between the leader and members of the 'out-group'. The latter are extended relationships based on mutual trust, rapport and negotiation between the leader and members of the 'in-group' [13]. The theory posits that leaders generally establish high-quality interactions with a few dependable followers who act as aides and confidants [28]. Whether subordinates become members of the 'in-group' or 'out-group' depends on how the delineation of leader-member responsibilities evolves as well as their personal appraisal of the risks and benefits of a closer leader-member relationship and the wider organisational responsibilities beyond the formal contract that this will entail [13, 26].

Future leader-member relationships are influenced by prior experiences as well as the temperament, beliefs and personal and professional skills of both parties in the dyad [13]. There is evidence that the quality of the leader-member exchange is established very soon after the relationship starts, with first impressions of whether the leader and follower can meet each other's expectations counting [31]. The likelihood of 'in-group' membership is enhanced by a positive atmosphere between leader and follower and political savviness on the follower's part [32]. As the relationship develops, leaders offer followers collaborative duties extending beyond formally-defined contractual obligations by way of a trial. If a follower's reaction to this is interpreted positively by the leader, further opportunities for closer collaboration will be presented; otherwise, the relationship will revert to formally defined norms [13, 33].

The quality of the leader-member exchange is continuously evolving and has been postulated to be influenced by: (1) perceived contribution to the exchange; (2) loyalty; (3) affect (perceived interpersonal attraction) [33]; and (4) professional respect [13, 34].

4.3 Developing Leader-Member Relationships

Graen and Uhl-Bien [35] delineated three phases in the development of high-quality, 'in-group' relationships as part of leadership making [13, 32].

1. Phase 1 ('role-taking' or 'stranger phase'): Mutual awareness of how respect is perceived and expected by leader and follower.
2. Phase 2 ('role-making' or 'acquaintance phase'): Deepening of trust as a basis for the relationship progressing and leader and follower shaping one another's views and actions.
3. Phase 3 ('role routinisation' or 'partner phase'): A high-quality pattern of leader-member exchange is now standard.

4.4 Effectiveness of Leader-Member Relationships

Compared to out-group members, who receive little attention or support, in-group members perform better and are more dedicated to organisational goals [36]. They are more content at work and less likely to resign. In-group members in a healthcare context displayed superior citizenship behaviours—such as working harder to perform their roles—that increase the likelihood of a leader conferring additional resources, e.g. a budget increase or pay rise [37]. Furthermore, in-group members in various industrial settings are more likely to display helpful workplace-related behaviours extending beyond the call of duty and role definition [38, 39].

In the banking sector, managers perceived as being superior in status were able to forge higher-quality relationships with employees [40]. Furthermore, better LMX relationships across a range of industries increased the transformational effectiveness of leaders, with positive implications for teamwork [41]. High quality LMX relationships were shown to facilitate emergent leadership in a telecommunications team with a shared vision, which in turn may enhance team performance [42]. Better LMX relationships among group leaders in the manufacturing sector resulted in improved communication about, and attention to, safety and fewer work-related accidents [39, 43].

5 Similarities and Differences

A common underpinning of both the psychodynamic leadership approach and LMX theory is their focus on leader-follower relationships. LMX theory is more akin to a range of other leadership theories in focussing on observable aspects of leadership that can potentially be subjected to qualitative and quantitative investigation. The psychodynamic approach can complement and add greater depth to the understanding of leader-follower relationships afforded by the LMX model. Its underlying precepts may be harder to recognise or demonstrate, however, limiting the ability to undertake empirical leadership studies. Furthermore, whether in practice, training or research applications, psychodynamic theory may (ironically) be met with greater resistance from potential participants who may be unprepared to confront the deeper understanding of their behaviour encouraged by this approach. Embarking on an in-depth psychodynamic understanding of oneself outside a longer-term individual psychotherapeutic relationship may entail risks for some participants that need to be considered in rolling out time-limited, psychodynamically-oriented leadership workshops. With this perspective in mind, it is worth considering more fully the advantaged and limitations of each approach.

6 Advantages and Limitations

Strengths of the psychodynamic approach include its attention to the intricacy of relationships, conscious and unconscious motives, how leaders originate and are maintained, and relevance to a range of organisations [13]. Furthermore, the psychodynamic approach aims to generate reflective practitioners by concentrating on leaders' and followers' insight into the factors motivating their relationship.

Coaching, case studies and 360-degree feedback can be employed to provide a meaningful understanding of individual and organisational behaviour [9].

Strengths of LMX theory are that it comprehensively describes leadership processes (particularly the difference between 'in-group' and 'out-group' relationships), draws attention to the importance of communication in leadership [44], and is validated by research findings in the public sector [13, 45].

Weaknesses of the psychodynamic approach include overlooking the situational context in which leadership occurs and application of subjective Freudian theory—which is hard to scientifically verify—to 'normal' leader-follower behaviour [9, 13]. As the route to psychodynamic change varies between individuals, a uniform training model may be difficult to achieve. Furthermore, important structural factors may be missed by an individually-focussed training program [9].

LMX theory also has limitations. A leader must simultaneously value all followers while fostering a closer relationship with a subset of aides in the interests of advancing organisational goals [28]. Accordingly, a better understanding of how differing dyadic relationships impact on each other and collective organisational efficacy is needed [46], along with more research regarding how LMX relationships develop and role-making occurs over time. The impact of situational factors (e.g. demographic, job and organisational variables) on LMX relationships also merits further investigation [28, 47].

7 Discussion

Although the Psychodynamic Leadership Approach may appear abstract and distant from the reality in which organisations operate, there is some literature describing its practical application in the business sector. For example, Kets de Vries et al. [48] described a psychodynamically-informed leadership development program for business executives incorporating predominantly group-based leadership coaching practices. Similarly, Bell and Huffington [49] outlined a systems psychodynamic approach to leadership coaching designed to promote self-awareness and attunement to organisational factors that can impede or facilitate the leadership role. Ward et al. [50] examined the psychotherapeutic modalities that psychodynamic group coaching draws upon, such as intensive short-term dynamic psychotherapy and group therapy, and concluded that each was separately effective. However, further research into the effectiveness of psychodynamic leadership coaching as an entity in its own right appears warranted.

Adapting psychodynamic leadership coaching for use among psychiatrist leaders and other senior mental health leaders and managers is an obvious possibility. However, psychodynamic leadership coaching, as implemented in the business world, is not entirely alien to psychoanalytically-informed approaches to supervision and professional development already in use within medicine and psychiatry in particular. 'Balint groups' are a well-established approach to fostering reflective practice among medical practitioners and other health professionals involved in direct patient care [51]. Balint groups allow doctors to meet regularly to discuss the cases of patients they perceive as a source of interpersonal difficulties in their day-to-day work. Balint group leaders are either psychoanalysts or professionals working in closely-related fields. Their role is to foster a tolerant and safe environment in which cases can be presented by participants and subsequently commented on and discussed without fear of criticism or interrogation by other group members [52, 53]. Groups are made up of 6 to 12 members and convene over several years at weekly to monthly intervals. Over time, this approach assists group members to gain a broader outlook on the initial challenges they encountered and promotes improved doctor-patient interactions [54]. The benefits of Balint groups in introducing psychiatry trainees to psychological processes [55] and serving as an introduction to more comprehensive psychotherapeutic practice [56] have been reported.

It is evident that the Balint group methodology could be adapted from a patient-oriented to a leadership-oriented focus to assist psychiatrists and other mental health professionals with leadership responsibilities to reflect on and better understand the factors underlying professional (e.g. establishing organisational goals) and personal (e.g. building good workplace relations) leadership behaviours they display at work [57]. Insights gained may be invaluable in overcoming complex interpersonal challenges encountered in the leadership role. In Australia, the Royal Australian and New Zealand College of Psychiatrists' (RANZCP) Continuing Professional Development (CPD) Program provides a structure under which mandatory peer review groups are constituted to support practicing psychiatrists in exercising their duties [58]. Peer review activities will also be a mandatory part of the new Continuing Education Program (CEP) of the Royal Australasian College of Medical Administrators (RACMA) that is due to be implemented fully in 2017 [59]. Although RANZCP peer review groups are not required to adopt an explicitly Balint-style methodology or focus exclusively on leadership issues, psychiatrists with an interest or involvement in leadership can draw upon both these perspectives in presenting topics and participating in related group discussions.

The approaching introduction of revalidation for doctors by the Medical Board of Australia is likely to entail increased rigour in the CPD requirements expected by specialist medical colleges of their individual members [60, 61]. In view of the evolving regulatory climate in which medicine is practiced, there may be dual benefits for psychiatrists in leadership roles to establish Balint-style groups focussed on leadership topics—both to assist them in exercising their leadership responsibilities more effectively but also to satisfy regulatory authorities as to their continued learning and reflective practice in this field. For such groups to function effectively, however, it is essential that material discussed remains confidential within the groups themselves (as opposed to group attendance records, which may be of relevance to regulators). Ensuring confidentiality in leadership-oriented Balint-style groups is even more important than in their patient-oriented prototypes, as effectively de-identifying discussions about workplace colleagues may be problematic in small organisations. Indeed, this may be a major barrier to effective group functioning if not handled appropriately. Groups should preferably be established with psychiatrist leaders outside one's own workplace to address this issue.

In the spirit of interprofessional learning and practice [55, 62], such groups need not be restricted to psychiatrists. Indeed, there may be advantages in constituting groups comprised of psychiatrists and other senior medical administrators from other areas of medicine, or psychiatrists and senior mental health service leaders and managers from non-medical disciplines. The former approach allows psychiatrists to compare and contrast their own leadership experiences with those of colleagues in other medical fields whereas the latter may assist in better understanding the experiences and perspectives of non-medical team members in services akin to those which they lead. In both cases, psychiatrists may offer unique insights to other group members in relation to the leadership challenges that are discussed.

Balint groups constituted for doctors and healthcare workers outside the psychiatric sector rely on the psychodynamic expertise of the group leader [52, 53] but do not entail specialised knowledge of psychodynamic theory or related terminology by group participants [51]. However, in constituting such groups for psychiatrists and other mental health service leaders, it may be assumed that this group of professionals is likely to bring along a greater understanding of the complexity of human interactions than the average healthcare professional. This may facilitate deeper engagement with the psychodynamic aspects of the leadership dilemmas that are brought to the group for consideration.

Furthermore, while Balint groups may have their origins in psychodynamic thinking, the similarities between the Psychodynamic Leadership Approach and LMX theory suggest that the latter may also provide a useful framework for understanding leader-follower relationships within a Balint-style group. The potential benefits of effective leader-member relationships for the multidisciplinary mental health teams that psychiatrists lead—and the patients that team members collectively care for—are manifold. For example, there is evidence from the manufacturing sector that high-quality leader-member exchanges may improve communication about safety, with an increased commitment to safety and fewer workplace accidents as a result [39, 43]. The scope to translate this research into improved risk management and patient safety in a multidisciplinary mental health setting is readily apparent.

Similarly, the finding that positive leader-member exchanges resulted in superior citizenship behaviours in a hospital setting [37, 39] has implications for psychiatrist leaders in the multidisciplinary mental health sector. An improved understanding of the disparity with which members of the 'in-group' or 'out-group' may be inadvertently approached by leaders can enable psychiatrists to reduce unjustified preferential treatment between multidisciplinary team members and promote better leader-member relationships with all clinicians whom they lead. In turn, this may strengthen team members' commitment to organisational goals and to the teamwork needed to achieve these, as has been demonstrated in the retail sector [36, 39].

There is also evidence that emergent leadership is more likely to occur in teams characterised by positive LMX relationships and a shared vision for the future

[39, 42]. It is important for psychiatrist leaders to be aware of and actively foster this process in leading clinical teams and services more broadly, in the interests of identifying and supporting potential future leaders in both the clinical and administrative domains. The need to sponsor effective mental health leaders is especially pertinent and urgent in the face of a looming shortage globally in the availability of psychiatrists and other specialist mental healthcare workers [63].

A further example of how the Psychodynamic Leadership Approach and LMX theory can be incorporated into existing approaches for leadership development is provided by 360-degree appraisal or multi-source feedback, a process for assessing doctors' performance and professional behaviours by systematically obtaining feedback from colleagues and patients [64–66]. The introduction of revalidation may see 360-degree appraisal become a more widespread CPD activity for Australian doctors, as is currently the case in the United Kingdom [67]. Indeed, RACMA has already introduced this technique as an optional CEP activity and, from 2017, it will be one of the ways in which the mandatory requirement for annual peer review activities can be met [59]. For psychiatrist leaders, the Psychodynamic Leadership Approach and LMX theory dovetail well with the methodology of 360-degree appraisal. Both theories can assist psychiatrist leaders to maximise the learning derived from this process by offering useful frameworks for reflecting on and making sense of feedback received.

8 Conclusion

The important role that psychiatrists can play in the effective leadership and management of psychiatric services is being increasingly recognised. Psychiatrists' existing clinical expertise does not represent a sufficient skill set on its own to prepare them for leadership positions and cannot supplant the need for specialised training in this field. Such training, however, should not be conducted in isolation from, and oblivious to, the wide-ranging expertise in human behaviour and relationships that psychiatrists can bring to the leadership arena.

In this paper, the Psychodynamic Leadership Approach and Leader-Member Exchange theory have been compared and contrasted to illustrate this point. The former represents a less well-known approach to leadership theory and practice whereas the latter is a widely familiar, conventional theory that is regularly taught in leadership courses. Both are underpinned by their emphasis on leader-follower relationships—and human relationships more broadly—and are intuitively appealing to psychiatrists endeavouring to understand aspects of organisational behaviour in the healthcare settings in which they work and lead. These theories can be applied in leadership-oriented Balint-style groups to reflect on and work through challenging leadership dilemmas. They can also be used by psychiatrists to make sense of feedback received about their leadership behaviours through 360-degree appraisal. It is hoped that this paper will serve to stimulate thought and discussion about how leadership training for future psychiatrists can be tailored to better harness a range of existing psychiatric competencies that are unique among medical practitioners and potentially applicable to leadership roles. Such an approach may be helpful in developing richer formative learning experiences and, ultimately, achieving superior leadership outcomes in mental healthcare.

Funding No sources of funding were received for this work.

Conflicts of Interest The author reports no conflicts of interest and is alone responsible for the content and writing of this paper.

References

1. Bhugra, D., S. Bell, and A. Burns. 2007. *Management for Psychiatrists*. 3rd ed. Champaign, IL: RCPsych Publications.
2. Bhugra, D., P. Ruiz, and S. Gupta. 2013. *Leadership in Psychiatry*. Somerset: Wiley.
3. Bhugra, D., and A. Ventriglio. 2016. Medical Leadership in the 21st Century. Australas Psychiatry 1039856216641308.
4. Callaly, T., and H. Minas. 2005. Reflections on Clinician Leadership and Management in Mental Health. *Australasian Psychiatry* 13: 27–32.
5. Christensen, T., and J.K. Stoller. 2016. Physician Leadership Development at Cleveland Clinic: A Brief Review. *Australasian Psychiatry* 24: 235–239.
6. Kyratsis, Y., K. Armit, A. Zyada, and P. Lees. 2016. Medical Leadership and Management in the United Kingdom. *Australasian Psychiatry* 24: 240–242.
7. Goodall, A.H. 2015. A Theory of Expert Leadership (TEL) in Psychiatry. Australas Psychiatry 1039856215609760.
8. Kets de Vries, M.F.R. 2006. *The Leader on the Couch: A Clinical Approach to Changing People and Organizations*. San Francisco: Jossey-Bass.
9. Kets de Vries, M.F.R., and A. Cheak. 2014. *Psychodynamic Approach*. France: Fontainebleau Cedex.
10. Zaleznik, A. 1981. Managers and Leaders: Are They Different? *The Journal of Nursing Administration* 11: 25–31.
11. Maccoby, M. 1976. *The Gamesman: The New Corporate Leaders*. New York: Simon and Schuster.
12. Freud, S. 1938. *The Basic Writings of Sigmund Freud (1995)*. Modern Library ed. New York: Modern Library.
13. Winkler, I. 2010. *Contemporary Leadership Theories: Enhancing the Understanding of the Complexity, Subjectivity and Dynamic of Leadership*. Berlin: Physica-Verlag.
14. Kets de Vries, M.F.R. 1979. Managers Can Drive Their Subordinates Mad. *Harvard Business Review*: 125–134.
15. ———. 2001. *Struggling with the Demon: Perspectives on Individual and Organizational Irrationality*. Madison, CT: Psychosocial Press.
16. Pervin, L.A. 1993. *Personality: Theory and Research*. 6th ed. New York: Wiley.
17. Ogden, T.H. 1979. On Projective Identification. *The International Journal of Psycho-Analysis* 60: 357–373.
18. ———. 1992. *The Matrix of the Mind Object Relations and the Psychoanalytic Dialogue*. London: Karnac.
19. Goethals, G.R. 2005. The Psychodynamics of Leadership: Freud's Insights and Their Vicissitudes. In *The Psychology of Leadership: New Perspectives and Research*, ed. D.M. Messick and R.M. Kramer, 99–116. Mahwah, NJ: Lawrence Erlbaum Associates.

20. Argyris, C. 1957. *Personality and Organization: The Conflict Between System and the Individual*. New York: Harper.
21. Berne, E. 1961. *Transactional Analysis in Psychotherapy: A Systematic Individual and Social Psychiatry*. New York: Grove Press.
22. Harris, T.A. 1969. *I'm OK, You're OK: A Practical Guide to Transactional Analysis*. New York: Harper & Row.
23. Freud, S. 1953. *The Standard Edition of the Complete Psychological Works of Sigmund Freud*. London: Hogarth Press.
24. Maccoby, M. 2000. Narcissistic Leaders: The Incredible Pros, the Inevitable Cons. *Harvard Business Review*: 68–77.
25. McCullough Vaillant, L. 1997. *Changing Character: Short-Term Anxiety-Regulating Psychotherapy for Restructuring Defenses, Affects, and Attachment*. New York: Basic Books.
26. Dansereau, F., G. Graen, and W.J. Haga. 1975. A Vertical Dyad Linkage Approach to Leadership within Formal Organizations: A Longitudinal Investigation of the Role Making Process. *Organizational Behavior and Human Performance* 13: 46–78.
27. Graen, G., and J.F. Cashman. 1975. A Role Making Model of Leadership in Formal Organizations: A Developmental Approach. In *Leadership Frontiers*, ed. J.G. Hunt and L.L. Larson, 143–165. Comparative Administration Research Institute, Graduate School of Business Administration, Kent State University: Kent, OH.
28. Yukl, G.A. 2013. *Leadership in Organizations Global Edition*. 8th ed. Harlow: Pearson.
29. Messick, D.M. 2005. On the Psychological Exchange Between Leaders and Followers. In *The Psychology of Leadership: New Perspectives and Research*, ed. D.M. Messick and R.M. Kramer, 83–98. Mahwah, NJ: Lawrence Erlbaum Associates.
30. van Breukelen, W., B. Schyns, and P. Le Blanc. 2006. Leader-Member Exchange Theory and Research: Accomplishments and Future Challenges. *Leadership* 2: 295–316. doi:10.1177/1742715006066023.
31. Liden, R.C., S.J. Wayne, and D. Sitwell. 1993. A Longitudinal Study on the Early Development of Leader-Member Exchanges. *The Journal of Applied Psychology*: 662–674.
32. DuBrin, A.J., C. Dalglish, and P. Miller. 2006. *Leadership*. 2nd Asia-Pacific ed. Australia, Milton, Queensland: John Wiley and Sons.
33. Dienesch, R.M., and R.C. Liden. 1986. Leader-Member Exchange Model of Leadership: A Critique and Further Development. *Academy of Management Review* 11: 618–634.
34. Liden, R.C., and J.M. Maslyn. 1998. Multidimensionality of Leader-Member Exchange: An Empirical Assessment Through Scale Development. *Journal of Management* 24: 43–72.
35. Graen, G.B., and M. Uhl-Bien. 1991. The Transformation of Professionals into Self-Managing and Partially Self-Designing Contributors: Toward a Theory of Leadership-Making. *Journal of Manufacturing Systems* 3: 25–39.
36. Klein, H.J., and J.S. Kim. 1998. A Field Study of the Influence of Situational Constraints Leader-Member Exchange, and Goal Commitment on Performance. *Academy of Management Journal* 41: 88–95.
37. Settoon, R.P., N. Bennett, and R.C. Liden. 1996. Social Exchange in Organizations: Perceived Organizational Support, Leader–Member Exchange, and Employee Reciprocity. *The Journal of Applied Psychology* 81: 219.
38. Tierney, P., and T.N. Bauer. 1996. A Longitudinal Assessment of LMX on Extra-Role Behavior. *Academy of Management Best Papers Proceedings* 8: 298–302.
39. DuBrin, A.J. 2016. *Leadership: Research Findings, Practice, and Skills*. 8th ed. Boston, MA: Cengage Learning.
40. Venkataramani, V., S.G. Green, and D.J. Schleicher. 2010. Well-Connected Leaders: The Impact of Leaders' Social Network Ties on LMX and Members' Work Attitudes. *The Journal of Applied Psychology* 95: 1071–1084. doi:10.1037/a0020214.
41. Piccolo, R.F., and J.A. Colquitt. 2006. Transformational Leadership and Job Behaviors: The Mediating Role of Core Job Characteristics. *Academy of Management Journal* 49: 327–340.

42. Zhang, Z., D.A. Waldman, and Z. Wang. 2012. A Multilevel Investigation of Leader–Member Exchange, Informal Leader Emergence, and Individual and Team Performance. *Personnel Psychology* 65: 49–78.
43. Hofmann, D.A., and F.P. Morgeson. 1999. Safety-Related Behavior as a Social Exchange: The Role of Perceived Organizational Support and Leader-Member Exchange. *Journal of Applied Psychology* 84 (2): 286–296.
44. Northouse, P.G. 1997. *Leadership: Theory and Practice*. Thousand Oaks, CA: SAGE Publications.
45. Graen, G.B., and M. Uhl-Bien. 1995. Relationship-Based Approach to Leadership: Development of Leader-Member Exchange (LMX) Theory of Leadership over 25 Years: Applying a Multi-Level Multi-Domain Perspective. *The Leadership Quarterly* 6: 219–247.
46. Henderson, D.J., R.C. Liden, B.C. Glibkowski, and A. Chaudhry. 2009. LMX Differentiation: A Multilevel Review and Examination of Its Antecedents and Outcomes. *The Leadership Quarterly* 20: 517–534. doi:10.1016/j.leaqua.2009.04.003.
47. Green, S.G., S.E. Anderson, and S.L. Shivers. 1996. Demographic and Organizational Influences on Leader–Member Exchange and Related Work Attitudes. *Organizational Behavior and Human Decision Processes* 66: 203–214.
48. Kets de Vries, M.F.R., E. Florent-Treacy, L. Guillen, and K. Korotov. 2010. The Proof is in the Pudding: An Integrative, Psychodynamic Approach to Evaluating a Leadership Development Program. In *The Coaching Kaleidoscope: Insights from the Inside*, ed. M. Kets de Vries, L. Guillen, K. Korotov, and E. Florent-Treacy, 3–19. Basingstoke: Palgrave MacMillan.
49. Bell, J., and C. Huffington. 2008. Coaching for Leadership Development: A Systems Psychodynamic Approach. In *Leadership Learning: Knowledge into Action*, ed. K. Turnbull James and J. Collins, 93–110. Basingstoke; New York: Palgrave Macmillan.
50. Ward, G., E. van de Loo, and S. ten Have. 2014. Psychodynamic Group Executive Coaching: A Literature Review. *International Journal of Evidence Based Coaching and Mentoring* 12: 63–78.
51. Balint, M. 1957. *The Doctor, his Patient and the Illness*. London: Pitman Medical.
52. Samuel, O. 1989. How Doctors Learn in a Balint Group. *Family Practice* 6: 108–113.
53. Johnson, A.H., D.E. Nease, L.C. Milberg, and R.B. Addison. 2004. Essential Characteristics of Effective Balint Group Leadership. *Family Medicine* 36: 253–259.
54. Van Roy, K., S. Vanheule, and R. Inslegers. 2015. Research on Balint Groups: A Literature Review. *Patient Education and Counseling* 98: 685–694. doi:10.1016/j.pec.2015.01.014.
55. Graham, S., L. Gask, G. Swift, and M. Evans. 2009. Balint-Style Case Discussion Groups in Psychiatric Training: An Evaluation. *Academic Psychiatry* 33: 198–203.
56. Fitzgerald, G., and M.D. Hunter. 2003. Organising and Evaluating a Balint Group for Trainees in Psychiatry. *Psychiatric Bulletin* 27: 434–436.
57. Mastrangelo, A., E.R. Eddy, and S.J. Lorenzet. 2004. The Importance of Personal and Professional Leadership. *Leadership and Organization Development Journal* 25: 435–451. doi:10.1108/01437730410544755.
58. The Royal Australian and New Zealand College of Psychiatrists. 2016. Continuing Professional Development (CPD) program. https://www.ranzcp.org/Membership/CPD-program.aspx. Accessed 23 Sep 2016.
59. The Royal Australasian College of Medical Administrators. 2016. Continuing Professional Development Standard. http://www.racma.edu.au/index.php?option=com_content&view=article&id=22&Itemid=185. Accessed 23 Sep 2016.
60. The Medical Board of Australia. 2016. Registration standard: Continuing professional development. http://www.medicalboard.gov.au/documents/default.aspx?record=WD16%2f19473&dbid=AP&chksum=r3KA%2fZ0ECm3jZ1M7CfEMSw%3d%3d. Accessed 23 Sep 2016.
61. ———. 2016. Options for revalidation in Australia: Discussion Paper. http://www.medicalboard.gov.au/documents/default.aspx?record=WD16%2f21163&dbid=AP&chksum=1AmBXmPS80XN5gNGp%2bQvlQ%3d%3d. Accessed 19 Sep 2016.

62. Freeth, D. 2014. Interprofessional Education. In *Understanding Medical Education: Evidence,Theory and Practice*, ed. T. Swanwick, 2nd ed., 81–96. Chichester, West Sussex: John Wiley & Sons.
63. World Health Organization. 2015. *Mental Health Atlas 2014*. Geneva: World Health Organization.
64. Whitehouse, A., A. Hassell, A. Bullock, et al. 2007. 360 Degree Assessment (Multisource Feedback) of UK Trainee Doctors: Field Testing of Team Assessment of Behaviours (TAB). *Medical Teacher* 29: 171–176. doi:10.1080/01421590701302951.
65. Dubinsky, I., K. Jennings, M. Greengarten, and A. Brans. 2010. 360-Degree Physician Performance Assessment. *Healthcare Quarterly* 13: 71–76.
66. General Medical Council. 2016. 4a – Revalidation: Guidance on Colleague and Patient Questionnaires: Annex A. http://www.gmc-uk.org/4a__Revalidation__Guidance_on_Colleague_and_Patient_Questionnaires___Annex_A.pdf_40022212.pdf. Accessed 23 Sep 2016
67. Mason, R., S. Power, J. Parker-Swift, and E. Baker. 2009. 360-Degree Appraisal: A Simple Pragmatic Solution. *Clinical Governance: An International Journal* 14: 295–300. doi:10.1108/14777270911007818.

An Approach of Non-Linear Systems Through Fuzzy Control Based on Takagi-Sugeno Method

Andreas Giannakis, Konstantinos Giannakis, and Athanasios Karlis

Abstract Today, the advanced technology is a part of the everyday's life. As a result, most of the applications used require a more complex system in order to achieve a better performance. These systems have a mathematic background indicating the need of a better mathematical tool to increase the reliability of them. One of the most significant problems coming up against these systems is undoubtedly the non-linearity of the equations governing them. Herein, a linearization method is proposed and studied through intelligent control. The transformation of a non-linear system into a linear is based on fuzzy logic and more specifically on Takagi-Sugeno technique. Firstly, it is analyzed in a theoretical level followed by two examples. The fuzzy model was developed through Matlab program. Finally, the efficiency of the above method was investigated setting up various values for the under study variables and comparing the results of them with the "actual" ones. The square error method was used for a better evaluation indicating that this method is a useful technique except from the applications where the high accuracy is mandatory.

Keywords Non-Linearity • Intelligent control • Fuzzy logic • Takagi-Sugeno technique • Matlab

1 Introduction

One of the most significant problems today technology is facing is the description of complex systems [1–4]. In this study, the systems' non-linearity issue is analyzed proposing a method of transforming this into linear system using an approximate technique based on intelligent control. Recently, several strategies have been studied. Some of them are presented in [5–8]. Intelligent control faces the processing

A. Giannakis (✉) • A. Karlis
Electrical and Computer Engineering Department, Democritus University of Thrace, University Campus Kimmeria, Building B, Xanthi, 67100, Greece
e-mail: andgian21@gmail.com; akarlis@ee.duth.gr

K. Giannakis
Department of Informatics, Ionian University, Corfu, 49100, Greece
e-mail: kgiann@ionio.gr

P. Vlamos (ed.), *GeNeDis 2016*, Advances in Experimental Medicine and Biology 988, DOI 10.1007/978-3-319-56246-9_9

control problems in a different way than the existing conventional techniques do [9]. The knowledge and the experience of the person-operator is the base of this method and it does not require specialized learning of the controlled process. Consequently, intelligent control can be considered as the first control technique without standard indicating a promising future. Generally, intelligent systems use a collection of experienced and non-experienced logical facts and other types of knowledge, in combination with other methods in order to draw conclusions with respect to the required control acts for the controlled process. One of the most significant advantages of this control method is their ability to draw conclusions from insufficient and uncertain data. The main aims of the intelligent control use are; (1) Energy saving, (2) Increase the productivity, (3) Reduction of production cost and (4) Better performance of the control process.

Intelligent controllers consist of the fuzzy logic controllers (FLCs) and the neural networks (NNs) [3]. The FLCs are based on the fuzzy logic theory [8, 10] and the NNs on the Trainee Artificial Neural Networks. Herein, an approach of non-linear systems via FLCs is analyzed and proposed. The purpose of this is to enrich or go beyond the standard models such the non-conventional computing operates.

2 Fuzzy Systems

Fuzzy systems begun to grow in 1965 by Zadeh and later, their use started into real control problems. Their goal is to describe effectively the fuzziness of the real world having the fuzzy logic and the fuzzy sets as their "tools." Fuzzy sets are generalization of the classic crisp sets providing a way of a qualitative description of the sizes where the fuzziness and the indeterminacy are by definition. The fuzzy sets theory handles these values as linguistic variables giving them linguistic values. Each value is described by a fuzzy set with a specific meaning. Each input signal is measured and depending on the crisp value of the signals, it can be expressed in terms of the degree of membership (μ) of the fuzzy sets [11]. Fuzzy logic is an effort for modeling the human's way of thinking, providing mechanisms of approximate reasoning and decision making, as the human brain tends to perform similar actions instead of accurate reasoning based on abundance data. It can be understandable that the simplicity and the effectiveness of fuzzy systems have established them as one of the most promising control methods. So, several applications since '80 until today use fuzzy systems. The main subjects are the image and the signal processing, the identification of patterns, telecommunication systems, the economic theory, sociology and bio-engineering systems. Fuzzy logic controllers are considered non-linear intelligent controllers. They can be very accurate with high performance, using the experience of the system operator. The design of FLC consists of three basic steps: (1) the determination of the inputs, (2) the setting up of the rules and (3) the defuzzification which is a designing method in order to convert the rules into a crisp output signal.

2.1 Fuzzy Rules and Membership Functions

The fuzzy rules are depicted through if/then rules. These rules are conditional statements and are described by the below equation.

$$\text{IF x is A THEN y is B} \tag{1}$$

where A and B are fuzzy sets of x and y which are defined from the X and Y spaces. The phases "x is A" and "y is B" are fuzzy propositions. On the left side of the rule, the "IF" is called pre-conditional part (or premise part or IF part) and includes the assumption of the rule. On the right side, the phrase "THEN ..." is called the consequent part (or THEN part) and includes the conclusion of the rule.

The membership of the x element into the fuzzy set A is given by the membership function $\mu_{A(x)}$. This function gets values between 0 and 1. A fuzzy set is thoroughly described by the membership functions. The four most used types are the triangular, the trapezoidal, the Gaussian and the generalized Bell.

2.2 Fuzzy Models

The need for the modeling process to be automated with the help of numeric data used for trainee has emerged. Recently, data-driven techniques have attracted the worldwide interest in order to create more flexible models. There are the intelligent systems among them. As mentioned earlier, these systems consist of the fuzzy and neural systems. In this study, the first type is presented. Moreover, there are also two methods for the fuzzy systems, named Takagi-Sugeno and Mamdani method. Of course, there are systems where a combination of these methods is used [12]. A comparative study between the two methods is presented in [2]. Takagi-Sugeno models are considered as a powerful "tool" for the engineering science in order to model complex systems. They provide effective and computer attractive solutions in a wide range of problems, indicating a multiple model structure which is capable of approaching non-linear dynamic systems. An implementation of this technique is given through an example in [13].

2.3 Fuzzy Models Based on Takagi-Sugeno

As mentioned above, one of the most used fuzzy method is the Takagi-Sugeno (and Kang) model, known as TS (or TSK) fuzzy model, proposed by Takagi, Sugeno (and Kang later). Their aim was to develop an approximated system for fuzzy rules development from a known in advance input-output data set. A such model is described by the following equation.

$$\text{IF x is A AND y is B THEN z} = \text{f}(\text{x}, \text{y}) \quad (2)$$

where A and B are the fuzzy sets in premise part while the z = f(x,y) is a crispy function in consequent part. Regularly, the f(x,y) is a polynomial, but it can be also any other function describing appropriate the model's output. When the f is a constant function, then the TS model is called zero-order TS model and when it is a first degree polynomial, the arising fuzzy system is called first-order TS fuzzy model. In contrast with other fuzzy models where they need a defuzzification system in order to convert the fuzzy output to crispy value, the TS fuzzy model uses the weighted average method avoiding the time consuming defuzzification process. This leads to the TS model being a particularly interesting and popular choice for fuzzy modeling based on samples data. The aim of this study is to present the solutions for non-linear dynamic systems via TS fuzzy models. Two examples of non-linear equations are given, which can be solved following the fuzzy model TS. Their base is the IF-THEN rules. The fuzzy model construction is the main and the most significant process in the approach of the non-linear system. Generally, there are two techniques for this construction, as illustrated in Fig. 1, the first one is called "Identification using input-output data" and the second one is the "Physical model," where the equations of system is a mandatory requirement to extract the fuzzy model.

The process in the first technique consists of two parts. Firstly, it is the identification of the structure part and then the identification of the parameters. This technique is suitable for systems for which it is difficult to extract analytical

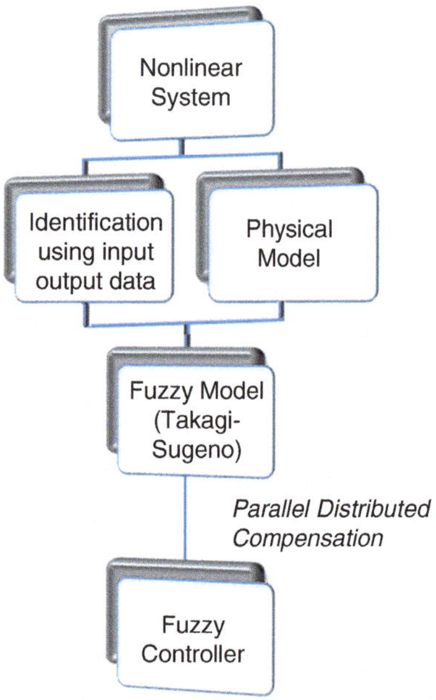

Fig. 1 Fuzzy controller construction

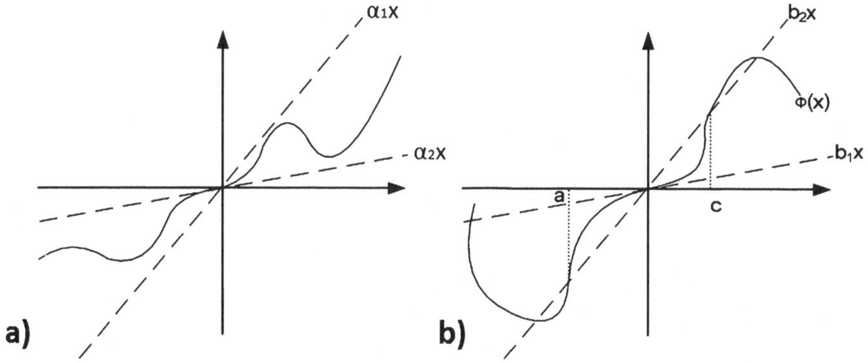

Fig. 2 Graphical representation of the expression of "non-linearity sector" (**a**) global and (**b**) local type

equations models. On the other hand, non-linear dynamics models for mechanical mainly systems can be represented by methods such as Lagrange and Newton-Euler, which are in line with the second technique. This technique uses the expressions "non-linearity sector" and "local estimators" or a combination of them. The first expression is based on the following idea: Considering a simple non-linear system, $dx/dt = f(x(t))$, where $f(0) = 0$. The goal is to be found a relation such that: $dx/dt = f(x(t))\ \varepsilon[\alpha_1\ \alpha_2]x(t)$. The next figure (Fig. 2) illustrates this approach. This method certifies the fuzzy model construction. However, in some cases it is difficult for such relations to be extracted. In these cases, a "local non-linearity sector" is considered. This is a very common approach due to the fact that the variables of the physical world are always within boundaries (Fig. 2b).

3 First Example

Subsequently, a first example of linearization is given and solved through Takagi-Sugeno fuzzy model. In the first stage, they are referred the problem's data and then the way of solution via Matlab is following. The non-linear equation system, which is a simple mathematical problem, is given:

$$\begin{pmatrix} \frac{dx_1}{dt} \\ \frac{dx_2}{dt} \end{pmatrix} = \begin{pmatrix} -x_{1(t)} + x_{1(t)}x^2{}_{2(t)} \\ -x_{2(t)} + \left(3 + x_{2(t)}\right)x^3{}_{1(t)} \end{pmatrix} \tag{3}$$

where $x_{1(t)}\ \varepsilon[-1,1]$ and $x_{2(t)}\ \varepsilon[-1,1]$. The above equation can be written as:

$$\frac{dx}{dt} = \begin{pmatrix} -1 & x_{1(t)}x^2{}_{2(t)} \\ \left(3 + x_{2(t)}\right)x^2{}_{1(t)} & -1 \end{pmatrix} x_{(t)} \tag{4}$$

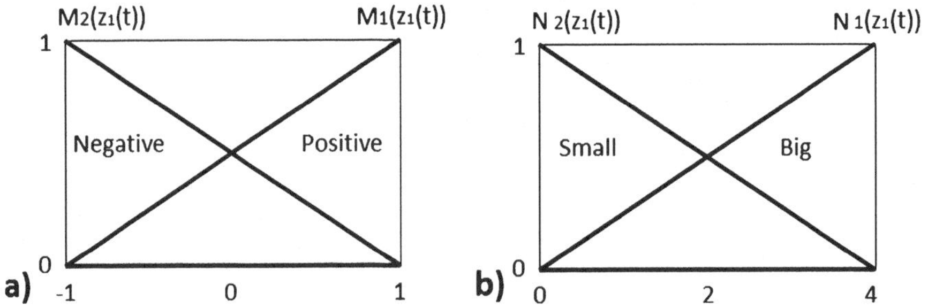

Fig. 3 Membership functions of (**a**) Mi and (**b**) Ni

t is defined that $z_{1(t)} = x_{1(t)}\, x^2_{2(t)}$ and $z_{2(t)} = (3 + x_{2(t)})\, x^2_{1(t)}$. The next step includes the minimum and the maximum calculations of the z_i functions. These are achieved through a code, developed in Matlab's environment. It is given that M_i and N_i ($i = 1,2$) are the membership functions of z_i for the minimum and the maximum values. The arising graphical depictions of these are illustrated in Fig. 3.

Considering the membership functions as "positive," "negative," "big" and "small" respectively and based on the four rules for every possible couple, M_1-N_1, M_1-N_2, M_2-N_1, M_2-N_2, the desired fuzzy model is extracted (i.e. if z_1 "positive" ($M_1 = 1$ and $M_2 = 0$) and z_2 "big" ($N_1 = 1$ and $N_2 = 0$) then $dx/dt = A_1x$). The output value is given by the following equation:

$$\frac{dx}{dt} = \sum_{i=1}^{4} h_{i(z(t))} A_i x_{(t)} \tag{5}$$

where $h_{1(z(t))} = M_{1(z1(t))} \times N_{1(z2(t))}$, $h_{2(z(t))} = M_{1(z1(t))} \times N_{2(z2(t))}$, $h_{3(z(t))} = M_{2(z1(t))} \times N_{1(z2(t))}$, $h_{4(z(t))} = M_{2(z1(t))} \times N_{2(z2(t))}$.

For the above, we consider the minimum of these elements in fuzzy logic. The efficacy of the fuzzy model was investigated by setting up various values for the variables x_1 and x_2 and then making a comparison with the "real" solutions for dx_1/dt and dx_2/dt. Specifically, setting up $x_1 = 0.9$ and $x_2 = 0.9$ in the original system, the arising results are $dx_1/dt = -0.2439$ and $dx_2/dt = 1.9431$, while running the program the results are given as $dx_1/dt = -0.3838$ and $dx_2/dt = 1.7207$. For these values, some significant results are depicted (Fig. 4).

In Tables 1 and 2 and Figs. 5 and 6, the efficiency of the proposed linearization system through fuzzy model compared to real responses is illustrated. Specifically, at first stage, the x_1 variable was set as steady with value 0.5 and the second variable, x_2 was set with various values. Subsequently, from the arising results, using the square error method in order to investigate the efficiency of the fuzzy model approach, the imported errors are 0.0392 and 0.0736 for the two outputs dx_1/dt and dx_2/dt, respectively.

a)

```
min1 =
   -1.0000

max1 =
   1

min2 =
   4.2971e-24

max2 =
   4
```

b)

```
M1 =
   0.8645

M2 =
   0.1355

N1 =
   0.7898

N2 =
   0.2102
```

c)

```
A1 =
   -1     1
    4    -1

A2 =
   -1.0000    1.0000
    0.0000   -1.0000

A3 =
   -1.0000   -1.0000
    4.0000   -1.0000

A4 =
   -1.0000   -1.0000
    0.0000   -1.0000
```

d)

```
h1 =
   0.7898

h2 =
   0.2102

h3 =
   0.1355

h4 =
   0.1355
```

e)

```
dxdt =
   -0.3838
    1.7207
```

Fig. 4 Arising results through Matlab using fuzzy model approach: (**a**) minimum and maximum values of z_i, (**b**) membership functions values, (**c**) A_i arrays, (**d**) intermediate results related to h_i and (**e**) final results

Table 1 Results for steady x_1 and various x_2 for both actual and TS fuzzy model used

$x_1 = 0.5$					
x_2	−1	−0.5	0	0.5	1
dx_1/dt	−1	−0.5625	−0.5	−0.4375	0
dx_1/dt (with TS)	−0.9	−0.5476	−0.5	−0.4565	−0.1667
dx_2/dt	1.25	0.8125	0.375	−0.0625	−0.5
dx_2/dt (with TS)	1.4	0.9762	0.5455	0.1087	−0.3333

Table 2 Results for steady x_2 and various x_1 for both actual and TS fuzzy model used

$x_2 = 0.5$					
x_1	−1	−0.5	0	0.5	1
dx_1/dt	0.875	0.4375	0	−0.4375	−0.875
dx_1/dt (with TS)	0.9	0.4565	0	−0.4565	−0.9
dx_2/dt	−4	−0.9375	−0.5	−0.0625	3
dx_2/dt (with TS)	−3.7	−1.1087	−0.5	0.1087	2.7

Similarly to the previous case with steady x_1 and various x_2, herein, it was investigated the efficiency of the TS fuzzy model changing the x_1 variable and keeping the x_2 steady. The arising results are illustrated in the below table and figure. It is mentioned that using the square error method, the imported errors for the two outputs dx_1/dt and dx_2/dt are 0.0089 and 0.0977 respectively.

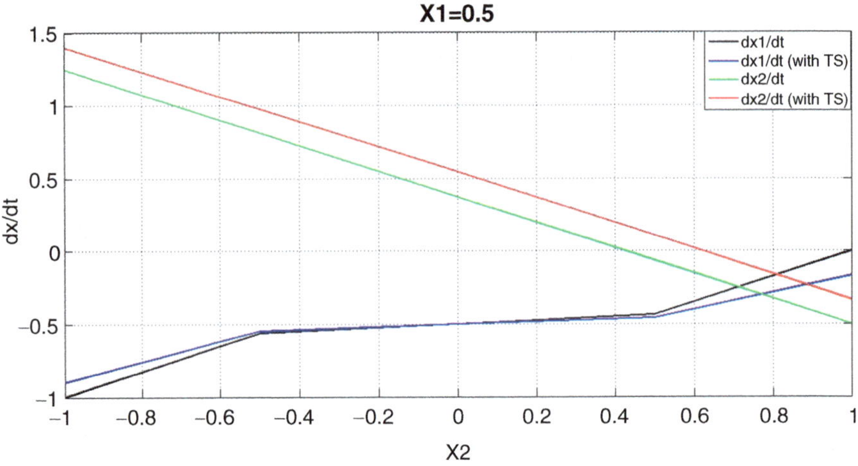

Fig. 5 Graphical depiction of the arising results for steady x_1 and various x_2

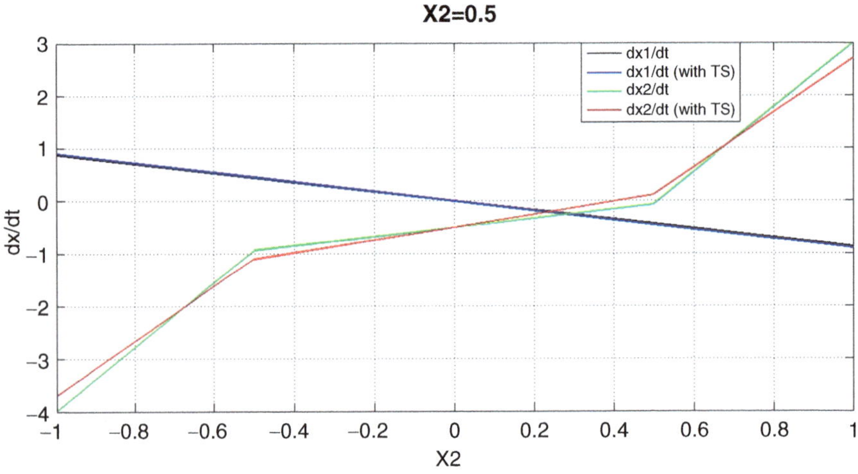

Fig. 6 Graphical depiction of the arising results for steady x_2 and various x_1

4 Second Example

For a better understanding, a second example is following. Firstly, it is given the below non-linear equations system. In contrary with the first example, these equations simulate a physical system. More specifically the motion of an inverted pendulum [14] is governed by the below set of equations.

$$\frac{dx_1}{dt} = x_{2(t)} \tag{6}$$

$$\frac{dx_2}{dt} = \frac{gsin\left(x_{1(t)}\right) - \frac{\alpha mlx^2{}_{2(t)}\sin\left(2x_{1(t)}\right)}{2} - \alpha cos\left(x_{1(t)}\right)u_{(t)}}{\frac{4l}{3} - \alpha mlcos^2\left(x_{1(t)}\right)} \quad (7)$$

where $x_{1(t)}$ is the angle of the pendulum from the vertical and $x_{2(t)}$ is the angular speed. Moreover, $g = 9.8$ m/s^2 the gravity standard, m is the maze of the pendulum, M the overall maze, $2\ l$ the length of the pendulum and u the force. Also, it is given that $\alpha = 1/(m + M)$ and $x_{1(t)}\ \varepsilon(-\pi/2,\pi/2)$ and $x_{2(t)}\ \varepsilon[-\alpha,\alpha]$. The following variables are defined:

$$z_{1(t)} = \frac{1}{\frac{4l}{3} - amlcos^2\left(x_{1(t)}\right)} \quad (8)$$

$$z_{2(t)} = \sin\left(x_{1(t)}\right) \quad (9)$$

$$z_{3(t)} = x_{2(t)}\sin\left(2x_{1(t)}\right) \quad (10)$$

$$z_{4(t)} = \cos\left(x_{1(t)}\right) \quad (11)$$

From the above equations, the arising system of equations is:

$$\begin{pmatrix} \frac{dx_1}{dt} \\ \frac{dx_2}{dt} \end{pmatrix} = \begin{pmatrix} 0 & 1 \\ gz_1z_2 & -\frac{aml}{2}z_1z_3 \end{pmatrix}\begin{pmatrix} x_1 \\ x_2 \end{pmatrix} + \begin{pmatrix} 0 \\ -az_1z_4 \end{pmatrix}\begin{pmatrix} u_1 & u_2 \end{pmatrix} \quad (12)$$

For $x_{1(t)} = -\pi/2$ and $x_{1(t)} = \pi/2$ the system becomes uncontrollable. For this reason, it is considered that $x_{1(t)}\ \varepsilon[-88^0,88^0]$. The first step includes the calculation of the minimum and maximum values of the z_i. In contrast with the previous example, herein the shape of the z_2 membership function defined as generalized bell which depicted in Fig. 7. In the right of Fig. 7 illustrates the local estimators with $b_1{*}x_1$ ($b_1 = 1$) and $b_2{*}x_2$ ($b_2 = 2/p$) as linearization sectors of z_2, according to the theory. The rest of the membership functions have the same shape with the example I. It is given that E_i (i = 1 “positive”, i = 2 “negative”), M_i (i = 1 “zero,” i = 2 “not zero”), N_i (i = 1 “positive”, i = 2 “negative”) and S_i (i = 1 “big”, i = 2 “small”) are the membership functions of the z_i (i = 1,2,3,4) for the minimum and maximum values.

Following, the arrays A_{ijkl} and B_{ijkl} are calculated such that dx/dt = Ax + Bu. The configuration of these arrays is based on the above consideration of 16 rules: A combination of the membership functions E_i-M_j-N_k-S_l (from 1111 to 2222) based on the IF-THEN logic (i.e. if z_1 “positive” ($E_1 = 1$ and $E_2 = 0$) and z_2 “not zero” ($M_1 = 0$ and $M_2 = 1$) and z_3 “negative” ($N_1 = 0$ and $N_2 = 1$) and z_4 “big” ($S_1 = 1$ and $S_2 = 0$) then dx/dt = A_{1221}x + B_{1221}u, and the arrays A_{1221} και B_{1221} are extracted). Then, calculating all the A_{ijkl} and B_{ijkl} arrays with the same way, the

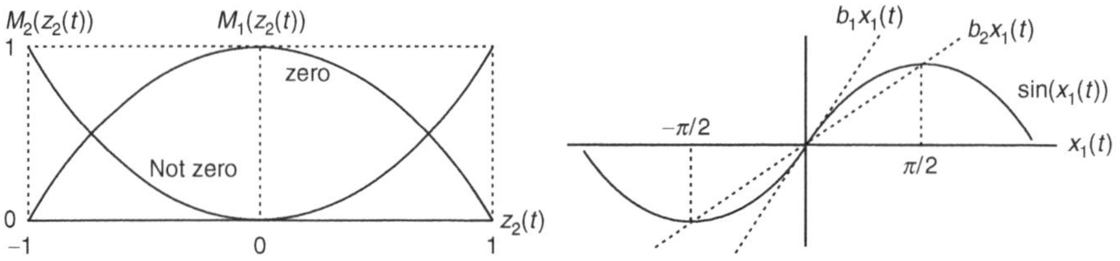

Fig. 7 Membership functions of $M_1(z2(t))$ and $M_2(z2(t))$ on the *left*. Function z_2 and its linearization sectors on the right

desired fuzzy model is extracted. Finally, the output value is given by the equations below:

$$\frac{dx}{dt} = \sum_{p=1}^{16} h_{p(z(t))} \left(A_p x_{(t)} + B_p u_{(t)} \right) \tag{13}$$

$$p = l + 2(k-1) + 4(j-1) + 8(i-1) \tag{14}$$

$$h_{p(z(t))} = E_{i(z_{1(t)})} \times M_{j(z_{2(t)})} \times N_{k(z_{3(t)})} \times S_{l(z_{4(t)})} \tag{15}$$

It is mentioned that for the above multiplication of two elements, it is the minimum of these element in fuzzy logic. Similarly to example I, the efficacy of the fuzzy model was investigated setting up various values for the variables x_1 and x_2 and then making a comparison with the "actual" solutions for dx_1/dt and dx_2/dt. More specifically, setting up $x_1 = 0.9$, $x_2 = 0.1$, $m = 1$, $M = 4$, $l = 1$, $u_1 = 1$ and $u_2 = 0$ in the original system, the arising results are $dx_1/dt = 0.1$ and $dx_2/dt = 6.1109$, while running the program (code in Matlab) through fuzzy model the results are given as $dx_1/dt = 0.1$ and $dx_2/dt = 5.9279$. For these values, in the above figures (Figs. 8 and 9), some significant results are illustrated.

In Tables 3 and 4 and Figs. 10 and 11, the efficiency of the proposed linearization system through fuzzy model compared to real responses for the pendulum example is illustrated. Initially, the x_1 variable was set as steady with value 57^0 and the second variable x_2 was set with various values. Subsequently, using the square error method in order to investigate the efficiency of the fuzzy model approach, the imported errors are 0.0 and 0.0273 for the two outputs dx_1/dt and dx_2/dt, respectively. It is noted that the rest of the parameters remained steady ($m = 1$, $M = 4$, $l = 1$ and $u = [1\ 0]$).

As in the previous case with steady x_1 and various x_2, herein, it was investigated the efficiency of the TS fuzzy model changing the x_1 variable and keeping the x_2 steady (=0.1). The arising results are illustrated in the below table and figure. It is mentioned that using the square error method, the imported errors for the two outputs dx_1/dt and dx_2/dt are 0.0 and 0.125 respectively.

Fig. 8 Arising results through Matlab: (**a**) minimum and maximum values of z_i, (**b**) membership functions values and (**c**) final results

```
max1 =

    0.8824

min1 =

    0.7501

max2 =

    0.9994

min2 =

   -0.9994

max3 =

    0.2000

min3 =

   -0.2000

max4 =

    1

min4 =

    0.0349
```

a)

```
E1 =

    0.3480

E2 =

    0.6520

M1 =

    0.6432

M2 =

    0.3568

N1 =

    0.7435

N2 =

    0.2565

S1 =

    0.6079

S2 =

    0.3921
```

b)

```
dxdt =

    0.1000
    5.9279
```

c)

5 Conclusion

In the above study, the fuzzy model approach for non-linear systems was investigated. More specifically, the Takagi-Sugeno technique was used in order to evaluate the efficiency of the proposed linearization method. For a better validation of the under-study scheme, two examples were examined using the square error method as an indicator of the performance. Both of them showed that the imported error in the solutions had a minimum value indicating probably that this method is useful for applications where the time consuming and complexity are more significant than the accuracy.

```
Ak(:, :,1) =

      0    1.0000
 8.6471   -0.0176


Ak(:, :,2) =

      0    1.0000
 8.6471   -0.0176


Ak(:, :,3) =

      0    1.0000
 8.6471    0.0176


Ak(:, :,4) =

      0    1.0000
 8.6471    0.0176


Ak(:, :,5) =

      0    1.0000
 5.5049   -0.0176


Ak(:, :,6) =

      0    1.0000
 5.5049   -0.0176


Ak(:, :,7) =

      0    1.0000
 5.5049    0.0176


Ak(:, :,8) =

      0    1.0000
 5.5049    0.0176


Ak(:, :,9) =

      0    1.0000
 7.3513   -0.0150


Ak(:, :,10) =

      0    1.0000
 7.3513   -0.0150


Ak(:, :,11) =

      0    1.0000
 7.3513    0.0150


Ak(:, :,12) =

      0    1.0000
 7.3513    0.0150


Ak(:, :,13) =

      0    1.0000
 4.6800   -0.0150


Ak(:, :,14) =

      0    1.0000
 4.6800   -0.0150


Ak(:, :,15) =

      0    1.0000
 4.6800    0.0150


Ak(:, :,16) =

      0    1.0000
 4.6800    0.0150


Bk =

 Columns 1  through 7

       0        0        0        0        0        0        0
 -0.1765  -0.0062  -0.1765  -0.0062  -0.1765  -0.0062  -0.1765

 Columns 7  through 14

       0        0        0        0        0        0        0
 -0.0062  -0.1500  -0.0052  -0.1500  -0.0052  -0.1500  -0.0052

 Columns 15  through 16

       0        0
 -0.1500  -0.0052
```

Fig. 9 Results for the 16 A_{ijkl} and B_{ijkl} arrays

Table 3 Results for steady x_1 and various x_2 for both actual and TS fuzzy model used

$x_1 = 57^0$					
x_2	−0.2	−0.1	0	0.1	0.2
dx_1/dt	−0.2	−0.1	0	0.1	0.2
dx_1/dt (with TS)	−0.2	−0.1	0	0.1	0.2
dx_2/dt	6.3804	6.3826	6.3833	6.3826	6.3804
dx_2/dt (with TS)	6.4337	6.4593	6.4186	6.4593	6.4337

Table 4 Results for steady x_2 and various x_1 for both actual and TS fuzzy model used

$x_2 = 0.1$					
x_1 (^{0})	−88	−44	0	44	88
dx_1/dt	0.1	0.1	0.1	0.1	0.1
dx_1/dt (with TS)	0.1	0.1	0.1	0.1	0.1
dx_2/dt	−7.352	−5.6515	−0.1765	5.4176	7.3416
dx_2/dt (with TS)	−7.4829	−5.1495	0	5.1495	7.4829

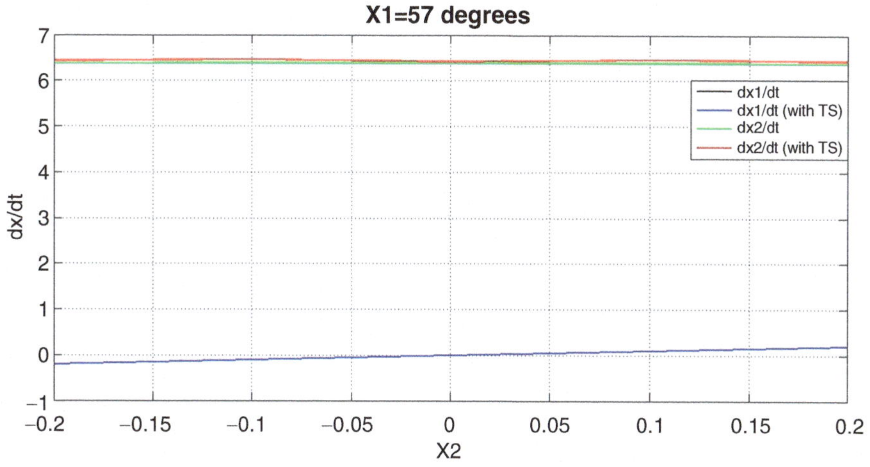

Fig. 10 Graphical depiction of the arising results for steady x_1 and various x_2

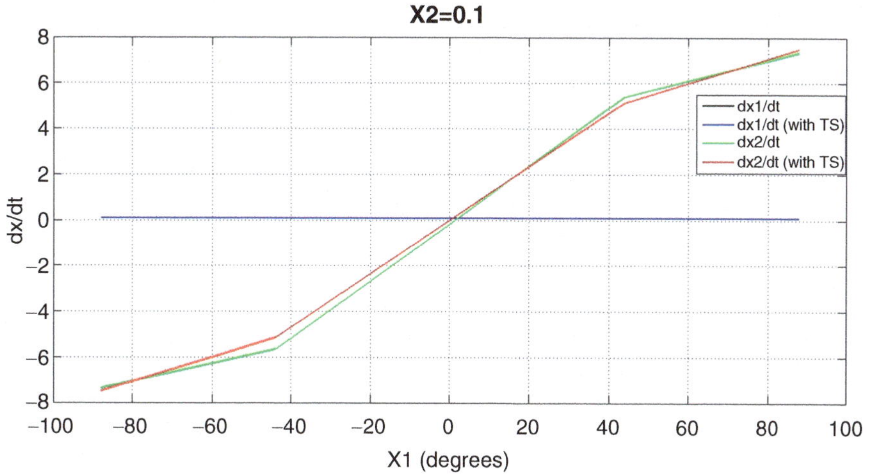

Fig. 11 Graphical depiction of the arising results for steady x_2 and various x_1

References

1. Agarwal, N., and R. Mukkamala. 2005. Design of Complex Systems: Issues and Challenges. Twenty fourth digital avionics systems conference, Vol. 2, p. 8. IEEE.
2. Bajpai, D., and A. Mandal. 2015. Comparative Analysis of t-Sugeno and Mamdani Type Fuzzy Logic Controller for pmsm Drives. *International Journal of Engineering Research and General Science* 3 (2): 495–507.
3. Berenji, H. 1990. Neural Networks and Fuzzy Logic in Intelligent Control. Proceedings of the 5th IEEE International Symposium on Intelligent Control, 1990, pp. 916–920. IEEE.

4. Califano, C., and C.H. Moog. 2014. The Observer Error Linearization Problem via Dynamic Compensation. *IEEE Transactions on Automatic Control* 59 (9): 2502–2508.
5. Denaï, M.A., F. Palis, and A. Zeghbib. 2007. Modeling and Control of Non-Linear Systems Using Soft Computing Techniques. *Applied Soft Computing* 7 (3): 728–738.
6. Dierks, T., and S. Jagannathan. 2011. Online Optimal Control of Nonlinear Discrete-Time Systems Using Approximate Dynamic Programming. *Journal of Control Theory and Applications* 9 (3): 361–369.
7. Jouan, P. 2003. Immersion of Nonlinear Systems Into Linear Systems Modulo Output Injection. *SIAM Journal on Control and Optimization* 41 (6): 1756–1778.
8. Wang, H.O., K. Tanaka, and M.F. Griffin. 1996. An Apprach to Fuzzy Control of Nonlinear Systems: Stability and Design Issues. *IEEE Transactions on Fuzzy Systems* 4 (1): 14–23.
9. Liu, T., E. Ko, and J. Lee. 1993. Intelligent Control of Dynamic Systems. *Journal of the Franklin Institute* 330 (3): 491–503.
10. Tanaka, K., and H.O. Wang. 2004. *Fuzzy Control Systems Design and Analysis: A Linear Matrix Inequality Approach*. New York: Wiley.
11. Mishra, S., Y. Mishra, F. Li, and Z. Dong. 2009. Ts-Fuzzy Controlled d g Based Wind Energy Conversion Systems. 2009 IEEE Power & Energy Society General Meeting, pp. 1–7. IEEE.
12. Salimifard, M., and A.A. Safavi. 2013. Nonlinear System Identi Cation Based on a Novel Adative Fuzzy Wavelet Neural Network. 2013 21st Iranian Conference on Electrical Engineering (ICEE), pp. 1–5. IEEE.
13. Singh, A., and M. Badoni. 2015. Design and Implementation of Takagi-Sugeno Fuzzy Logic Controller for Shunt Compensator. *Journal of The Institution of Engineers (India): Series B*: 1–11.
14. Prasad, L.B., H.O. Gupta, and B. Tyagi. 2011. Intelligent Control of Nonlinear Inverted Pendulum Dynamical System with Disturbance Input Using Fuzzy Logic Systems. 2011 International Conference on Recent Advancements in Electrical, Electronics and Control Engineering (ICONRAEeCE), pp. 136–141. IEEE.

Chronic Lymphocytic Leukemia Patient Clustering Based on Somatic Hypermutation (SHM) Analysis

Eleftheria Polychronidou, Aliki Xochelli, Panagiotis Moschonas, Stavros Papadopoulos, Anastasia Hatzidimitriou, Panayiotis Vlamos, Kostas Stamatopoulos, and Dimitrios Tzovaras

Abstract Somatic Hypermutation (SHM) load in the immunoglobulin heavy variable (IGHV) gene of the clonotypic B cell receptor immunoglobulin (BcR IG) is one of the most important prognostic markers in CLL, segregating patients into two distinct categories, with contrariwise disease course. Over the last years, immunogenetic studies have identified that ~30% of CLL patients carry (quasi)identical BcR IG and thus can be assigned to different subsets with

---

E. Polychronidou (✉)
Centre for Research and Technology Hellas, Information Technologies Institute, Thessaloniki, 57001, Greece

Department of Informatics, Ionian University, Corfu, Greece
e-mail: epolyc@iti.gr

A. Xochelli • K. Stamatopoulos
Centre for Research and Technology Hellas, Institute of Applied Biosciences, Thessaloniki, 57001, Greece

Science for Life Laboratory, Uppsala University, Uppsala, Sweden

P. Moschonas (✉) • D. Tzovaras
Centre for Research and Technology Hellas, Information Technologies Institute, Thessaloniki, 57001, Greece
e-mail: moschona@iti.gr

S. Papadopoulos
Centre for Research and Technology Hellas, Information Technologies Institute, Thessaloniki, 57001, Greece

Department of Electrical and Electronic Engineering, Imperial College London, SW7 2AZ, London, UK

A. Hatzidimitriou
Centre for Research and Technology Hellas, Institute of Applied Biosciences, Thessaloniki, 57001, Greece

P. Vlamos
Department of Informatics, Bioinformatics and Human Electrophysiology Laboratory, Ionian University, Corfu, Greece
e-mail: vlamos@ionio.gr

P. Vlamos (ed.), *GeNeDis 2016*, Advances in Experimental Medicine and Biology 988, DOI 10.1007/978-3-319-56246-9_10

distinct clinicobiological profiles. This characterization was achieved by applying rules mainly concerning the diversity of the VH complementarity determining region 3 (CDR3). Following, studies have also identified subset-specific somatic hypermutation further highlighting antigen selection in disease ontogeny and evolution. In this study, an innovative attempt to explore possible associations amongst SHMs in different CLL patients is implemented and also the potential correlations with VH CDR3 stereotypy is examined, leading to a new classification algorithm implicating both SHM and CDR3 patterns. All results are classified to a ground level analysis, focusing on the most frequent SHMs, their paired associated amino acid changes and the formation of subgroups sharing the same VH CDR3 pattern, the latter being used as a similarity metric. In addition, all results are compared to established VH CDR3 patterns of the well-known CLL subsets in order to confirm the validity of our findings.

Keywords Chronic Lymphocytic Leukemia (CLL) • Somatic Hypermutation (SHM) • CDR3 patterns • Mutation Association • Visual Analytics

1 Introduction

Immunogenetic studies have provided strong evidence for antigen implication in Chronic Lymphocytic Leukemia (CLL) ontogeny and evolution by revealing repertoire skewing as well as patterns of somatic hypermutation (SHM) typical of antigen receptors selected by antigen [1–5].

SHM load is one of the most important prognostic markers in CLL, segregating patients into two distinct categories, with contrariwise disease course. More specifically, patients with no SHMs or those carrying a low SHM load (unmutated CLL; U-CLL) usually follow a more aggressive disease course whereas those carrying a heavier SHM load (mutated CLL; M-CLL) are characterized by a more indolent disease course [3, 6].

Yet, the strongest evidence for antigen selection in CLL emerged from the finding that ~30% of CLL patients carry identical or quasi-identical immunoglobulin B-cell receptor (Ig BcR) and thus can be assigned to different subsets with distinct clinicobiological profiles [7]. The identification of stereotypy in CLL involved a complex exhaustive search algorithm that included rules concerning VH CDR3 diversity. In more details, rules concerned amino acid and amino acid properties identity (50% and 70% respectively), the usage of different yet phylogenetically related IGHV genes, and also criteria related to the tertiary protein structure, thus clustering only sequences with identical VH CDR3 lengths [7]. All patterns shared the same offset.

Parallel and subsequent studies [18, 19] have demonstrated that in some cases, stereotyped subsets were also characterized by subset and disease specific SHMs, further highlighting antigen implication in disease course [17].

Prompted by the aforementioned findings, an attempt to explore possible associations amongst SHMs in different CLL patients was applied and also examined the potential correlations with VH CDR3 stereotypy aiming to a new classification algorithm implicating both SHM and VH CDR3 patterns. All results were set to a ground level (level 0) of SHM analysis by establishing a threshold of recurrent SHMs (associated amino acid changes). To this level the selected were linked to known "stereotyped" amino acid changes [5] and to subgroups of the already proved CLL stereotyped subset groups [7].

2 Methods

The current study was accomplished through the combination of a density based clustering algorithm with a force-directed graph drawing visualization. Case specific parameters were set related to the study requirements and there will be explained briefly in the following paragraphs. CLL patients were grouped based on IGHV gene usage, into three phylogenetic clans namely Clan I; Clan II and Clan III [8]. At that point, a region weight factor was set based on the physicochemical properties of the amino acids and the entropy of the SHMs in the complementarity determining regions (CDRs) and the framework regions (FRs). The Euclidean distance between the sequences was defined as a combination of the Blosum Matrix and the region weight factor, following by the implementation of DBSCAN clustering algorithm. Finally, through a force-directed graph drawing visualization the associations between SHMs and CLL groups was underlined using the VH CDR3 patterns as similarity metric.

2.1 Position Score

An evolutionary divergence sequence based analysis was accomplished for every amino acid sequence. To efficiently perform the analysis, germ line sequences were obtained from IMGT/Gene-DB and used as a benchmark for comparisons. Sequences held in average more than 80% identity to the corresponding germlines constituting the Blosum80 matrix [8] as the optimum divergence score. The process output concerned newly formed sequences where all amino acids were transformed to numbers corresponding to a score with respect to alignment outliers and gaps.

The SHM variance of each position was calculated through the position entropy. Shannon entropy [9] expresses the expected value of each SHM in each codon i.e. the position of the amino acid in the sequence, and is defined as:

$$H_2(X) = -\sum_{i=1}^{n} \frac{m_i}{N} \log_2 \frac{m_i}{N} \quad (1)$$

where m_i/N is the frequency of the SHM i in every codon of the IGHV genes sequence. The position entropy was normalized to the length of CDRs and FRs

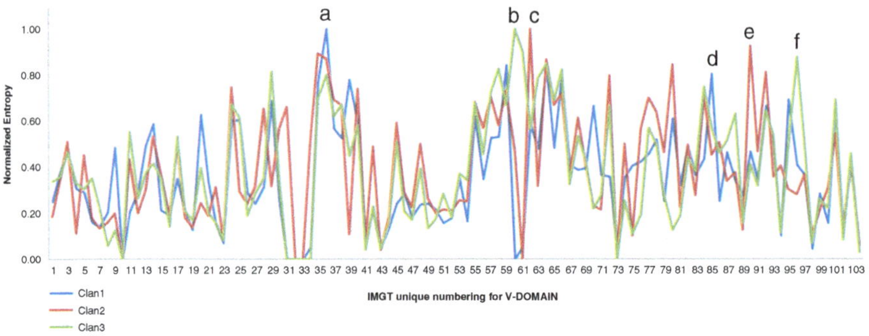

Fig. 1 Normalized SHM variation per position for the three phylogenetic clans. The points a,b,c underline the maximum entropy in CDRs while d,e,f in FRs for ClanI, ClanII, ClanIII respectively

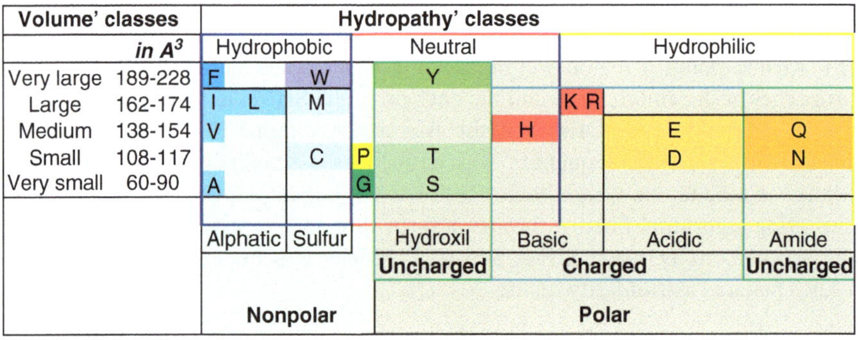

Fig. 2 The 11 "IMGT Physicochemical" classes, for the 20 amino acids"

accordingly. The normalized entropy revealed a greater SHM variance in CDR regions (Fig. 1). Due to the fact that evolutionary divergence score described by the Blosum matrix cannot fully evaluate the position SHM significance of the immunoglobulin regions (FRs and CDRs), a more complex factor needed to be defined. Accordingly, the information derived of position entropy was combined with the physicochemical similarity so that a multivariable position weight could be applied to each position score defined by Blosum80.

The physicochemical similarity was defined according to the 11 "IMGT Physicochemical" classes, for the 20 amino acids" [10]. The 11 IMGT 'Physicochemical' classes were designated based on the 'Hydropathy', 'Volume', 'Chemical', 'Charge', 'Hydrogen donor or acceptor atoms' and 'Polarity' properties of the side chains (or R- groups) into the following classes: (1) aliphatic (A, I, L, V), (2) basic (R, H, K), (3) sulfur (C, M), (4) hydroxyl (S, T), (5) contain acidic (D, E), (6) amide (N, Q), and (7–11) contain a single amino acid: G, F, P, W, Y. The description of each class is present in Fig. 2.

The aforementioned physicochemical weight was defined to characterize the expected value of the SHMs in codons. More explicitly, the position SHM entropy

Table 1 Multivariable physicochemical weight

	Similar physicochemical class	Dissimilar physicochemical class	Clan
FRs	0.35	0.8	Clan1
CDRs	0.4	1	
FRs	0.3	0.9	Clan2
CDRs	0.6	1	
FRs	0.35	0.85	Clan3
CDRs	0.5	1	

was measured and grouped into two categories, those presenting in FRs regions and those in CDRs. Each group was normalized based on the length of the region, according to the IMGT unique numbering for V-DOMAIN [11]. Next, for each group the maximum and minimum entropy values were computed, and all the values were normalized into the range [0, 1]. The total region weight [20] was calculated based on the following formula:

$$W_{region} = \sum_{i=1}^{n} \frac{H_i}{N} \tag{2}$$

where, H_i is the entropy position value of the region FR or CDR. To point out the significance of the physicochemical class alterations, the region weight was set to the maximum normalized entropy (Fig. 1). A summary of the physicochemical weight is forgathered in Table 1.

2.2 *SHM Clustering*

Regarding the distance metrics, the SHM distance between pairs of sequences was calculated based on the Euclidean distance:

$$Distance\left(S_i, S_j\right) = \sqrt{\sum_{i=1}^{n} \left(W_i q_i - W_j q_j\right)^2} \tag{3}$$

where q_i and q_j are the position scores based on Blosum80 and W_i W_j are the position normalized physicochemical weights. All the possible pairs of distances where stored in a distance matrix in a format that can be utilized by a dimensionality reduction algorithm. At that point, classic distance-based methods of multidimensional scaling were applied to facilitate data visualization for the selection of the more adequate clustering method for further analysis [12].

According to the data distribution in the two dimensional space, the need of an algorithm robust to outliers, that does not a priori request the definition of the number of clusters, led to the selection of a density cluster algorithm. DBSCAN algorithm [13] was implemented to the data. The algorithm required the predefinition of two parameters: (1) the radius from a given point p that defines the circular area where neighbouring points are reachable from p, ε (eps) and (2) the minimum number of points required to form a dense region, (minPts). Data two-dimensional representation demonstrated that the majority of points in all datasets laid within 20 units from their nearest neighbour, for this reason the optimal ε was set at 20. Additionally, the minimum number of neighbor points that could form a cluster was set to 20 for Clan I and Clan III and 30 for Clan II. Eventually, the clustering parameters set to (20, 20) for groups Clan I and Clan III and (20, 30) for Clan II. Clusters consistency was evaluated by the application of Davies–Bouldin index (DBI) a common metric for evaluating clustering algorithms [3]. Finally the comparison to the established subsets was defined by the use of Jaccard Index.

3 Results

A total of 19,635 IGHV-IGHD-IGHJ gene rearrangement sequences from CLL patients were examined. Based on IGHV gene usage, all cases were subgrouped into three phylogenetic clans namely Clan I|IGHV1/5/7), n = 5439; Clan II|IGHV2/4/6, n = 5092 and Clan III|IGHV3, n = 9104. Each group was studied separately following the same process.

The density based algorithm revealed two clusters for clan I, three clusters for clan II and two clusters for clan III. The sequences included in each cluster are summarized in Table 2.

The best intra-cluster similarity and inter-cluster differences were identified in the dataset group of Clan II with Davies–Bouldin index (DBI) equal to 1.08. The high consecration in the first cluster of each group is due to the high similarity of sequences belonging to evolutionary related subgroup genes [1]. The application of most severe clustering rules could produce adverse sequence group segmentation. The clustering results were plotted in two dimensional scatterplots (Fig. 3).

Associated SHM distribution in each cluster was as follows: (1) for the Clan I: n = 49,214 for the Cluster1 and n = 10,411 for the Cluster2, (2) Clan II: n = 130,181 for the Cluster1, n = 21,834 for the Cluster2 and n = 8294 for the Cluster3 and finally (3) for Clan III n = 217,733 for the Cluster1 and n = 36,541 for the Cluster2. The greatest number of associated SHMs could be identified among sequences utilizing IGHV genes belonging to subgroup 3 (Clan III).

Table 2 Clustering result

	Cluster 1	Cluster 2	Cluster 3	Outliers
Gr. clan I	4496	766	n/a	218
Gr. clan II	3992	824	171	107
Gr. clan III	7671	1092	n/a	342

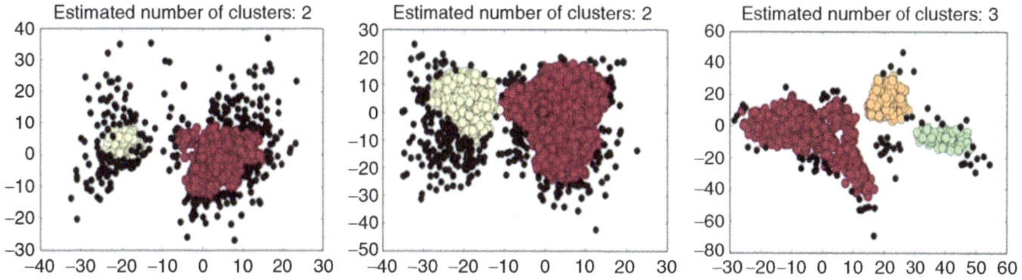

Fig. 3 Density based algorithm clustering results for Clan I, Clan II, Clan III respectively, based on Multidimensional Scaling results

The aforementioned associations across every cluster where represented through an in-house developed force-directed graph drawing. This visualization method is a network diagram that aims to position the nodes in the two dimensional space so that the system comes to a mechanical equilibrium state; i.e., their relative positions do not change anymore from one iteration to the next [14]. More explicitly, the nodes of the graph represented SHM associations and patients' IDs and while forces applied to the nodes, the nodes where pulled closer or pushed further apart. The force at a vertex v is described by

$$F(u) = \sum_{(u,v)\in E} (p_u - p_v) \tag{4}$$

A graph is an ordered pair G = (V,E) comprising a set V of vertices or nodes or points together with a set E of edges or arcs or lines. The attributes used for the graph implementations was the set I = {Patients}, and the set A = {Mutation Associations}. According to graph parameter definition [16] the set of vertices V is the union of the attributes V ≡ I ∪A while the set E of edges is the subgroup E ⊆ I × A. The positions of the nodes in this equilibrium depict the relations between the SHM associations and the patients based on the frequency of the first to the second.

In order to manipulate the large amount of information, a threshold of SHM association significance was set in the minimum graph vertices. More precisely, level 0 of graph analysis was set to minimum 20 vertices in the small clusters and 50 vertices to the bigger ones. In terms of associations, this level involved only SHM associations that were connected with at least 50 different patients (large patient subsets) in the first clusters of all groups and above 20 connections (small patient subsets) for the rest. The aforementioned threshold resulted to 106 associations that conformed to the established "stereotyped" amino acid changes [5] and expanded them to SHM associations. An example of the known SHM S to T in the subgroup gene IGHV4-34 sequences: IMGT VH FR2, codon 40 that is common in Subsets 4 & 16 of patients was discovered through the visualization tool as an associate to the SHM I to M in codon 78 and to S to N in codon 92 and is included in Fig. 4.

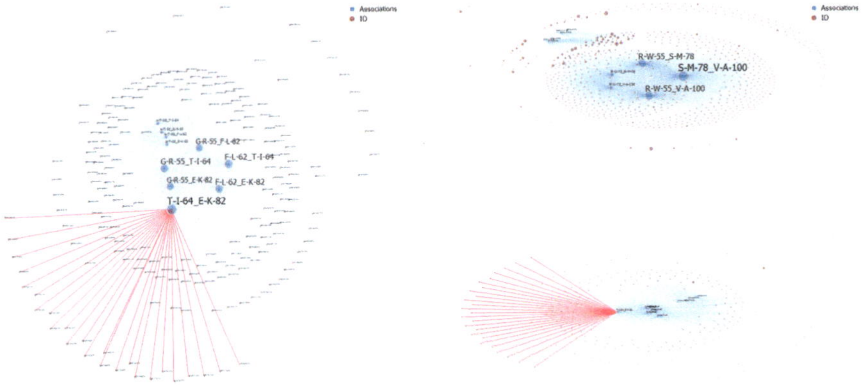

Fig. 4 Relation between SHM associations and CLL patients in Cluster1 of Clan I. On the right side there is a total representation of the SHMs associated to patients under the ground level analysis for ClanI. On the left side there is the ability of specific view, where a preferred area can be projected in larger scale. In this mode a more detailed study of the area is available. Here a detailed view for the T-I-64_E-K-82 SHM association was selected that is linked to 25 patients. The *red lines* stand for selected parameters and are interactive in both modes. *Red dots* represent the patients while *blue dots* the SHMs' associations

The format of the SHM associations that was created for the data visualization is based on the format: G-M-C_AG-AM-AC where: G: Germline amino acid, M: Mutated amino acid, C: Codon of the mutation, AG: Associated Germline amino acid, AM: Associated Mutated amino acid and, AC: Associated Codon.

The ground level of the SHMs association analysis named S-M-78_V-A-100 as the most common SHM in Clan I (related to approximately 700 patients). Although Clan II appeared as more mutated, the most common SHM associations (P-A-46_V-I-30, P-A-46_Y-R-55, P-A-46_Y-T-58 & P-S-16_C-Y-103) do not exceed the relation to 100 patients. Similar to Clan II, Clan III presented four SHMs associations that are linked to individual groups of approximately 100 patients (K-R-72_K-3-84, S-N-36_A-V-96, S-N-38_S-G-62, T-A-68_G-D-82).

By the threshold application, SHM associations formed mainly independent groups analyzed in terms of VH CDR3 to discover the link between SHMs and patterns. The results revealed ten patterns linked to eight CLL established subsets.

Simultaneously, each group pattern was related to at least one SHM association. Groups linked to more than one SHM were Subset 59 that was characterized by two and Subsets 1 and 28A with five. The summarized information about the discoveries that establish the significance of the SHMal status of IGHV genes in forming prognostic subgroups of chronic lymphocytic leukemia (CLL) is represented in Table 3.

For each case, Jaccard similarity was defined in order to demonstrate the similarity to the original subset. The higher similarity was underlined in subsets 28A and 77 while the lower in subset 4.

Table 3 Somatic hypermutations (SHMs) associated with particular VH CDR3

Group No	Jaccard index	Subset	Associations	VH CDR3 pattern	Sequence logo
Group A-Clan1_Cluster1	0.841	#28A	S-M-78_V-A-100	ARxxxGxxYYYYYGMDx	
Group B-Clan1_Cluster1	0.317	#1	R-W-55_V-A-100, G-D-36_V-A-100, G-D-36_R-W-55, R-W-55_S-M-78	ARx[NQ]W[AVLI]xxxxFDx	
Group B-Clan1_Cluster1	0.493	#28A	R-W-55_V-A-100, G-D-36_V-A-100, G-D-36_R-W-55, M-I-39_R-W-55, R-W-55_S-M-79	ARxxxGxxYYYYYGMDx	
Group C-Clan1_Cluster1	0.384	#59	A-T-38_G-R-55, F-L-62_T-I-64	AxxxDFWSGxxx	
Group D-Clan2_Cluster1	0.067	#4	S-T-40_P-S-45	[AVLI]RGxxxxxxx[KRH]RYYYYGx[DE]x	
Group E-Clan2_Cluster1	0.330	#148B	S-T-31_N-D-59	(*not officially defined*)	
Group F-Clan2_Cluster1	0.250	#14	P-A-46_Y-T-58	x[KRH]GGxWxFDx	

(contined)

Table 3 (continued)

Group No	Jaccard index	Subset	Associations	VH CDR3 pattern	Sequence logo
Group G-Clan2_Cluster1	0.240	#14	S-T-40_I-M-78	x[KRH]GGxWxFDx	
Group H-Clan2_Cluster1	0.400	#77	P-S-16_C-Y-103	[AVLI]RGxxx[ST]GWxxxxx	
Group I-Clan3_Cluster1	0.292	#31	S-E-38_S-G-62	ARxxxxxxxxxxxYYYxMDx	

VH CDR3 amino acid patterns of different subsets were visualized using WebLogo (http://weblogo.berkeley.edu/). Each logo consists of stacks of symbols, and the size of the letter represents its frequency. CDR3 are shown from IMGT positions 105 to 117 [10].

4 Discussion

In the present study, we analyzed 19,635 unique IGHV-IGHD-IGHJ gene rearrangement sequences from CLL patients. Sequences were grouped in three datasets according to the phylogenetic relationship of the IGHV genes used. The cohort was clustered conforming to density based algorithm. At this point, an agglomerative approach of clustering was tested. The comparison of the two methods (density based and hierarchical clustering) revealed a better cluster consistency on the density based results. Additionally the agglomerative method had a bigger error factor in comparison to the well-established CLL subsets.

The density based clustering methodology formed seven clusters in the total patient cohort and 474,208 somatic hypermutations SHMs with at least once interactively connection underlined.

The visualization of the associated SHMs for different patients was accomplished through a force-directed graph-drawing algorithm that achieved a detailed description of each cluster's SHM status. A threshold of (50, 20) was set to highlight the most frequent relations between SHMs across different patients. Defining the ground level of analysis in the 106 main SHM associations, solid groups of patients where formed in the two dimensional space linked to specific SHMs. All groups were analyzed based on the patients' VH CDR3 pattern -which is the primary specificity characteristic [15] and their possible relation to the established CLL patterns of each subset [7].

Frequently observed SHMs in IGHV leading to amino acid changes (replacement) that were characterized as "stereotyped" [5] and were linked to specific subsets, based on our results, are now upgraded into subset specific associated SHMs.

In conclusion a novel classification algorithm that combines both SHM and CDR3 patterns was demonstrated and the results were presented through new associations only between pairs of SHMs.

Next level analysis will include associations between more than two SHMs. SHM patterns in the VH region will then be associated with CDR3 patterns leading to the identification of newly recognized and yet uncharacterized group of patients.

Acknowledgment This work is supported by the EU funded project MEDGENET (Grant agreement No: 692298).

References

1. Damle, R., et al. 1999. Ig V Gene SHM Status and CD38 Expression as Novel Prognostic Indicators in Chronic Lymphocytic Leukemia. *Blood* 94 (6): 1840–1847.
2. Darzentas, N., et al. 2010. A Different Ontogenesis for Chronic Lymphocytic Leukemia Cases Carrying Stereotyped Antigen Receptors: Molecular and Computational Evidence. *Leukemia* 24 (1): 125–132.
3. Darzentas, N., and K. Stamatopoulos. 2013. The Significance of Stereotyped B-cell Receptors in Chronic Lymphocytic Leukemia. *Hematology/Oncology Clinics of North America* 27 (2): 237–250.
4. Ester, M., et al. 1996. A Density-Based Algorithm for Discovering Clusters in Large Spatial Databases with Noise. Second International Conference on Knowledge Discovery and Data Mining (KDD-96), 226–231. AAAI Press.
5. Lisitsyn, S., et al. 2013. Tapkee: An Efficient Dimension Reduction Library. *Journal of Machine Learning Research* 14: 2355–2359.
6. Chiorazzi, N., et al. 2005. Chronic Lymphocytic Leukemia. *New England Journal of Medicine* 352 (8): 804–815.
7. Agathangelidis, A., et al. 2012. Stereotyped B-cell Receptors in One-Third of Chronic Lymphocytic Leukemia: A Molecular Classification with Implications for Targeted Therapies. *Blood* 119 (19): 4467–4475.
8. Hamblin, T., et al. 1999. Unmutated Ig V(H) Genes are Associated with a More Aggressive Form of Chronic Lymphocytic Leukemia. *Blood* 94 (6): 1848–1854.
9. Pommie, C., et al. 2004. IMGT Standardized Criteria for Statistical Analysis of Immunoglobulin V-REGION Amino Acid Properties. *Journal of Molecular Recognition* 17 (1): 17–32.
10. Pramanik, S., et al. 2015. 3D Structure, Dimerization Modeling, and Lead Discovery by Ligand-Protein Interaction Analysis of p60 Transcription Regulator Protein (p60TRP). *Molecular Informatics* 35 (3–4): 99–108.
11. Kirkham, P., et al. 1992. Immunoglobulin VH Clan and Family Identity Predicts Variable Domain Structure and May Influence Antigen Binding. *EMBO Journal* 11 (2): 603–609.
12. Lefranc, M.P., et al. 2003. IMGT Unique Numbering for Immunoglobulin and T Cell Receptor Variable Domains and Ig Superfamily V-like Domains. *Developmental and Comparative Immunology* 27 (1): 55–77.
13. Dighiero, G. 2005. CLL Biology and Prognosis. *Hematology. American Society of Hematology. Education Program*: 278–284.
14. Bikos, V., et al. 2012. Over 30% of Patients with Splenic Marginal Zone Lymphoma Express the Same Immunoglobulin Heavy Variable Gene: Ontogenetic Implications. *Leukemia* 26 (7): 1638–1646.
15. Davies, D.L., and D.W. Bouldin. 1979. A Cluster Separation Measure. *IEEE Transactions on Pattern Analysis and Machine Intelligence* 1 (2): 224–227.
16. Brandenburg, F.J., et al. 1996. An Experimental Comparison of Force-Directed and Randomized Graph Drawing Algorithms. *Graph Drawing* 1027: 76–87.
17. Davies, D.R., et al. 1990. Antibody-Antigen Complexes. *Annual Review of Biochemistry* 59: 439–473.
18. Hadzidimitriou, A., et al. 2009. Evidence for the Significant Role of Immunoglobulin Light Chains in Antigen Recognition and Selection in Chronic Lymphocytic Leukemia. *Blood* 113 (2): 403–411.
19. Murray, F., et al. 2008. Stereotyped Patterns of Somatic hyperSHM in Subsets of Patients with Chronic Lymphocytic Leukemia: Implications for the Role of Antigen Selection in Leukemogenesis. *Blood* 111 (3): 1524–1533.
20. Shannon, C.E. 1951. Prediction and Entropy of Printed English. *Bell System Technical Journal* 30 (1): 50–64.

Development of Novel Indole Molecules for the Screening of Anti-Inflammatory Activity

Konda Swathi, Cidda Manasa, and Manda Sarangapani

Abstract In the present work, some new 5-[2(3)-dialkylamino alkoxy] Indole 2,3-diones were prepared from 5-hydroxy isatin. A mixture of 5-hydroxy isatin, dialkylamino alkylhalide in alcoholic potassium hydroxide was stirred at room temperature for 6 h to get the 5-[2(3)-dialkylamino alkoxy] Indole 2,3-diones. The structures of the products were characterized by IR, NMR, MASS Spectral studies. All the compounds were evaluated for anti-inflammatory activity. Some of these compounds showed good anti-inflammatory activity compared with standard compound Indomethacin.

Keywords Synthesis • 5-[2(3)-dialkyl amino alkoxy] Indole 2,3-diones • Antiinflamatory activity

1 Introduction

Isatin is an endogenous compound isolated in 1998 and reported [1] to possess a wide range of central nervous system activities. Pandeya and Senthil Raja [2] reported the synthesis and anticonvulsant activity of some novel n-methyl/acetyl, 5-(un)-substituted isatin-3-semicarbazones. In the last few years, Isatin derivatives have been discovered which show potential hypnotic [3], antibacterial [4–6] and MAO inhibitory [7], antihistaminic [8] activity. It is evident from the literature survey that Isatin derivatives dialkylamino alkyl derivatives showing more promising antiinflamatory activities. Keeping in view of these two molecular moieties viz., 5-hydroxy isatin and dialkylamino alkyl (Resembles diphenhydramine), it is our endeavor to bring such important moieties into a single molecular frame as a model for molecular conjunction by appropriate synthetic routes and to screen them for anti-inflammatory activity.

We are reporting in the present communication the synthesis and characterization of some new compounds: 5-[2(3)-dialky amino alkoxy] Indole 2, 3-diones.5-

K. Swathi (✉) • C. Manasa • M. Sarangapani
Medicinal Chemistry Laboratory, U.C.P.Sc., Kakatiya University, Wararagal, 506009, Telangana, India
e-mail: kswathi84@yahoo.co.in

P. Vlamos (ed.), *GeNeDis 2016*, Advances in Experimental Medicine and Biology 988, DOI 10.1007/978-3-319-56246-9_11

Hydroxyisatin condensed with dialkylamino alkyl halide by using Williamson synthesis to prepare the 5-[2(3)-dialkylamino alkoxy] Indole 2, 3-dione derivatives. All the compounds of the series have been screened for antiinflammatory activity and the structures of these compounds were identified by IR, NMR and Mass Spectrums.

2 Materials and Methods

The compounds were mostly synthesized by conventional methods and described in experimental selection and also by the methods established in our laboratory.

2.1 Chemicals

Caragenan, Indomethacin, Dialkyl amino alkylhalides purchased from Sigma-Aldrich Chemicals Private Limited, Hyderabad, India. p-Amino phenol, Hydroxylamine hydrochloride, Sodium sulfate were purchased from Merck Chemicals Private Limited, Hyderabad, India.

2.2 Chemistry

Solvents were dried or distilled before use. Melting points were obtained on a Thoshniwall melting point apparatus in open capillary tubes and are uncorrected. The purity of the compounds were ascertained by TLC on silica gel –G plates(Merck). Infrared spectra(IR) were recorded with KBR pellet on a Perkin-Elmer BX series, Infrared spectrophotometer. Mass spectra were recorded by the direct inlet method on Thadmam-mass-quantam API 400H mass spectrophotometer. ^{1}H NMR spectra were recorded on Brucker spectrospin 400 MHz spectrophotometer in DMSO-d_6.

5-hydroxy Isatin was synthesized from p- amino phenol by using Sandmayer[8] method(Scheme-1). It consists in the reaction of aniline with chloral hydrate and hydroxylamine hydrochloride in aqueous sodium sulfate to form an isonitrosoacetanilide(I), which after isolation, when treated with concentrated sulfuric acid, furnishes isatin (II)in >75% overall yield (Fig. 1).

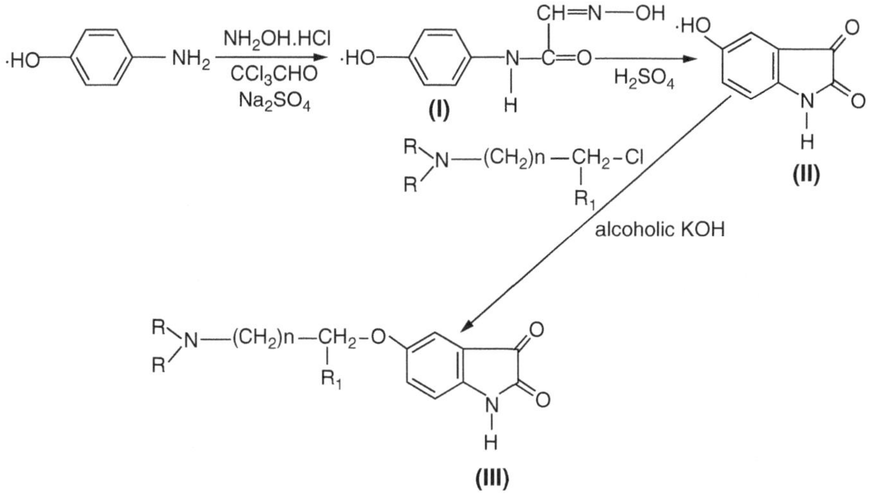

Fig. 1 Scheme 1

2.3 *Preparation of 5-[2(3)-Dialkyl Amino Alkoxy] Indole 2,3 Dione(III) Derivatives*

A mixture of 5-hydroxyisatin (I)(0.01 moles) and dialkylamino alkylhalide (0.01 moles) placed in 10% alcoholic potassium hydroxide and this mixture was stirred at room temperature for 6 h (Scheme-1). The alcohol was reduced to half of its volume and cooled. The product separated was filtered, washed with small portions of cold alcohol repeatedly and dried. It was purified by recrystalisation from hydro alcoholic mixtures to get a crystalline solid. Similarly other 5-Hydroxy Isatin derivatives were prepared and their melting points were determined in Open capillary tubes using Toshniwall melting point apparatus and are uncorrected. Purity of the compounds was checked by TLC (Fig. 2).

The physical data of the title compounds were presented in Table 1. The compounds were characterized by spectral data.

2.4 *Spectral Data*

The compounds have been characterized by the spectral data IR, PMR and Mass.

IR spectrum (KBr) of compound **(III)** exhibited absorption bands (cm-1) 3421.47 (OH), 1630.08 (C=O), 1548 (Ar,C=C), 1282(C–O–C), 883.85–579.8 (Ar). Its PMR spectrum (DMSO, **III**) showed characteristic peaks at (d ppm) 300 MHz 13.3 (s, 1H, OH), 10.36(s, 1H,–CONH), 6.65–7.29(m, 3 H, Ar-H). Mass spectrum of compound III showed molecular ion(M+) base peak at m/z (164.1).

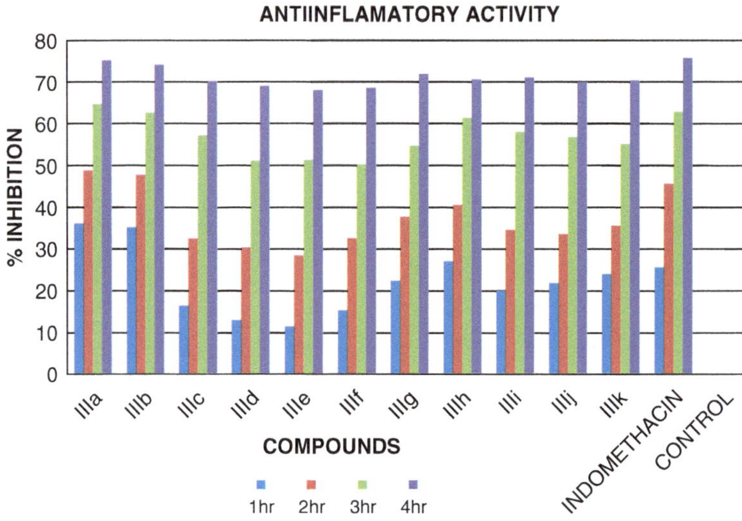

Fig. 2 Graph showing Antiinflammatory activity of 5-[2(3)-dialkylamino alkoxy] Indole 2,3-dione derivatives

Table 1 Characterization data of 5-[2(3)-dialkylamino Alkoxy] Indole 2,3-Diones

S. No.	Compound	R	R_1	n	X	M.F	% YEILD	M.P	M.Wt
1	IIIa	CH_3	H	1	O	$C_{12}H_{14}N_2O_3$	91%	>320	234
2	IIIb	C_2H_5	H	1	O	$C_{14}H_{18}N_2O_3$	86%	>320	252
3	IIIc	CH_3	CH_3	1	O	$C_{13}H_{16}N_2O_3$	93%	>320	248
4	IIId	C_2H_5	CH_3	1	O	$C_{15}H_{20}N_2O_3$	85%	>320	276
5	IIIe	CH_3, H_3C > CH_2–	H	1	O	$C_{16}H_{22}N_2O_3$	81.8%	>320	290
6	IIIf	CH_3	H	2	O	$C_{13}H_{16}N_2O_3$	93%	>320	248
7	IIIg	C_2H_5	H	2	O	$C_{15}H_{20}N_2O_3$	75%	>320	276
8	IIIh	CH_3, H_3C > CH_2–	H	2	O	$C_{17}H_{24}N_2O_3$	74%	>320	304
9	IIIi	CH_3	H	0	O	$C_{11}H_{12}N_2O_3$	85%	>320	220
10	IIIj	C_2H_5	H	0	O	$C_{13}H_{16}N_2O_3$	90%	>320	238
11	IIIk	CH_3, H_3C > CH_2–	H	0	O	$C_{15}H_{22}N_2O_3$	75%	>320	276

Compound (**IIIa**) showed characteristic IR peaks at 3276(NH), 1651.96 (C=O), 1569.82 (Ar,C=C), 1276(C–O–C), 1080(C–N), 2860(C–C), 807.93(Ar). Its PMR spectrum (DMSO, **IIIa**) showed characteristic peaks at (d ppm) 300 MHz 10.36(s, 1H,–CONH), 7.21(d, H,Ar-H), 7.26(d, H,Ar-H), 7.01(s, H,Ar-H), 3.2 (t,2H,O–CH_2), 2.9 (t,2H,N–CH_2), 1.36(s,6H,N–$(CH_3)_2$). Mass spectrum of compound **IIIa** showed molecular ion (M+) base peak at m/z 234 (100%). It also shows peak at m/z (72) may be due to the fragmentation of the alkyl chain from the molecule ion.

Compound (**IIIb**) showed characteristic IR peaks at 3274(NH), 1681.53 (C=O), 1570.21(Ar,C=C), 1243(C–O–C), 1084(C–N), 2890(C–C), 845.51(Ar). Its PMR spectrum (DMSO, **IIIb**) showed characteristic peaks at (d ppm) 300 MHz 10.25(s, 1H,–CONH), 7.22(d, H,Ar-H), 7.26(d, H,Ar-H), 7.11(s, H,Ar-H), 2.99(t,2H,O–CH_2), 2.72(t,2H,N–CH_2), 1.24(s,4H,N–$(CH_2–C)_2$), 1.22(s,6H,(N–C–$CH_3)_2$). Mass spectrum of compound **IIIb** showed molecular ion (M+) base peak at m/z 252 (100%). It also shows peak at m/z (90) may be due to the fragmentation of the alkyl chain from the molecule ion.

Compound (**IIIc**) showed characteristic IR peaks at 3274(NH), 1651.96(C=O), 1579.72(Ar,C=C), 1266(C–O–C), 1095(C–N), 2898(C–C), 805.91(Ar). Its PMR spectrum (DMSO, **IIIc**) showed characteristic peaks at (d ppm) 300 MHz 10.46(s, 1H,–CONH), 7.22(d, H,Ar-H), 7.26(d, H,Ar-H), 7.11(s, H,Ar-H), 2.84(m,H,O–CH), 2.51(d,3H, R_1=CH_3), 2.48(d,2H,N–CH_2), 1.25(s,6H,N–$(CH_3)_2$).

Mass spectrum of compound **IIIc** showed molecular ion (M+) base peak at m/z 248 (100%). It also shows peak at m/z (86) may be due to the fragmentation of the alkyl chain from the molecule ion.

Compound (**IIId**) showed characteristic IR peaks at 3257(NH), 1679.64 (C=O), 1546.86(Ar,C=C), 1245(C–O–C), 1180(C–N), 2960(C–C), 812.71(Ar). Its PMR spectrum (DMSO, **IIId**) showed characteristic peaks at (d ppm) 300 MHz 10.51(s, 1H,–CONH), 7.22(d, H,Ar-H), 7.26(d, H,Ar-H), 7.11(s, H,Ar-H), 2.76(m,2H,O–CH), 2.45 (t,3H, R_1=CH_3), 2.48(d,2H,N–CH_2), 1.24(s,4H,N–$(CH_2–C)_2$), 1.22(s,6H,(N–C–$CH_3)_2$). Mass spectrum of compound **IIId** showed molecular ion (M+) base peak at m/z 276 (100%). It also shows peak at m/z (114) may be due to the fragmentation of the alkyl chain from the molecule ion.

Compound (**IIIe**) showed characteristic IR peaks at 3257(NH), 1689.46 (C=O), 1576.34(Ar,C=C), 1228(C–O–C), 1170(C–N), 2870(C–C), 814.53(Ar). Its PMR spectrum (DMSO, **IIIe**) showed characteristic peaks at (d ppm) 300 MHz 10.26(s, 1H,–CONH), 7.22(d, H,Ar-H), 7.26(d, H,Ar-H), 7.11(s, H,Ar-H), 2.96 (t,2H,O–CH_2), 2.82(t,2H,N–CH_2), 1.35(s, 2H,N–$(CH)_2$), 1.21 (d,12H,N–C–$(CH_3)_2$). Mass spectrum of compound **IIIe** showed molecular ion (M+) base peak at m/z 290 (100%). It also shows peak at m/z (128) may be due to the fragmentation of the alkyl chain from the molecule ion.

Compound (**IIIf**) showed characteristic IR peaks at 3286(NH), 1651.96 (C=O), 1566.82 (Ar,C=C), 1266(C–O–C), 1150(C–N), 2910(C–C), 808.93(Ar). Its PMR spectrum (DMSO, **IIIf**) showed characteristic peaks at (d ppm) 300 MHz 10.46(s, 1H,-CONH), 7.22(d, H,Ar-H), 7.26(d, H,Ar-H), 7.11(s, H,Ar-H), 3.2(t,2H,O–CH_2), 2.9(t,2H,N–CH_2), 3.01(m,2H,C–CH_2–C), 1.36(s,6H,N–$(CH_3)_2$). Mass spectrum of compound **IIIf** showed molecular ion (M+) base peak at m/z 248 (100%). It also shows peak at m/z (86) may be due to the fragmentation of the alkyl chain from the molecule ion.

Compound (**IIIg**) showed characteristic IR peaks at 3274(NH), 1681.53 (C=O), 1570.21 (Ar,C=C), 1243(C–O–C), 1210(C–N), 2885(C–C) 845.51(Ar). Its PMR spectrum (DMSO, **IIIg**) showed characteristic peaks at (d ppm) 300 MHz

10.25(s, 1H,-CONH), 7.22(d, H,Ar-H), 7.26(d, H,Ar-H), 7.11(s, H,Ar-H), 2.99 (t,2H,O–CH_2), 3.04(m,2H,C–CH_2–C), 2.72 (t,2H,N–CH_2),1.23 (s,4H,N–(CH_2–C)$_2$), 1.21 (s,6H,(N–C–CH_3)$_2$). Mass spectrum of compound **IIIg** showed molecular ion (M+) base peak at m/z 276 (100%). It also shows peak at m/z (114) may be due to the fragmentation of the alkyl chain from the molecule ion.

Compound (**IIIh**) showed characteristic IR peaks at 3257(NH), 1689.46 (C=O), 1576.34 (Ar,C=C), 1228(C–O–C), 1280(C–N), 2970(C–C) 814.53(Ar). Its PMR spectrum (DMSO, **IIIh**) showed characteristic peaks at (d ppm) 300 MHz 10.26(s, 1H,–CONH), 7.22(d, H,Ar-H), 7.26(d, H,Ar-H), 7.11(s, H,Ar-H), 2.96 (t,2H,O–CH_2), 3.06(m,2H,C–CH_2–C), 2.82 (t,2H,N–CH_2), 1.35 (s, 2H,N–(CH)$_2$), 1.21 (d,12H,N–C–(CH_3)$_2$). Mass spectrum of compound **IIIh** showed molecular ion (M+) base peak at m/z 304 (100%). It also shows peak at m/z (142) may be due to the fragmentation of the alkyl chain from the molecule ion.

Compound (**IIIi**) showed characteristic IR peaks at 3276(NH), 1651.96 (C=O), 1569.82 (Ar, C=C), 1276(C–O–C), 1089(C–N), 2865(C–C) 807.93(Ar). Its PMR spectrum (DMSO, **IIIi**) showed characteristic peaks at (d ppm) 300 MHz 10.36(s, 1H,–CONH), 7.21(d, H,Ar-H), 7.26(d, H,Ar-H), 7.01(s, H,Ar-H), 2.8 (s,2H,N–CH_2–O), 1.36 (s,6H,N–(CH_3)$_2$). Mass spectrum of compound **IIIi** showed molecular ion (M+) base peak at m/z 220 (100%). It also shows peak at m/z (58) may be due to the fragmentation of the alkyl chain from the molecule ion.

Compound (**IIIj**) showed characteristic IR peaks at 3274(NH), 1681.53 (C=O), 1570.21. (Ar,C=C), 1243(C–O–C), 1180(C–N), 2940(C–C), 845.51(Ar). Its PMR spectrum (DMSO, **IIIj**) showed characteristic peaks at (d ppm) 300 MHz 10.25(s, 1H,–CONH), 7.22(d, H,Ar-H), 7.26(d, H,Ar-H), 7.11(s, H,Ar-H), 2.78 (s,2H,N-CH_2–O)1.24 (s,4H,N–(CH_2–C)$_2$), 1.22 (s,6H,(N–C–CH_3)$_2$).

Mass spectrum of compound **IIIj** showed molecular ion (M+) base peak at m/z 238 (100%). It also shows peak at m/z (76) may be due to the fragmentation of the alkyl chain from the molecule ion.

Compound (**IIIk**) showed characteristic IR peaks at 3257(NH), 1689.46 (C=O), 1576.34 (Ar,C=C), 1228(C–O–C), 1165(C–N), 2970(C–C), 814.53(Ar). Its PMR spectrum (DMSO, **IIIk**) showed characteristic peaks at (d ppm) 300 MHz 10.26(s, 1H,–CONH), 7.22(d, H,Ar-H), 7.26(d, H,Ar-H), 7.11(s, H,Ar-H), 2.76 (s,2H,N–CH_2–O), 1.35 (s, 2H,N–(CH)$_2$), 1.21 (d,12H,N–C–(CH_3)$_2$). Mass spectrum of compound **IIIk** showed molecular ion (M+) base peak at m/z 276 (100%). It also shows peak at m/z (114) may be due to the fragmentation of the alkyl chain from the molecule ion.

3 In Vivo Experiments

3.1 Anti-Inflammatory Activity

Carrageenan - induced rat paw edema method [9] was employed for evaluating the anti-inflammatory activity of the synthesized compounds by in vivo model. Wister Albino rats of either sex weighing approx 200–300 gm, were housed in clean polypropylene cages and kept under room temperature (25 ± 2 °C), and relative humidity 40–50% in a 12 h light-dark cycle. Food was withdrawn 12 h before and during experimental hours. In this study, the animals were divided into groups as shown in the Table 2. Acute inflammation was produced by sub plantar injection of 0.1 ml of 1% suspension of Carrageenan in normal saline, in the right hind paw of the rats. After intra peritoneal administration of the test compounds, the paw volume was measured Plethysmometrically at 1, 2, 3, and 4 h intervals. Indomethacin 5 mg/kg in normal saline was used as standard drug.

In Vitro TMPD Assay Method for Anti-Inflammatory Activity [10] was employed for evaluating the anti-inflammatory activity of the synthesized compounds by in vitro method. Cayman's Colorimetric COX (ovine) Inhibitor Screening Assay Kit was used to carry out the in vitro anti-inflammatory activity. Principle involved in this method is the enzyme activity was measured using chromogenic assay based on oxidation of N, N, N′, N′-tetramethyl-*p*-phenylenediamine (TMPD) during the reduction of prostaglandin G_2 to prostaglandin H_2 by COX-2 enzyme. The Colorimetric COX Inhibitor Screening Assay measures the peroxidase component of cyclooxygenases. The peroxidase activity is assayed colorimetrically by monitoring the appearance of oxidized N,N,N′,N′-tetramethyl-p-phenylenediamine (TMPD) at 590 nm. Inhibition of COX

Table 2 Antiinflammatory activity of 5-[2(3)-dialkylamino alkoxy] Indole 2, 3-dione derivatives by caragenan induced rat paw edema method

Compound	1 h	% Inhibition	2 h	% Inhibition	3 h	% Inhibition	4 h	% Inhibition
IIIa	0.54	36.29	0.43	48.91	0.51	64.7	0.43	75.4
IIIb	0.54	35.6	0.51	47.9	0.48	62.9	0.48	74.3
IIIc	0.72	16.6	0.56	32.6	0.55	57.6	0.53	70.5
IIId	0.74	13.1	0.68	30.6	0.63	51.5	0.55	69.4
IIIe	0.76	11.7	0.70	28.5	0.63	51.5	0.55	68.4
IIIf	0.72	15.4	0.68	32.6	0.65	50.45	0.56	68.8
IIIg	0.63	22.5	0.58	37.9	0.56	54.9	0.45	72.1
IIIh	0.62	27.2	0.48	40.8	0.50	61.5	0.43	71.1
IIIi	0.67	20.1	0.53	34.7	0.53	58.2	0.46	71.4
IIIj	0.67	22	0.65	33.6	0.56	56.9	0.48	70.3
IIIk	0.67	24	0.63	35.7	0.58	55.3	0.51	70.6
Indomethacin	0.63	25.8	0.53	45.9	0.48	63.07	0.43	76.1
Control	0.85	–	0.98	–	1.3	–	1.8	–

Table 3 Anti-inflammatory activity of 5-[2(3)-dialkylamino alkoxy] Indole 2,3-dione derivatives by TMPD method

S. No.	Compound	% Inhibition of COX-2 at 100 μM
1	IIIa	75.00
2	IIIb	66.13
3	IIIc	50.00
4	IIId	38.71
5	IIIe	10.48
6	IIIf	NA
7	IIIg	39.52
8	IIIh	36.13
9	IIIi	33.87
10	IIIj	2.42
11	IIIk	24.31
12	Etoricoxib	67.74
13	Celecoxib	58.06

activity, measured by TMPD oxidation, by a variety of selective and nonselective inhibitors, shows changed potencies similar to those observed with other in vitro methods. The Cayman Colorimetric COX Assay is a time saving tool for screening vast numbers of inhibitors. In this method compound IIIa-k are compared with standard drugs Celicoxib and Eetoricoxib (Table 3).

4 Results and Discussions

Physical data TLC, IR, 1H NMR and mass spectra confirmed the structures and purity of the synthesized compounds. All the title compounds de-composed before melting. All the synthesized compounds were evaluated for their in vivo antiinflamatory activity. Among the compounds were subjected to antiinflamatory activity and it was observed that compounds **IIIa** and **IIIb** showed more promising activity compared to standard drug Indomethacin in caragenan induced rat paw edema method (Table 2). Compounds **IIIg, IIIh, IIIi, IIIk, IIIj and IIIc** were found to be next in the order of reducing the duration of inflamation. **IIIc** and **IIId** compounds also exhibited moderatere protective activity against carrageenan induced inflammation. In Vitro TMPD Assay Method for Anti-Inflammatory Activity showed compound IIIa, IIIb and IIIc showed good COX inhibition compared to standard drugs Celicoxib and Etoricoxib.

5 Conclusion

A new series of five 5-[2(3)-dialkyl amino alkoxy] Indole 2,3 dione derivatives were synthesized by reacting 5-hydroxyindole 2,3 dione with 2-N,N di alkylamino alkyl halides. Evaluation of these compounds as antiinflamatory activity by invivo rat paw edema method and invitro TMPD assy method. Compounds with a dimethyl (**IIIa**), di ethyl(**IIIb**) amino ethyl chain derivatives was found to be relatively superior in antiinflamatory activity and other compounds (**IIIg, IIIh, IIIi, IIIk, IIIj and IIIc**) are next in the order of activity.

Acknowledgments The First author would like to thank the UGC, New Delhi for providing financial support. Authors are thankful to Principal University College of Pharmaceutical Sciences, Kakatiya University, Warangal for providing facilities.

References

1. Bhattacharya Salil, K., et al. 1991. *Journal of Psychopharmacology* 5 (202): 1991.
2. Pandeya, S.N., and A. Senthil Raja. 2002. *Journal of Pharmaceutical Sciences* 5 (3): 275.
3. Pandeya, S.N., et al. 2000. *European Journal of Medicinal Chemistry* 35: 879–886.
4. Padhyet, A.K., et al. 2004. *Indian Journal of Chemistry* 43B: 971.
5. Raviraj, A., et al. 2004. *Journal of Chemical Sciences* 116 (5): 265.
6. Gupta, S., Raman Asian, et al. 2004. *Journal of Chemistry* 16 (2): 779–783.
7. Gringberg, B., et al. 1990. *Chemija* 2: 87.
8. Marvel, C.S., and G.S. Heirs. 1941. *Organic synthesis Collect* 1: 327.
9. Winter, C.A., et al. 1962. *Proceedings of the Society for Experimental Biology and Medicine* 111: 544.
10. Kulmacz, R.J., and W.E.M. Lands. 1983. *Prostaglandins* 25: 531–540.

Short Review on Quantum Key Distribution Protocols

Dimitris Giampouris

Abstract Cryptographic protocols and mechanisms are widely investigated under the notion of quantum computing. Quantum cryptography offers particular advantages over classical ones, whereas in some cases established protocols have to be revisited in order to maintain their functionality. The purpose of this paper is to provide the basic definitions and review the most important theoretical advancements concerning the BB84 and E91 protocols. It also aims to offer a summary on some key developments on the field of quantum key distribution, closely related with the two aforementioned protocols. The main goal of this study is to provide the necessary background information along with a thorough review on the theoretical aspects of QKD, concentrating on specific protocols. The BB84 and E91 protocols have been chosen because most other protocols are similar to these, a fact that makes them important for the general understanding of how the QKD mechanism functions.

Keywords Quantum key distribution • Quantum key • Key distribution • Quantum key distribution protocol • Quantum cryptography • Quantum computer • Physical review letter • Quantum key transmission • Pseudo random sequence • Quantum repeater • Raw key • Short review • Key agreement protocol • Decoy pulse • Quantum game • Quantum channel • Quantum key distribution system • Practical quantum key distribution

1 Introduction

Processing power is growing fast. A 20–30% increase has been observed over a period of every 2 years, as stated in the famous Moore's law [1]. Transistor sizes have been shrinking exponentially over the last decades, but this is about to come to an end in the next few years [2]. One potential step regarding the evolution of computing, in general, could be the implementation of the quantum computer. As

D. Giampouris (✉)
Department of Informatics, Ionian University, Tsirigoti Square 7, Corfu 49100, Greece
e-mail: p14giab@ionio.gr

P. Vlamos (ed.), *GeNeDis 2016*, Advances in Experimental Medicine and Biology 988, DOI 10.1007/978-3-319-56246-9_12

it has been shown, problems that have been considered hard to solve with classical computers, are not that difficult to solve with the use of a quantum computer [3].

The problem that arises is that most of the current public key cryptographic models rely on those problems, something that makes them vulnerable when opposed to a quantum computer. However, not all the existing classes of cryptographic systems are vulnerable when opposed to a quantum computer. Hash-based, code-based, lattice-based, secret-key and multivariate-quadratic-equations cryptographic systems are believed to have resilience against this kind of adversary [4]. Although these system can serve as replacement for public key cryptography, quantum key distribution serves as a replacement that in some cases can guarantee absolute security. Given the power of quantum computers, the next step in order to protect key transmissions and sensitive information conveyance is quantum cryptography. Quantum cryptography is vastly different from the classical cryptography, because instead of using difficult to solve mathematical problems, it relies on physics laws as a basis for establishing security. In the case of quantum key transmission and more specifically the BB84 protocol, security is based (amongst other laws) upon the Heisenberg's uncertainty principle, as an eavesdropper, by trying to measure photons, will disturb their state and cause hopefully detectable errors [5].

The goal with quantum key distribution is not to protect the transmission of big amounts of data over a quantum channel, but to securely distribute a key which upon the data will be encrypted. After the encryption, data can be transferred with the use of a classical channel.

This paper aims to give a quick start to new researchers attempting to understand the field of quantum cryptography. Quantum computing in general can be a daunting field in the eyes of someone who has only been researching classical computing. This paper is written with balance between technical terms and as easy as possible explanations in mind. The structure is as follows: first is the current literature and advancements in the field, then are the main protocols and important research concerning each one of them and finally there is a short general discussion on the matter.

2 Current Literature and Advancements in the Field of QKD

Despite the fact that quantum key transmission is theoretically an efficient solution for the transmission of a key without leaks, there are economical and technical obstacles for the actual implementation of such mechanisms. The first physical problem for these systems is the noise. Recently, it has been shown that there is progress concerning the speed of the transmission related to distance. This is achieved with the use of division-multiplexing wavelength, the use of the colorless interferonic method and the use of specific quantum key transmission and key distillation hardware. More specifically, a key generation rate 200-kbps was achieved, with 14.5-db transmission loss at the distance of 45-km, with the perspective of 1-mbps for less than 10-db transmission loss [6] in 2012. In 2015 a compact and autonomous QKD system was presented with the capability of distributing secure cryptographic keys over 307-km of optical fibre [7].

There are various approaches regarding the study and the design of efficient and trustworthy protocols. Some of them include game-theoretic notion. Games and cryptography are strongly connected and various studies have explored this observation, i.e. [8, 9]. Quantum games are well-studied in recent years, see for example the works in [10–12]. Also, in [13], the authors showed the existence of a relation between specific games and automata variants (in particular quantum periodic automata of [14]).

Besides, the computational cost, there are more barriers (at least for the maintenance of absolute security). One example was the transmission of videos, which was considered to be very difficult until 10 years ago through quantum key transmission and the use of one time pad (since there was a need for coding for every data bit) was necessary. However, after the relevant progress at this field, we have the ability to transmit a video between nodes with a rate of 128–1024-kb/s without considering any overhead data, with the only disadvantage of the delay for a few seconds [15].

On the other hand, in the technical aspect of the barriers, we mainly encounter the problem of the relay of photons or other particles [16]. This fact makes the quantum key transmission possible only at the distances of a few hundred kilometers. The idea behind the embodiment of a quantum repeater is to have a unit which regenerates the original bits every one kilometer. So far there has been no practical application of a quantum repeater. However, theoretical models and methodologies that investigate the theoretical foundation for a future achievement, are constantly developed [17, 18].

Additionally, another significant problem that emerges is the device independence. It means that if the quantum device that the quantum cryptographic protocol is utilized upon, are created by a malicious creator, will there be a way for the protocol to function securely? This question was answered with the use of a slightly modified version of E91 and the proof was based on the understanding of the monogamous nature of quantum co-relations [19].

3 Cryptographic Protocols

3.1 BB84

The BB84 protocol was proposed by Bennet and Brassard in 1984. It is based on three basic principles:

- The no-cloning theorem, which makes quantum states impossible to duplicate, something that hinders an eavesdropper (Eve) from intercepting the quantum channel of communication, copying the states in order to create a key, and sending the original states to the original recipient (Bob).
- Measurements lead to the collapse of the quantum states. An important fact in quantum key distribution (QKD) is the use of different bases for creating series

of bits. When one of the given bases is measured, the state of the other base collapses, resulting in completely random measurements. Consequently, when we extract information from the cryptosystem, we agitate its state.
- Measurements are non reversible [20].

Let's assume that Alice has a source of solitary photons. The spectral properties of photons are strictly defined so that the only degree of freedom is polarization. Alice and Bob align the polarizers and agree to use the horizontal or vertical position $(+)$, or the complimentary base of linear polarizations $+45/-45(\times)$. Specifically, the bits are encoded as below:

$|H\rangle$, corresponds to 0_+
$|V\rangle$, corresponds to 1_+
$|+45\rangle$, corresponds to $0_\times$
$|-45\rangle$, corresponds to $1_\times$

We observe that both bit values (0 and 1) can be encoded with two possible ways, or more specifically to non-orthogonal states because:

$$|\pm 45\rangle = (1/\sqrt{2})(|H\rangle \pm |V\rangle)$$

Given the above, the BB84 protocol is utilized as follows:

- Alice prepares a photon in one of the for aforementioned states and sends it to Bob using a quantum channel. Bob makes measurements using either the $+$ or the $\times$ base. This step is repeated N times. As a result, Alice and Bob have a list of N pairs (bits, basis).
- Alice and Bob communicate with a classic channel and compare the basis value for every object, deleting every case that both of them have used different bases. This procedure is called sifting. Finally, Alice and Bob have a list $N/2$ bits, with the promise that for every pair, the encoding used by Alice is matched by Bob's measurements. This list is called raw key.
- Alice and Bob disclose a random sample of bits that are contained in the raw keys and estimate the error rate in the quantum channel. In the case that there are no errors and the raw key is the same for Alice and Bob, the raw key is considered the secret key. In case there are errors, Alice and Bob must correct them and delete the information that a passive enemy could potentially obtain. Both of these procedures can be conducted in a classic channel of communication. This part of the protocol is called classical post-processing. When this part is over Alice and Bob share a truly secret key or nothing at all (in the case that the enemy intercepted a big part of the transmitted information) [21].

3.1.1 Proof of Security and the PNS Attack

In theory the BB84 has been proven secure given some conditions, but this is not the case when one party inadvertently sends two or more identical copies of it's qubit instead of just one copy [22]. A multi-photon part of a signal creates a security hole which can be exploited by the photon number splitting attack (PNS). Let's assume that an active enemy exists (Eve). Eve splits one photon off each multi-photon signal. In order to do that, she projects the state onto subspaces characterised by n, which is the total photon number, which can be measured via a quantum nondemolition (QND) measurement.

The projection into these subspaces does not modify the polarization of the photons. Then she performs a polarization-preserving splitting operation, for example, by an interaction described by a Jaynes-Cummings Hamiltonian or an active arrangement of beam-splitters combined with further QND measurements. She keeps one photon and sends the other $(n - 1)$ photons to Bob. We assume (conservatively) that Bob's detector cannot resolve the photon number of arriving signals. When receiving the data regarding the basis, Eve measures her photon and obtains full information. Each signal containing more than one photon in this way will yield its complete information to an eavesdropper [23].

3.1.2 The SAGR04 Protocol

This protocol uses the same four states and the same measurements (from Bob's side) as the BB84 protocol, but the bit is encoded on the basis and on the state (base X equals to zero and base Y to one). Bob has to choose the basis with $1/2$ probability. The creation of the raw key is more complex in comparison to the BB84 protocol. If we assume that Alice sends $|+x\rangle$: in an error free case, if Bob measures X, the result will be $s_b = +$, else if he measures Y, the result will be $s_b = +/-$ with the same probability.

In the sifting phase Bob discloses the s_b and Alice indicates to Bob to accept if he had prepared a state where $s_a \neq s_b$, in which case Bob accepts the bit that corresponds to the base he did not use. The reason will become clear after the next example: in an error free case, $s_b = -$ a wrong base is excluded. The SAGR04 was invented to be applied when attenuated laser are used, because it is more resistant to the PNS attack that the BB84 protocol [21].

3.1.3 Decoy States

As mentioned before, a multi-photon part of a signal creates an opportunity for the PNS attack to be implemented. However, there have been proposed some methods in order to mediate the problem. One of these methods involves the use of decoy-pulses. In this setup, a legitimate user intentionally and randomly replaces signal pulses by multi-photon pulses (decoy pulses). Then they check the yield of the decoy pulses. If the yield of decoy pulses is abnormally higher than that of other signal pulses, the whole protocol is aborted. Otherwise, to continue the protocol, they estimate the yield of signal multi-photon pulses based on that of decoy pulses. This estimation can be done with an assumption that the two losses have similar values [24].

3.1.4 Quantum Key Agreement

A key agreement protocol is one whereby two or more parties agree upon a key over insecure communication channels based on their exchanges messages. In contrast to the key distribution, where one party decides the key and then distributes it to the other parties, each party in a key agreement protocol contributes its part to the shared key and the share key should not be determined fully by any party alone [25]. There have been QKA protocols which are based on the BB84 protocol, for example the one presented by Song-Kong Chong and Tzonelih Hwang [25], which claims three basic advantages over the QKD:

1. The outcome of the protocol is influenced by both parties: no one can determine the shared key alone.
2. The protocol has 50% qubit efficiency after the random sampling discussion.
3. It provides the unconditional security.

3.2 *E91*

This protocol is based on the quantum entanglement phenomenon. Let's assume that we have a Bell state, where Alice has one part of the EPR pair and Bob has the other. Let's also assume that the state of the EPR pair is:

$$|\beta_{00}\rangle = \frac{|00\rangle + |10\rangle}{\sqrt{2}}$$

Considering the above, we know that Alice and Bob will have measurement results completely co-related. In contrast, if we use the state:

$$|\beta_{01}\rangle = \frac{|01\rangle + |10\rangle}{\sqrt{2}}$$

then Alice and Bob will have measurement results completely non-related. Alice and Bob will then have to measure each qubit with randomly selected bases. When the measurements are complete, they have to communicate by using a classical channel, in order to verify the bits that they measured with the same bases.

From the fact that the measurement results are either completely co-related or completely non-related, it is easy for Alice and Bob to discover any possible eavesdropper. The errors that occur just by default and not because of some enemy, can be corrected to a degree with various techniques. In all cases of QKD, a distilled key is a result from a procedure known as privacy amplification [21, 26].

3.3 Important Research Concerning E91

3.3.1 Proof of Security and the Detection Loophole

The E91 protocol can be summarized as follows: If Eve provides separable states to Alice and Bob, then Eve can get information on the key. However, in this case the separable states cannot violate Bell's inequality in the checking phase, thus Eve is detected. For example, let us assume that Eve provides either $|0\rangle|1\rangle$ or $|1\rangle|0\rangle$ with equal probability after recording which state she sends at each instance. In the normal phase nothing unexpected happens here. In the checking phase, however, the samples cannot violate Bell's inequality and thus the attack by Eve is detected.

On the contrary, if Eve provides the legitimate Bell state, she can pass the checking phase. However, in this case Eve has no information on the key generated at Alice's and Bob's sites. If Eve provides a partially entangled state, then she will get partial information on the key. However, Eve can utilize the free-choice loophole in the E91 protocol. Let's assume that a manufacturer designs a QKD system such that each device chooses spin-measurement directions according to a pseudo-random sequence that is installed in the device beforehand. Here the pseudo-random sequences in the two devices are independent. The pseudo-random sequence is one that appears to be random but actually is not. The QKD system is also designed such that one device contains an algorithm for generating the pseudo-random sequence of the other device.

Thus, effectively, one device has information about the choices on the spin-measurement direction of the other device. Therefore, the locality condition:

$$S_A(p) = f(p, \lambda), S_B(q) = g(q, \lambda)$$

can be effectively violated in the QKD system provided by the manufacturer. In this case, users, after careful inspections, will become aware of a problem in the devices and that the measurement choices claimed to be random are not really the measurement choices claimed to be random, are not really random, of course. However, it is impractical for many users to perform such a careful inspection, and moreover, it is a very difficult task to identify a pseudo-random sequence.

In other respects, the design of the devices is inline with that the hidden variable λ is encoded on the timing of the pulses carrying qubits. The devices can read out and make use of the classical information thus encoded. Therefore, Eve can successfully eavesdrop by adopting an effectively nonlocal hidden variable model that simulates the Bell state $|\Psi^-\rangle$ [27].

4 General Discussion

Quantum key distribution does not require the use of a quantum computer. It is realized with commercially available technologies [28]. There are already some well known and operational quantum key distribution networks like the DARPA quantum network, which is a 10 node network operating since 2004, and the Tokyo QKD network, which involves commercial and academic entities.

Also, QKD systems are commercially available by companies like Toshiba or MagiQ Technologies, meaning that although a general purpose, every day use quantum computer is years ahead from implementation, quantum key distribution is already a reality. Although, QKD is heading to commercialisation, it is not yet that of a general use type. Quantum repeaters don't exist yet, something that constricts the range of a QKD network to that of a few hundred kilometers. The realization of the quantum repeater seems like the most important barrier towards global QKD networking.

Acknowledgements I would like to thank Theodore Andronikos and Konstantinos Giannakis for their suggestions, guidance, and motivation they provided me in order to complete this review. I would also like to thank my colleague and fellow student Alexander Kalavitis for his valuable assistance and comments.

References

1. Moore, G.E., et al. 1975. Progress in digital integrated electronics. In *Electron Devices Meeting* 21:11–13.
2. Colwell, R. 2013. The chip design game at the end of Moore's law. In Hot Chips 25 Symposium (HCS), 2013 IEEE, 1–16. IEEE.
3. Shor, P.W. 1999. Polynomial-time algorithms for prime factorization and discrete logarithms on a quantum computer. *SIAM Review* 41(2):303–332.
4. Bernstein, D.J., J. Buchmann, and E. Dahmen. 2009. *Post-quantum cryptography*. New York: Springer Science & Business Media.
5. Bennett, C.H., and G. Brassard. 2014. Quantum cryptography: Public key distribution and coin tossing. *Theoretical Computer Science* 560:7–11.
6. Tanaka, A., M. Fujiwara, K.i. Yoshino, S. Takahashi, Y. Nambu, A. Tomita, S. Miki, T. Yamashita, Z. Wang, M. Sasaki, et al. 2012. High-speed quantum key distribution system for 1-mbps real-time key generation. *IEEE Journal of Quantum Electronics* 48(4):542–550.
7. Korzh, B., C.C.W. Lim, R. Houlmann, N. Gisin, M.J. Li, D. Nolan, B. Sanguinetti, R. Thew, and H. Zbinden. 2015. Provably secure and practical quantum key distribution over 307 km of optical fibre. *Nature Photonics* 9(3):163–168.
8. Katz, J. 2008. Bridging game theory and cryptography: Recent results and future directions. In Theory of Cryptography Conference, 251–272. Springer.
9. Kol, G., and M. Naor. 2008. Cryptography and game theory: Designing protocols for exchanging information. In Theory of Cryptography Conference, 320–339. Springer.
10. Buhrman, H., N. Chandran, S. Fehr, R. Gelles, V. Goyal, R. Ostrovsky, and C. Schaffner. 2014. Position-based quantum cryptography: Impossibility and constructions. *SIAM Journal on Computing* 43(1):150–178.
11. Cooney, T., M. Junge, C. Palazuelos, and D. Pérez-García. 2015. Rank-one quantum games. *Computational Complexity* 24(1):133–196.
12. Guo, H., J. Zhang, and G.J. Koehler. 2008. A survey of quantum games. *Decision Support Systems* 46(1):318–332.
13. Giannakis, K., C. Papalitsas, K. Kastampolidou, A. Singh, and T. Andronikos. 2015. Dominant strategies of quantum games on quantum periodic automata. *Computation* 3(4):586–599.

14. Giannakis, K., C. Papalitsas, and T. Andronikos. 2015. Quantum automata for infinite periodic words. In 2015 6th international conference on information, intelligence, systems and applications (IISA), 1–6. IEEE.
15. Mink, A., X. Tang, L. Ma, T. Nakassis, B. Hershman, J.C. Bienfang, D. Su, R. Boisvert, C.W. Clark, C.J. Williams. 2006. High speed quantum key distribution system supports one-time pad encryption of real-time video. In Defense and security symposium, 62,440M–62,440M. International Society for Optics and Photonics.
16. Ouellette, J. 2004. Quantum key distribution. *Industrial Physicist* 10(6):22–25.
17. Yuan, Z.S., Y.A. Chen, B. Zhao, S. Chen, J. Schmiedmayer, and J.W. Pan. 2008. Experimental demonstration of a bdcz quantum repeater node. *Nature* 454(7208):1098–1101.
18. Zhao, Z., T. Yang, Y.A. Chen, A.N. Zhang, and J.W. Pan. 2003. Experimental realization of entanglement concentration and a quantum repeater. *Physical Review Letters* 90(20):207,901.
19. Vazirani, U., and T. Vidick. 2014. Fully device-independent quantum key distribution. *Physical Review Letters* 113(14):140,501.
20. McMahon, D. 2007. *Quantum computing explained*. New York: Wiley.
21. Scarani, V., H. Bechmann-Pasquinucci, N.J. Cerf, M. Dušek, N. Lütkenhaus, and M. Peev. 2009. The security of practical quantum key distribution. *Reviews of Modern Physics* 81(3):1301.
22. Shor, P.W., and J. Preskill. 2000. Simple proof of security of the bb84 quantum key distribution protocol. *Physical Review Letters* 85(2):441.
23. Brassard, G., N. Lütkenhaus, T. Mor, and B.C. Sanders. 2000. Security aspects of practical quantum cryptography. In International conference on the theory and applications of cryptographic techniques, 289–299. New York: Springer.
24. Hwang, W.Y. 2003. Quantum key distribution with high loss: Toward global secure communication. *Physical Review Letters* 91(5):057,901.
25. Chong, S.K., and T. Hwang. 2010. Quantum key agreement protocol based on bb84. *Optics Communications* 283(6):1192–1195.
26. Deutsch, D., A. Ekert, R. Jozsa, C. Macchiavello, S. Popescu, and A. Sanpera. 1996. Quantum privacy amplification and the security of quantum cryptography over noisy channels. *Physical Review Letters* 77(13):2818.
27. Hwang, W.Y. 2005. Bell's inequality, random sequence, and quantum key distribution. *Physical Review A* 71(5):052,329.
28. Elliott, C. 2002. Building the quantum network. *New Journal of Physics* 4(1):46.

Objective Structured Clinical Examination (OSCE) in Psychiatry Education: A Review of Its Role in Competency-Based Assessment

Christos Plakiotis

Abstract Over the last two decades, Objective Structured Clinical Examination (OSCE) has become an increasingly important part of psychiatry education and assessment in the Australian context. A reappraisal of the evidence base regarding the use of OSCE in psychiatry is therefore timely. This paper reviews the literature regarding the use of OSCE as an assessment tool in both undergraduate and postgraduate psychiatry training settings. Suitable articles were identified using the search terms 'psychiatry AND OSCE' in the ERIC (educational) and PubMed (healthcare) databases and grouped according to their predominant focus: (1) the validity of OSCEs in psychiatry; (2) candidate preparation and other factors impacting on performance; and (3) special topics. The literature suggests that the OSCE has been widely adopted in psychiatry education, as a valid and reliable method of assessing psychiatric competencies that is acceptable to both learners and teachers alike. The limited evidence base regarding its validity for postgraduate psychiatry examinations suggests that more research is needed in this domain. Despite any shortcomings, OSCEs are currently ubiquitous in all areas of undergraduate and postgraduate medicine and proposing a better alternative for competency-based assessment is difficult. A critical question is whether OSCE is sufficient on its own to assess high-level consultancy skills, and aspects of professionalism and ethical practice, that are essential for effective specialist practice, or whether it needs to be supplemented by additional testing modalities.

Keywords Objective structured clinical examination • OSCE • Psychiatry education • Medical education • Competency-based assessment

C. Plakiotis (✉)
Monash Ageing Research Centre (MONARC), Monash University, Melbourne, VIC, Australia

Department of Psychiatry, School of Clinical Sciences at Monash Health, Monash University, Melbourne, VIC, Australia

Aged Persons Mental Health Service, Monash Health, Melbourne, VIC, Australia

Aged Psychiatry Academic Unit (Monash University), c/o Medical Administration, Kingston Centre, Warrigal Road, Cheltenham, VIC, 3192, Australia
e-mail: Chris.Plakiotis@monash.edu

P. Vlamos (ed.), *GeNeDis 2016*, Advances in Experimental Medicine and Biology 988, DOI 10.1007/978-3-319-56246-9_13

1 Background

Clinical competence, and its reliable and valid assessment, has increasingly become a focus for medical educators and a range of stakeholders with legitimate interests in the clinical competence of medical students and postgraduate medical trainees. In this context, it is essential that methods for assessing clinical competence effectively discriminate between adequate and inadequate performers [1].

The field of competency assessment in medical education was revolutionised in the 1970s with the introduction of the first objective structured clinical examination (OSCE), which evaluated the performance of professional behaviours through actors and choreographed scenarios [2, 3]. Candidates involved in OSCE rotate sequentially through several 'stations' at which specific tasks are undertaken in relation to structured cases. Tasks are usually clinically-focussed, e.g. taking a history or performing a physical examination or procedural skill. A specified time is allocated for each station and marking schemes are structured and pre-determined. OSCEs may vary with regard to the time allocated to individual stations, the use of mannequins or standardised patients, the choice of examiners (e.g. clinicians or standardised patients), and the marking method used (e.g. checklist or rating scale). However, the notion that all candidates must undertake the same task in the same amount of time and be marked using identical criteria is essential to all OSCEs [1].

In a seminal paper in the field of psychiatry education, Hodges [4] provided a comprehensive guide for developing, monitoring and improving the quality of psychiatry OSCEs that involved the following key elements:

1. Planning (budgeting, funding, team set-up, sourcing standardised patients)
2. Creation (blueprinting, station development, measurement instruments)
3. Preparation (recruiting and training standardised patients and examiners, venue location)
4. Conduct (preparation, useful documents, reporting results)
5. Quality control (monitoring and improvement, data analysis, standard setting)
6. Research.

In a commentary on the OSCE guide developed by Hodges [4], Davis [5] argued that medical education had stalled educationally and depended on 'attestation' as the means for assuring competence and quality for far too long. The focus on core competencies in medical education, and the use of OSCE to assess them, was identified as a potential driver of educational innovation. The scope of OSCEs to both educate and test was emphasised. A strategic selection of cases across the spectrum of competence was recommended to ensure the validity and reliability of the process. OSCEs' ability to evaluate performance in acute care situations (that would otherwise be practically and ethically difficult) and low-frequency but high-stakes situations (such as suicide risk assessment and management) was emphasised. However, a need to demonstrate a relationship between OSCE performance and clinical outcome was identified.

Over the last two decades, the OSCE has become an increasingly important part of psychiatry education and assessment in the Australian context. At an undergraduate (medical student) level, psychiatry stations frequently feature in OSCEs conducted jointly with other medical disciplines. The OSCE has also been a key component of the examination for Fellowship of the Royal Australian and New Zealand College of Psychiatrists. A review of the evidence base regarding the use of OSCE in psychiatry is therefore timely to evaluate the positive contributions it has made to psychiatry education and training as well as to consider its potential shortcomings.

2 Objectives

This paper reviews the literature regarding the use of OSCE as an assessment tool in both undergraduate and postgraduate psychiatry training settings. Suitable articles from English-speaking countries primarily were identified using the search terms 'psychiatry AND OSCE' in the ERIC (educational) and PubMed (healthcare) databases. Papers were grouped according to their predominant focus: (1) the validity of OSCEs in psychiatry; (2) candidate preparation and other factors impacting on performance; and (3) special topics. Key findings from the literature are subsequently discussed in light of the author's own experience as an undergraduate and postgraduate OSCE candidate and examiner. Given the increasing emphasis on interprofessional education and assessment for postgraduate health practitioners, the use of OSCE in the related fields of clinical psychology and neuropsychology is also briefly considered.

3 The Validity of OSCEs in Psychiatry

Papers focused predominantly on the validity of OSCEs in psychiatry are reviewed below and summarised in Table 1.

Loschen [6] reported on the use of OSCEs to assess the clinical skills of second- and fourth-year psychiatry residents at the Southern Illinois University School of Medicine since 1985. A separate examination was developed for each trainee group. Stations generally incorporated a written component. Each OSCE was comprised of six 40-min stations, allowing eight residents to complete it over 6 h. The OSCE format fared favourably as a method of formative assessment. However, marked differences between global OSCE ratings and clinical ratings of residents' performance by supervisors and unit heads prompted the authors to avoid its use for summative assessment until more acceptable performance standards could be established. A tendency to base OSCE scenarios on unusual or rare cases was discouraged as a marked departure from the goal of testing clinically relevant skills was felt to compromise the method's validity.

Table 1 Papers regarding the validity of OSCEs in psychiatry

Author year	Participants and OSCE characteristics	Key findings
Loschen [6]	• Second-year (18) and fourth-year (25) psychiatry residents • Six 40-min stations	• OSCE fared favourably as a formative assessment method • Differences between global OSCE ratings and clinical ratings of residents' performance by supervisors and unit heads prompted avoidance of its use for summative assessment, pending better performance standards • Scenarios based on unusual or rare cases are best avoided
Hodges et al. [7]	• Medical students (192) • Ten 12-min stations	• OSCE viably and reliably assessed high-level psychiatric skills in a format appreciated by students and teachers
Hodges and Lofchy [8]	• Medical students (42) • 'Mini-OSCE' (four 15-min stations)	• 'Mini-OSCE' demonstrated good interstation reliability and was well-received by students and academics • Benefits conferred without expense of longer OSCEs
Hodges et al. [9]	• Medical students (33) and psychiatry residents (17) • Eight 12-min stations	• Rankings on global scores, but not checklist scores, were accurately predicted by communication instructors • Performance was accurately predicted by supervisors (as measured by checklist scores but not global ratings) • Residents obtained significantly higher mean OSCE scores than medical students on global ratings only • Scenarios rated by residents as highly realistic
Loschen [10]	• Three stations for medical students (20–30-min simulated patient interview, 15–20-min testing) • Five to six stations for psychiatry residents	• OSCE was a cost-effective means of performance-based clinical skills assessment • Practical application of knowledge was evaluated (unlike traditional pen-and-paper knowledge-based tests)

Park et al. [12]	• Third-year medical students (286) • Nine-station psychiatry OSCE (including five 15-min standardised patient interviews)	• History taking, interpersonal skills and physical examination scores in a Clinical Skills Examination reflected scores for same skill set in an OSCE • Acceptable construct validity for both checklist and global process scores, supporting their continued use
Walters et al. [13]	• Fourth-year medical students (128) • Fifteen 'active' 6-min stations and two to three 'rest' stations	• Perceived face and content validity were high and overall reliability was moderate to good • OSCE was practically viable and well-received by students, simulated patients and examiners • Array of modalities enabled testing of many students and broad topic inclusion without cost of simulated patients
Marwaha [14]	• Debate paper	• OSCEs might be 'gold standard' in medical student clinical assessment, but their validity, authenticity and educational impact are open to challenge in postgraduate psychiatry • An inability to test complex real life scenarios and high-level psychiatric reasoning might compromise the standing of the profession and future patient care

Hodges et al. [7] assessed the feasibility, reliability, and validity of an OSCE for University of Toronto medical students undertaking their psychiatry rotation. A ten-station OSCE was developed in two parallel forms and administered three times. Performance was assessed via a checklist (content) and a global-rating scale (process) for each station. Additionally, each student's performance was graded as 'pass,' 'borderline' or 'fail' (by the examiner) and comments were recorded (by the examiner and standardised patient). Ninety-four students were examined with the first form (mean score 70.47%) and 98 with the second (mean score 67.66%). The findings suggested the viability and reliability of OSCE for assessing high-level psychiatric skills in a format appreciated by students and teachers.

In an effort to encourage psychiatry educators to adopt the OSCE in place of oral exams with inferior psychometric properties, Hodges and Lofchy [8] developed a four-station 'mini-OSCE' and trialled its use among 42 medical students undertaking their psychiatry clinical placement at the University of Toronto. Marks were assigned for content and process and the two domains equally weighted to derive a final score. Feedback was provided based on a global judgment of performance (as 'pass,' 'borderline,' or 'fail'). Exam scores ranged from 56 to 86%, with a mean of 74%. The mini-OSCE demonstrated good interstation reliability and was well-accepted by students and academic staff, benefits conferred without the expense associated with longer OSCE formats.

Hodges et al. [9] examined the validity of a psychiatry OSCE comprised of eight 12-min stations conducted among 33 medical students on their psychiatry rotation and 17 psychiatry residents at the University of Toronto. Medical students' rankings on global scores, but not checklist scores, were accurately predicted by communication instructors. Prediction of medical students' OSCE performance by faculty supervisors, as measured by checklist scores but not global ratings, was reasonably accurate. Residents obtained significantly higher mean OSCE scores than medical students on global ratings only. Scenarios were rated by residents as highly realistic. Based on these findings, the authors suggested the OSCE to be a valid tool for assessing medical students' clinical competence.

Loschen [10] described the role of OSCE in both resident evaluation and medical student teaching at the Southern Illinois University School of Medicine. This examination was composed of 5–6 stations and evolved over the years to include more comprehensive tasks than the circumscribed ones reported in an earlier paper [6]. A 3-station OSCE exam—after a format developed by Barrows et al. [11]—was used to test senior medical students' clinical skills competency at the conclusion of their psychiatry rotation. Each station involved a 20–30-min outpatient psychiatric evaluation of a standardised patient followed by 15–20-min of computerised short-answer and multiple choice questions. The author concluded that OSCE represented a cost-effective means of performance-based clinical skills assessment that allowed evaluation of the practical application of knowledge in a manner not possible with traditional pen-and-paper knowledge-based tests.

Park et al. [12] analysed aggregated archival data from two classes of third-year medical students (286 in total) at the St Louis University School of Medicine to assess the construct validity of the checklist and global process scores for an

OSCE in psychiatry. Binary checklists were used to evaluate clinical skills and numeric rating scales to assess standardised patients' perception of interpersonal skills. History taking, interpersonal skills and physical examination scores in a separate Clinical Skills Examination reflected scores for the same skill set in the psychiatry OSCE. The pattern of relationships that emerged confirmed acceptable construct validity for both the checklist score and global process score especially, supporting their continued use in psychiatry OSCEs.

Walters et al. [13] evaluated the validity and reliability of a multimodal OSCE developed on a limited budget for undergraduate psychiatry teaching at the Royal Free and University College Medical School in London. Four OSCEs were evaluated, comprised of 15–18 stations each and administered to 128 fourth-year medical students. Station types included: (1) history taking 'interactive' stations; (2) communication skills 'interactive' stations; (3) telephone communication with colleague; (4) video mental state examination; (5) written case vignette; and (6) written problem orientated vignette with visual prompt. Perceived face and content validity were high and overall reliability was moderate to good. The OSCE was practically viable and well-received by students, simulated patients and examiners. Integrating an array of modalities enabled the testing of many students and broad topic inclusion while avoiding the cost of simulated patient payments.

The Clinical Assessment of Skills and Competencies (CASC) is an OSCE used as a clinical examination gateway, granting access to becoming a senior psychiatrist in the United Kingdom (UK). In a debate paper, Marwaha [14] examined the utility of the CASC from the viewpoint of a senior psychiatrist. The author argued that while OSCEs might be the 'gold standard' in medical student clinical assessment, their validity, authenticity and educational impact are open to challenge in postgraduate professional psychiatry examinations. It was suggested that an inability to test complex real life scenarios and high level psychiatric reasoning might compromise the standing of the profession and future patient care.

4 Candidate Preparation and Other Factors Impacting on Performance

Papers regarding candidate preparation and other factors impacting on psychiatry OSCE performance are reviewed in this section and summarised in Table 2.

Blaskiewicz et al. [15] analysed aggregated archival OSCE performance data (in the form of a 25-item binary content checklist) from a class of 141 third-year medical students at Saint Louis University School of Medicine to investigate the impact of testing context and rotation order on student performance on a station common to both the obstetrics and gynaecology (O&G) and psychiatry OSCEs. Regardless of rotation order, students were less likely to address O&G issues in the psychiatry OSCE and psychiatric issues in the O&G OSCE. The authors concluded that testing context may bias students' collection and interpretation of patient

Table 2 Papers regarding candidate preparation and other factors impacting on performance

Author year	Participants and OSCE characteristics	Key findings
Blaskiewicz et al. [15]	• Third-year medical students (141) • OSCE station common to both O&G (1 of 10) and psychiatry (1 of 6) OSCEs	• Testing context may bias students' collection and interpretation of patient information • OSCEs may better simulate clinical reality by incorporating scenarios with diagnostic options not restricted to most recently completed rotation
Park et al. [16]	• Third-year medical students (869) • Psychiatry OSCE included five 15-min standardised patient interviews	• No association between performance and rotation timing • No association between performance trends and potential performance moderators (student preference for rotation order and specialty choice)
Robinson et al. [17]	• Psychiatry residents (26) • Monthly 2-h OSCE training session • Department-wide level OSCE (number of stations equal to year of training)	• Residents participating in OSCE training sessions (in which they authored and examined new stations) passed the psychiatry specialty examination of the Royal College of Physicians and Surgeons of Canada on their first attempt
Goisman et al. [18]	• Fourth-year medical students (136) • Nine stations (one 15-min psychiatry station)	• Psychiatry clerkship completion was associated with significantly better OSCE performance on phenomenology, mental state examination and differential diagnosis • Performance on history taking, communication skills and treatment planning was not affected
Morreale et al. [19]	• Medical students (active learning for 102 and standard lectures for 108) • Institutional-based OSCE (not graded)	• Active learning gave rise to superior student satisfaction and sense of preparedness for OSCE than standard lectures • Active learning included videotaped patient interviews, small group formulation of mental state examination and differential diagnosis, and discussion of these components and treatment planning with the clerkship director
Griswold et al. [20]	• Third-year medical students (integrated curriculum for 59 out of 922) • Comprehensive fourth-year OSCE (one psychiatry station)	• OSCE scores demonstrated effectiveness of an integrated curriculum incorporating didactic and clinical teaching, mentoring, psychotherapy experience, pharmacology instruction and immersive acute psychiatry exposure • Increased uptake of psychiatry as a career choice resulted

information and that OSCEs may better simulate clinical reality by incorporating scenarios with diagnostic options not restricted to the most recently completed rotation.

Park et al. [16] analysed archival data from six classes of third-year medical students (869 in total) at the St Louis University School of Medicine to see whether completion of the psychiatry rotation early in the clerkship cycle adversely affected student performance. No association was found between performance and rotation timing or between performance trends and potential performance moderators (student preference for rotation order and specialty choice).

Robinson et al. [17] described a training program to prepare psychiatry residents at the University of Western Ontario for the OSCE component of the Royal College of Physicians and Surgeons of Canada's (RCPSC) psychiatry specialty examination. Each month, residents attended a 2-h OSCE training session in which they acted as examiners for new stations they had authored themselves. A moderator facilitated the conduct of these training stations in a group setting. Once refined through a group feedback and discussion process, stations were incorporated in a bank of training stations. Residents also participated in a formal departmental OSCE twice a year. The OSCE training program was a key drawcard of the University's residency program and all 26 residents who completed it in the 5 years since its inception passed the RCPSC examination on their first attempt.

A psychiatry OSCE station completed by 136 fourth-year medical students at Harvard Medical School as part of a nine-station OSCE was used to compare the performance of students who had (46%) or had not (51%) completed a core psychiatry clerkship and determine which areas distinguished the two groups [18]. Psychiatry clerkship completion was associated with significantly better OSCE performance on phenomenology and mental state examination (six items) and differential diagnosis (three items). History taking, communication skills and treatment planning did not differ between groups.

Morreale et al. [19] demonstrated that, relative to conventional lectures, an active learning curriculum gave rise to superior student satisfaction and sense of preparedness for the OSCE at Wayne State University School of Medicine. Active learning seminars involved viewing video vignettes of patient interviews, formulating the mental state examination and differential diagnosis in small groups, and discussion of these components and treatment planning with the clerkship director.

A longitudinal, integrated, third-year psychiatry curriculum implemented within the Harvard Medical School-Cambridge Integrated Clerkship was demonstrated over 8 years to provide effective learning through OSCE scores, National Board of Medical Examiners shelf-exam scores, written work, and observed clinical work [20]. In addition to didactic and clinical teaching, the program incorporated longitudinal mentoring, psychotherapy experience, psychopharmacology instruction and immersive acute psychiatry exposure. An increased uptake of psychiatry as a career choice was a further benefit.

5 Special Topics

In this section, papers focussed on a range of special topics related to psychiatry OSCEs are reviewed. To further orient the reader, these papers are summarised in Table 3.

5.1 Attitudes of Learners and Teachers Towards Psychiatry OSCEs

Sauer et al. [21] evaluated the experience and views of psychiatry senior house officers in the UK on the inherent ability of OSCE to assess their clinical skills, using a modified version of the questionnaire developed by Hodges et al. [22] to assess Canadian psychiatry residents' attitudes to the OSCE. The authors conducted a mock OSCE based on Royal College of Psychiatrists' (RCPsych) OSCE format (twelve 7-min stations), but each station had three junior trainees (one candidate and two observers, 36 in total) and each candidate was examined at four stations instead of twelve. Trainees evaluated the OSCE positively, with 86% considering it fair and 89% appropriate in assessing clinical ability, and found it preferable to the individual patient assessment which it replaced in the RCPsych membership examination.

In a survey of 111 final-year medical students at University College Dublin, administered after the final psychiatry examination but prior to results being released, OSCE was the most highly-rated teaching and examination method [23]. The utility of an OSCE-based methodology in both teaching clinical skills and subsequently assessing the effectiveness of learning may have underpinned students' preference for this technique.

5.2 Standardised Patients in Psychiatry OSCEs

Sadeghi et al. [24] examined the views of 21 final-year psychiatry residents and 24 board-certified psychiatrist examiners regarding standardised patients' competence in simulating psychiatric disorders, and the accuracy of their performances, in an OSCE setting in Iran. The OSCE was comprised of eight 12-min stations and the standardised patients were university students or Tehran Institute of Psychiatry employees rather than professional actors. Both psychiatry residents and examiners found standardised patients acceptable, with both groups considering them competent to depict complex scenarios and accurate in their portrayal of psychiatric illness. The higher ratings provided by examiners in relation to competence may have been due to their involvement in writing case scenarios.

Table 3 Papers focussing on special topics in psychiatry OSCEs

Author year	Topic	Participants and OSCE characteristics	Key findings
Sauer et al. [21]	Attitudes of learners and teachers	• Psychiatry senior house officers (36) • Twelve 7-min stations	• Trainees evaluated OSCE positively, with 86% considering it fair and 89% appropriate in assessing clinical ability • Trainees preferred OSCE to the individual patient assessment it replaced in the RCPsych membership exam
Jabbar et al. [23]	Attitudes of learners and teachers	• Final-year medical students (111)	• OSCE was the most highly-rated teaching and examination method • The utility of OSCE in teaching and assessing clinical skills may have underpinned this preference
Sadeghi et al. [24]	Standardised patients	• Final-year psychiatry residents (21) and board-certified psychiatrist examiners (24) were surveyed • Eight 12-min stations	• Psychiatry residents and examiners found standardised patients acceptable, competent and accurate in their roles • Examiners (who wrote scenarios) rated competence more highly
Whelan et al. [25]	OSCE examiners	• Psychiatry trainees (55) • Eleven or 12 stations	• Moderate correlation between examiner and standardised patient scores for communication skills and overall performance • Examiners' scores were more strongly correlated • Including standardised patient scores in postgraduate psychiatry OSCEs required caution due to poor concurrent validity
Hodges and McNaughton [26]	OSCE examiners	• Commentary paper	• OSCE scoring by examiners involves a subjective, evaluative dimension • Emotionally-laden roles might affect standardised patients' perceptions of interviewers' competence • Research into suitability of OSCE examiners should incorporate broader educational and sociopolitical factors

(contined)

Table 3 (continued)

Author year	Topic	Participants and OSCE characteristics	Key findings
O'Connor et al. [28]	OSCE examiners	• Final year medical students (163) • Four five-min stations with simulated patients and two written stations	• Self-assessment of empathy was higher among female students • Higher concurrent validity between simulated patient (versus clinical examiner) assessment and self-assessment of empathy
Hanson et al. [29]	Child and adolescent psychiatry	• Medical students (402) • One of four 12-min child psychiatry stations in psychiatry clerkship OSCE	• Despite requiring considerable financial and staffing resources, substantial educational benefits were derived • Mean scores were 68–86% for content and 69–76% for process • Station reliability and examiner feedback were satisfactory
Hanson et al. [30]	Child and adolescent psychiatry	• Adolescent standardised patients in a medical student OSCE (39)	• Job skill acquisition and satisfaction of contributing to society identified as benefits • High stress psychosocial (substance abuse) role negatively received • No long-term adverse effects • Time invested in developing and pilot testing OSCE scenarios may reduce risks to adolescent standardised patients
Hung et al. [31]	Suicide risk assessment	• Psychiatry residents (26) and clinical psychology interns (5) • Single 40-min station	• A competency-assessment instrument for suicide risk-assessment (CAI-S) showed good internal consistency, reliability, and interrater reliability in an OSCE setting • Senior trainees performed better than junior trainees, supporting its concurrent validity
Matthews et al. [32]	Addiction psychiatry	• Third-year medical students • Two observed clinical interviews (OSCE format)	• An addiction psychiatry interclerkship training program improved knowledge, attitudes and confidence in this field • Skills in substance abuse assessment and intervention, as assessed by OSCE, were significantly improved

Padilla et al. [33]	Assessing cultural competence	• Psychiatry residents and fellows (17) • Observed clinical interview (OSCE format), one of two cases	• OSCE-based training familiarised trainees with culturally competent interviewing (centred on DSM-5 Cultural Formulation Interview) • Training incorporated a cultural formulation presentation and feedback
Moss [34]	Comprehensive psychiatric evaluation (single-station OSCE)	• Eleven first-year family medicine residents (five psychiatry training seminar participants) • Single observed clinical interview (OSCE format)	• Scenario revolved around a 47-year old separated woman with fatigue, abdominal pain and depressed mood • Superior OSCE performance (including diagnosis of major depression) among seminar participants
McLay et al. [35]	Comprehensive psychiatric evaluation (single-station OSCE)	• Third-year medical students (82) • Forty-five-minute interview, 15-min written test	• Performance on an OSCE simulating a detailed psychiatric interview correlated with students' essay examination, ward grades and scores on the National Board of Medical Examiners psychiatry examination • Students rated time availability, usefulness and believability of the process favourably
Chandra et al. [36]	OSCE as a teaching tool	• Thirty-four individual sessions over 6 months (15-min interview and 15-min group discussion and feedback)	• Objective structured clinical assessment with feedback (OSCAF) was convenient, economical and required few resources • Improved scoring, feedback and evaluation of effectiveness were recommended
Hodges [37] Hodges [3]	Contextual factors impacting on performance	• Theoretical papers	• Approaches to assessing OSCE validity may themselves be invalid • OSCEs identified as contextual, formative social experiences influenced by culture, economics and power relations • Qualitative research (cross-cultural approaches and analysis of sociological variables behind doctors' behaviours) needed to explain 'contextual fidelity'

5.3 Who Should Serve as Psychiatry OSCE Examiners?

In anticipation of the possibility that standardised patients might be called on to contribute to the scoring of postgraduate psychiatry OSCEs in the UK in future, Whelan et al. [25] measured the degree of agreement between scores given by examiners and standardised patients in two consecutive postgraduate mock OSCEs for 55 psychiatry trainees on a London psychiatry rotation. Standardised patients only allocated marks for communication skills and overall performance, whereas examiners also scored other skills and technical domains. Examiner and standardised patient scores for communication skills and overall performance were moderately correlated. Correlation between examiners' scores on these two domains was stronger. The authors concluded that including standardised patient scores in postgraduate psychiatry OSCE marking schemes required caution, as their moderate correlation with examiner scores was indicative of poor concurrent validity.

In a commentary on the paper by Whelan et al. [25], Hodges and McNaughton [26] argued that it is important to look beyond an OSCE's psychometric properties in considering who is most suited to the examiner role. A subjective, evaluative dimension is clearly present in OSCEs in which examiners complete global ratings or overall judgments of competence, and use of binary checklists to assess complex phenomena which exist on a continuum, such as empathy, rapport, and problem-solving, may be invalid and create an illusion of impartiality [27]. Furthermore, the process of scoring an OSCE performance is not entirely objective from a simulated patient perspective either. Some emotionally-laden roles (e.g. a patient with borderline personality disorder experiencing abandonment) might affect standardised patients' perceptions of interviewers' competence. Economic incentives were identified as a possible contributor to the popularity of standardised patient examiners in some settings. Incorporation of broader educational and sociopolitical factors into research regarding the appropriateness of OSCE examiners was recommended to ensure the fair appraisal of doctors' competence.

O'Connor et al. [28] compared 163 final-year medical students' self-assessment of empathy prior to their psychiatry OSCE at University College Dublin with assessments of empathy by clinical examiners and simulated patients during the OSCE itself. Self-assessment of empathy was significantly higher among female students. Concurrent validity was higher between simulated patient assessment (rather than clinical examiner assessment) and self-assessment of empathy, highlighting the potential validity of simulated patients as assessors of medical student empathy in an OSCE setting.

5.4 Child and Adolescent Psychiatry in the OSCE Setting

Hanson et al. [29] reported on the development and integration of four child psychiatry stations into a medical student psychiatry clerkship OSCE, designed to assess skills in recognising four common conditions. Child psychiatrists with experience in OSCE prepared the case scenarios and marking checklists and supervised the training of standardised patients. Despite considerable financial and staffing resources being required, the educational benefits derived from the success of this initiative were substantial. Following the examination of 402 students, the mean scores were 68–86% for content and 69–76% for process, and station reliability and examiner feedback were satisfactory.

Recognising the potential for adolescent standardised patients to experience adverse simulation effects in psychiatry OSCEs, Hanson et al. [30] evaluated an adolescent standardised patient selection method and simulation effects resulting from low- and high-stress roles. A two-component standardised patient selection method excluded 21% of the 83 applicants (7% on employment and 14% on psychological grounds). Selected applicants were randomly assigned to a low-stress medical role (abdominal pain of infectious origin), a high-stress psychosocial role (substance abuse) or wait list control group. Thirty-nine standardised patients participated in a medical student OSCE at the University of Toronto. Acquisition of job skills and the satisfaction of contributing to society were identified as benefits of OSCE participation by standardised patients. Conversely, an unexpectedly strong negative perception of the substance abuse role emerged. No long-term adverse effects were noted. The authors concluded that time invested in developing and potentially pilot testing high-stress OSCE case scenarios may be worthwhile in reducing risks posed to adolescent standardised patients.

5.5 Suicide Risk Assessment in the OCSE Setting

Hung et al. [31] developed a competency-assessment instrument for suicide risk-assessment (CAI-S) and evaluated its use in an OSCE involving 31 trainees (26 psychiatry residents and five clinical psychology interns) at the University of California. The OSCE included a 15-min standardised patient interview; 15 min to write a progress note; a 10-min oral presentation, including an assessment summary and suicide risk management plan; CAI-S completion by the faculty assessor; and a 25-min discussion and feedback session for trainees. Good internal consistency, reliability, and interrater reliability were demonstrated for the CAI-S and the better performance of senior compared to junior trainees in the OSCE setting supported its concurrent validity.

5.6 Addiction Psychiatry in the OSCE Setting

In order address the problem of inadequate substance abuse teaching in the medical student curriculum, Matthews et al. [32] examined the immediate and delayed effects of an intensive 1 or 2-day substance abuse interclerkship program. The program integrated several teaching modalities (with a focus on small-group teaching) and was delivered to third-year medical students at the University of Massachusetts Medical School between standard clerkship blocks. Improvements in knowledge, attitudes and confidence were demonstrated in pre- to post-interclerkship assessments. Skills in substance abuse assessment and intervention, as assessed by two OSCE interviews with simulated patients (one with and one without active substance abuse problems) at the end of a 6-week psychiatry clerkship, were significantly improved where the interclerkship had previously been completed.

5.7 Assessing Cultural Competence in Psychiatry OSCEs

OSCE-based training centred on the DSM-5 Cultural Formulation Interview was shown to be effective in familiarising 17 psychiatry trainees at the University of Massachusetts with culturally competent interview skills [33]. The training incorporated the presentation of a cultural formulation by trainees on which feedback was provided.

5.8 Comprehensive Psychiatric Evaluation in a Single-Station OSCE Format

In order to improve the teaching of psychiatric history taking, diagnosis, and management to first-year family medicine residents at the Sunnybrook Medical Centre (University of Toronto teaching hospital), Moss [34] developed and delivered a series of weekly, 1-h psychiatry seminars to six residents during their four-month Family Medicine Rotation. The program's effectiveness was evaluated via an OSCE conducted among 11 residents, five of whom had not participated in the seminar program. The single OSCE scenario revolved around a 47-year old separated woman with fatigue, abdominal pain and a depressed mood. Performance on history taking, communications skills and diagnostic and management skills was rated using a 100-point marking system that emphasised the importance of accurately diagnosing a major depressive episode. OSCE performance was significantly better among residents who participated in the seminar.

An OSCE designed to simulate a detailed psychiatric interview was used to assess 52 medical students at Tulane University School of Medicine [35]. Students were provided with the patient's chief complaint, a brief medical history and vital signs prior to undertaking a 45-min simulated patient interview. A written task encompassing differential diagnosis, safety risks and treatment planning was completed over a further 15 min. Students were graded using a 36-item content checklist and patient perception scale (both completed by the standardised patients) and a written examination component (completed by a psychiatrist). Checked independently by three fourth-year medical students who watched the interviews on videotape (content checklist and patient perception scale) or an independent grader (written exam component), all three assessments correlated with students' essay examination and ward grades and scores on the National Board of Medical Examiners psychiatry examination. Students rated the time availability, usefulness and believability of the process favourably.

5.9 The OSCE as a Teaching Tool in Psychiatry

Chandra et al. [36] described the implementation of objective structured clinical assessment with feedback (OSCAF), an adaptation of OSCE for teaching purposes, at the National Institute of Mental Health and Neurosciences in Bangalore, India. OSCE adaptation involved language and cultural modification, use of supervised role play rather than standardised patients, and development of a 14-item checklist to guide feedback. While OSCAF was convenient, economical and required few resources, improvements in scoring and feedback, and evaluation of effectiveness, were recommended to establish its place in postgraduate psychiatry education.

5.10 Evaluation of Contextual Factors Impacting on Psychiatry OSCE Performance

In a theoretical paper regarding the validity of OSCE, Hodges [37] raised the possibility that approaches to assessing validity may themselves be invalid. OSCEs were identified as distinctly contextual and strongly formative social experiences that are significantly influenced by culture, economics and power relations. Sophisticated qualitative research, incorporating cross-cultural approaches and analysis of sociological variables underpinning doctors' behaviours, was suggested to be necessary to explain their 'contextual fidelity.' These themes were elaborated upon in a further paper by the same author [3].

6 Discussion

Review of the literature suggests that the OSCE has been widely adopted in psychiatry education in the English-speaking world. Available studies indicate that it can be a valid and reliable method of assessing competencies in psychiatry and is acceptable to both learners and teachers alike. Some shortcomings are readily apparent, however. Given its wide uptake in postgraduate (specialist) examinations, it is striking that much of the available research has been undertaken in undergraduate (medical student) psychiatry teaching. The validity of the OSCE as a postgraduate psychiatry examination has been directly challenged [14]. For this reason, the recommendation that OSCEs be used in conjunction with other methods of clinical skills assessment [38] is particularly pertinent in the postgraduate setting. Problems in using OSCE as a form of summative assessment in residency training were identified since its inception [6]. While subsequent refinements in technique may have led to improvements in its validity as a summative assessment tool for medical students, it is unclear that this issue has been satisfactorily addressed in the postgraduate psychiatry domain. OSCE may be less prone to challenge when used exclusively as a tool for teaching [36] and providing formative assessment [6].

The aim of OSCE to examine practical competencies that are necessary for effective professional practice is undoubtedly well-intentioned. However, despite being labelled 'objective' and 'structured,' there is inherently greater subjectivity in the assessment of OSCE performance relative to other forms of assessment such a multiple choice questions, where absolute consensus regarding what constitutes a correct or incorrect answer is more easily achieved. In other words, highly objective but practically uninformative tests of knowledge have been replaced by the clinically relevant but potentially more subjective OSCE. Hodges [3, 37] identified a need to assess broader sociocultural and other contextual factors that may impact on performance in the psychiatry OSCE. Despite this call over a decade ago, qualitative research in this area is still lacking. Psychiatrists will be among the first to recognise the impact of 'unseen' psychodynamic factors [39] and sociocultural influences on human interactions outside the examination setting. The same variables are likely to come into play in the complex, high-stakes, emotionally-charged OSCE setting. However, subjecting these variables to scientific scrutiny may be far more complicated than merely identifying their potential existence.

With the above limitations in mind, the importance of adequately preparing both candidates and examiners for this form of assessment must be emphasised. OSCEs are very resource intensive to conduct and this may pose challenges in providing candidates and examiners with adequate practice for their respective roles. Furthermore, OSCE scenarios represent an idealised view of clinical practice and expecting candidates to acquire proficiency in performing them solely through observing and participating in busy clinical work schedules (as opposed to dedicated clinical skills teaching sessions) may be unrealistic. In the author's experience, groups of candidates will often establish their own OSCE preparation groups outside the formal teaching curriculum. Ideally, this should be complemented by formally instituted training sessions involving both candidates and examiners akin to those described by Robinson et al. [17].

It has been noted that creating high-quality OSCE stations requires time and effort [1]. It is important that OSCE stations are fit for the purpose for which they are designed. Relatively brief stations with checklist-like marking sheets may be suitable for examining discrete tasks in medical students (such as enquiring about manic symptoms) but may be inadequate for examining more sophisticated scenarios at a postgraduate training level (such as arranging assistance for an emotionally distressed junior colleague). Potentially longer stations and more sophisticated marking schemes are necessary for the latter. At the same time, however, it is important from the candidates' perspective—and in the interests of fairness—that transparency about what is expected of them is maintained in presenting OSCE scenarios. A tendency to base stations on unusual or rare cases was one of the earliest identified challenges to the validity of psychiatry OSCEs [6] and remains pertinent to this day.

The benefits of 'reverse engineering' or 'deconstructing' OSCE stations became apparent to the author in the process of preparing for his own postgraduate psychiatry OSCE. This approach was based on the notion that a better understanding of how OSCE stations were developed increased the chances of exam success, and has recently been espoused in an entire textbook on the topic [40]. However, deconstruction may be ineffective if a station is poorly conceptualised and its required tasks not clearly presented. An understanding of the standard-setting method used for determining the OSCE pass mark may also be advantageous to both candidates and examiners [1].

OSCE may not be optimally suited to truly evaluating ethics and professionalism, as socially-expected responses may easily be learned and portrayed by candidates in simulated examination settings. The same limitation may be true, however, of other formal examination modalities. An essential, albeit subjective, adjunct to the appraisal of these domains may be supervisors' observation of students' and trainees' conduct in day-to-day clinical activities [41].

In light of the fatigue that traditional multi-station OSCE formats may cause in candidates and examiners [38, 42, 43], the question is raised of whether developing OSCEs with small numbers of in-depth stations would be advantageous. This approach might help to avoid sudden, repeated shifts in thinking, but its sampling of a smaller area of the curriculum may be a disadvantage. Two studies included in this review [34, 35] reported on the use of single-station OSCE formats involving lengthier patient interviews that were reminiscent of older-style clinical interview-based exams [44] but potentially better 'operationalised' in terms of their marking schedules. However, whether such single-station examinations should be regarded as OSCEs in the traditional sense—which imply a circuit of multiple stations—or classed in a category of their own remains open to debate, due to their sampling of a limited area of the curriculum.

Psychiatry, like medicine as a whole, has taken a leading role in adopting OSCE to assess clinical competencies. There is limited emerging evidence, however, of its adoption in the related field of psychology. Cramer et al. [45] recently proposed core competencies and an integrated training framework for suicide risk assessment training in doctoral psychology programs. At Monash University, with which the author is affiliated, both Clinical Psychology and Neuropsychology students are required to participate in two formative OSCEs and one summative OSCE during their Doctor of Psychology (DPsych) program, in keeping with recommendations for competency assessment by the Australian Psychology Accreditation Council (APAC) [46]. Sharma et al. [47] recently took the OSCE even further, describing their development of a Team OSCE (TOSCE) to promote interprofessional learning among psychiatrists, clinical psychologists and social workers in a mental health setting.

Despite any shortcomings, OSCEs are currently ubiquitous in all areas of undergraduate and postgraduate medicine and proposing a better alternative (other than returning to and refining former assessment methods) is difficult. The limited evidence base regarding the validity of OSCE in postgraduate psychiatry examinations suggests that more research is needed in this domain. A critical question is whether OSCE is sufficient on its own to assess high-level consultancy skills, and aspects of professionalism and ethical practice, that are essential for effective specialist practice, or whether it needs to be supplemented by additional testing modalities.

Funding No sources of funding were received for this work.

Conflicts of Interest The author reports no conflicts of interest and is alone responsible for the content and writing of this paper.

References

1. Boursicot, K.A.M., T.E. Roberts, and W.P. Burdick. 2014. Structured Assessments of Clinical Competence. In *Understanding Medical Education: Evidence, Theory and Practice*, ed. T. Swanwick, 2nd ed., 293–304. Chichester, West Sussex: Wiley.
2. Harden, R.M., and F.A. Gleeson. 1979. Assessment of Clinical Competence Using an Objective Structured Clinical Examination (OSCE). *Medical Education* 13: 39–54.
3. Hodges, B. 2003. OSCE! Variations on a Theme by Harden. *Medical Education* 37: 1134–1140.
4. ———. 2002. Creating, Monitoring, and Improving a Psychiatry OSCE: A Guide for Faculty. *Academic Psychiatry* 26: 134.
5. Davis, G.C. 2002. The Hodges Psychiatry OSCE Guide and Emerging Trends in Assessment. *Academic Psychiatry* 26: 184–186.
6. Loschen, E.L. 1993. Using the Objective Structured Clinical Examination in a Psychiatry Residency. *Academic Psychiatry* 17: 95–100.
7. Hodges, B., G. Regehr, M. Hanson, and N. McNaughton. 1997. An Objective Structured Clinical Examination for Evaluating Psychiatric Clinical Clerks. *Academic Medicine* 72: 715–721.
8. Hodges, B., and J. Lofchy. 1997. Evaluating Psychiatric Clinical Clerks with a Mini-Objective Structured Clinical Examination. *Academic Psychiatry* 21: 219–225.
9. Hodges, B., G. Regehr, M. Hanson, and N. McNaughton. 1998. Validation of an Objective Structured Clinical Examination in Psychiatry. *Academic Medicine* 73: 910–912.
10. Loschen, E.L. 2002. The OSCE Revisited: Use of Performance-Based Evaluation in Psychiatric Education. *Academic Psychiatry* 26: 202–204.
11. Barrows, H.S., R.G. Williams, and R.H. Moy. 1987. A Comprehensive Performance-Based Assessment of Fourth-Year Students' Clinical Skills. *Journal of Medical Education* 62: 805–809.
12. Park, R.S., J.T. Chibnall, R.J. Blaskiewicz, et al. 2004. Construct Validity of an Objective Structured Clinical Examination (OSCE) in Psychiatry: Associations with the Clinical Skills Examination and Other Indicators. *Academic Psychiatry* 28: 122–128.
13. Walters, K., D. Osborn, and P. Raven. 2005. The Development, Validity and Reliability of a Multimodality Objective Structured Clinical Examination in Psychiatry. *Medical Education* 39: 292–298. doi:10.1111/j.1365-2929.2005.02091.x.

14. Marwaha, S. 2011. Objective Structured Clinical Examinations (OSCEs), Psychiatry and the Clinical Assessment of Skills and Competencies (CASC) Same Evidence, Different Judgement. *BMC Psychiatry* 11: 1.
15. Blaskiewicz, R.J., R.S. Park, J.T. Chibnall, and J.K. Powell. 2004. The Influence of Testing Context and Clinical Rotation Order on Students' OSCE Performance. *Academic Medicine* 79: 597–601.
16. Park, R.S., J.T. Chibnall, and A. Morrow. 2005. Relationship of Rotation Timing to Pattern of Clerkship Performance in Psychiatry. *Academic Psychiatry* 29: 267–273.
17. Robinson, D.J. 2009. A Training and Examination Program to Prepare Psychiatric Residents for OSCE-Style Exams. *Academic Psychiatry* 33: 331.
18. Goisman, R.M., R.M. Levin, E. Krupat, et al. 2010. Psychiatric OSCE Performance of Students with and without a Previous Core Psychiatry Clerkship. *Academic Psychiatry* 34: 141–144.
19. Morreale, M., C. Arfken, P. Bridge, and R. Balon. 2012. Incorporating Active Learning Into a Psychiatry Clerkship: Does It Make a Difference? *Academic Psychiatry* 36: 223–225.
20. Griswold, T., C. Bullock, E. Gaufberg, et al. 2012. Psychiatry in the Harvard Medical School—Cambridge Integrated Clerkship: An Innovative, Year-Long Program. *Academic Psychiatry* 36: 380–387.
21. Sauer, J., B. Hodges, A. Santhouse, and N. Blackwood. 2005. The OSCE has Landed: One Small Step for British Psychiatry? *Academic Psychiatry* 29: 310–315.
22. Hodges, B., M. Hanson, M.N. McNaughton, and G. Regehr. 1999. What do Psychiatry Residents Think of an Objective Structured Clinical Examination? *Academic Psychiatry* 23: 198–204.
23. Jabbar, F., P. Casey, and B.D. Kelly. 2014. Undergraduate Psychiatry Students' Attitudes Towards Teaching Methods at an Irish University. *Irish Journal of Medical Science*: 1–4. doi:10.1007/s11845-014-1211-3.
24. Sadeghi, M., A. Taghva, G. Mirsepassi, and M. Hassanzadeh. 2007. How do Examiners and Examinees Think About Role-Playing of Standardized Patients in an OSCE Setting? *Academic Psychiatry* 31: 358–362.
25. Whelan, P., L. Church, and K. Kadry. 2009. Using Standardized Patients' Marks in Scoring Postgraduate Psychiatry OSCEs. *Academic Psychiatry* 33: 319–322.
26. Hodges, B.D., and N. McNaughton. 2009. Who Should Be an OSCE Examiner? *Academic Psychiatry* 33: 282–284.
27. Norman, G. 2005. Checklists vs. Ratings, the Illusion of Objectivity, the Demise of Skills and the Debasement of Evidence. *Advances in Health Sciences Education* 10: 1–3. doi:10.1007/s10459-005-4723-9.
28. O'Connor, K., R. King, K.M. Malone, and A. Guerandel. 2014. Clinical Examiners, Simulated Patients, and Student Self-Assessed Empathy in Medical Students During a Psychiatry Objective Structured Clinical Examination. *Academic Psychiatry* 38: 451–457. doi:10.1007/s40596-014-0133-8.
29. Hanson, M., B. Hodges, N. McNaughton, and G. Regehr. 1998. The Integration of Child Psychiatry Into a Psychiatry Clerkship OSCE. *Canadian Journal of Psychiatry* 43: 614–618.
30. Hanson, M., R. Tiberius, B. Hodges, et al. 2002. Adolescent Standardized Patients: Method of Selection and Assessment of Benefits and Risks. *Teaching and Learning in Medicine* 14: 104–113. doi:10.1207/S15328015TLM1402_07.
31. Hung, E.K., S.R. Fordwood, S.E. Hall, et al. 2012. A method for Evaluating Competency in Assessment and Management of Suicide Risk. *Academic Psychiatry* 36: 23–28.
32. Matthews, J., W. Kadish, S.V. Barrett, et al. 2002. The Impact of a Brief Interclerkship About Substance Abuse on Medical Students' Skills. *Academic Medicine* 77: 419–426.
33. Padilla, A., S. Benjamin, and R. Lewis-Fernandez. 2016. Assessing Cultural Psychiatry Milestones Through an Objective Structured Clinical Examination. *Academic Psychiatry* 40: 600–603. doi:10.1007/s40596-016-0544-9.
34. Moss, J.H. 1990. Evaluating a Seminar Designed to Improve Psychiatry Skills of Family Medicine Residents. *Academic Medicine* 65: 658–660.

35. McLay, R.N., P. Rodenhauser, M.D.S. Anderson, et al. 2002. Simulating a Full-Length Psychiatric Interview with a Complex Patient. *Academic Psychiatry* 26: 162–167.
36. Chandra, P.S., S.K. Chaturvedi, and G. Desai. 2009. Objective Standardized Clinical Assessment with Feedback: Adapting the Objective Structured Clinical Examination for Postgraduate Psychiatry Training in India. *Indian Journal of Medical Sciences* 63: 235. doi:10.4103/0019-5359.53391.
37. Hodges, B. 2003. Validity and the OSCE. *Medical Teacher* 25: 250–254. doi:10.1080/01421590310001002836.
38. Barman, A. 2005. Critiques on the Objective Structured Clinical Examination. *Annals Academy of Medicine Singapore* 34: 478–482.
39. Malan, D.H. 1979. *Individual Psychotherapy and the Science of Psychodynamics*. London: Butterworths.
40. Harding, D. 2014. *Deconstructing the OSCE*. Oxford: Oxford University Press.
41. Marrero, I., M. Bell, L.B. Dunn, and L.W. Roberts. 2013. Assessing Professionalism and Ethics Knowledge and Skills: Preferences of Psychiatry Residents. *Academic Psychiatry* 37: 392–397.
42. Rutala, P.J., D.B. Witzke, E.O. Leko, et al. 1990. Student Fatigue as a Variable Affecting Performance in an Objective Structured Clinical Examination. *Academic Medicine* 65: S53–S54.
43. Reznick, R., S. Smee, A. Rothman, et al. 1992. An Objective Structured Clinical Examination for the Licentiate: Report of the Pilot Project of the Medical Council of Canada. *Academic Medicine* 67: 487–494.
44. Flynn, B. 2015. The Abandonment of the RANZCP OCI – and Why We May Live to Regret It. *Australasian Psychiatry* 23: 303–305. doi:10.1177/1039856215581295.
45. Cramer, R.J., S.M. Johnson, J. McLaughlin, et al. 2013. Suicide Risk Assessment Training for Psychology Doctoral Programs: Core Competencies and a Framework for Training. *Training and Education in Professional Psychology* 7: 1–11. doi:10.1037/a0031836.
46. Monash University. 2016. Psych: Graduate Research Info Guide – 10. DPsych OSCE. http://www.med.monash.edu.au/psych/students/current/hdr-info-guide/dpsych-osce.html. Accessed 8 Aug 2016
47. Sharma, M.K., P.S. Chandra, and S.K. Chaturvedi. 2015. Team OSCE: A Teaching Modality for Promotion of Multidisciplinary Work in Mental Health Settings. *Indian Journal of Psychological Medicine* 37: 327. doi:10.4103/0253-7176.162954.

The Effectiveness of Neurofeedback Training in Algorithmic Thinking Skills Enhancement

Antonia Plerou, Panayiotis Vlamos, and Chris Triantafillidis

Abstract Although research on learning difficulties are overall in an advanced stage, studies related to algorithmic thinking difficulties are limited, since interest in this field has been recently raised. In this paper, an interactive evaluation screener enhanced with neurofeedback elements, referring to algorithmic tasks solving evaluation, is proposed. The effect of HCI, color, narration and neurofeedback elements effect was evaluated in the case of algorithmic tasks assessment. Results suggest the enhanced performance in the case of neurofeedback trained group in terms of total correct and optimal algorithmic tasks solution. Furthermore, findings suggest that skills, concerning the way that an algorithm is conceived, designed, applied and evaluated are essentially improved.

Keywords Neurofeedback • Learning disabilities • Alpha waves • Beta waves • Algorithmic thinking • Neuroeducation

1 Introduction

In this study, a method is proposed in order to check algorithmic thinking ability with the use of a screener presented in an interactive way enhanced with neurofeedback elements. Besides generalized test concerning dyscalculia, there are several skill games, mainly commercial releases that are related to creativity and algorithmic thinking. However, the motivation of this case study is the evaluation of users' efficiency in algorithmic problem solving using interactive environment with the use of cognitive and neuroscience aspects within the frame of neuroeducational studies. The term Neuroeducation is referred to a recent interdisciplinary field that carries together researchers of developmental cognitive neuroscience, educational psychology, and other related disciplines. In this context, neurosciences, and

A. Plerou (✉) • P. Vlamos
Department of Informatics, Bioinformatics and Human Electrophysiology Laboratory, Ionian University, Corfu, Greece
e-mail: tplerou@ionio.gr; vlamos@ionio.gr

C. Triantafillidis
Department of Informatics, Ionian University, Corfu, Greece

P. Vlamos (ed.), *GeNeDis 2016*, Advances in Experimental Medicine and Biology 988, DOI 10.1007/978-3-319-56246-9_14

cognitive educational theories combine the perception of brain function in order to improve the educational process. The goals of neuroeducation are to enhance the perception of the mind, brain and learning interrelationship, through natural science investigation of educational issues [1]. Neuroeducation could encompass a wide variety of learning disabilities, allowing educators to improve teaching methods to students with mild to more severe mental handicaps. This is a reformation that would have innumerable benefits [2].

Research in Neuroeducation is based on tools and technologies of brain imaging in order to explore cognitive functions and improve educational practices. Considering the fact that genetic factors affect brain function and cognitive abilities, mathematical perception, learning difficulties in mathematics and algorithmic thinking abilities are evaluated within this frame with the use of imaging techniques. In particular brain regions and functions associated with the mathematical perception and the ability to algorithmic thinking are evaluated with the frame of knowledge of molecular biology and cognitive science related to brain function. The researchers of educational neuroscience are studying the neural mechanisms and the interaction of biological processes with learning and namely in processes such as reading, arithmetic, attention and learning difficulties, dyslexia, ADHD and dyscalculia (difficulties in perception of number notion, in spatial perception, in computational procedures, in calculus and algorithmic processing) [3].

2 Participants

Participants were selected from the proper population according to sampling theory have to deal with algorithmic tasks in order to the study their brain activity during their engaging in the given problems and evaluating their ability of algorithmic lesion stimuli. The sample of the present research was parted from 182 participants and was randomly selected (convenience sample). From them, 91 participants were allocated to neurofeedback trained (NFB) group and 91 participates were allocated to control group. In the case of the control group (CG) elements of HCI, color, narration and neurofeedback elements were excluded. Subjects were randomly assigned to one of the two groups. None of the patients used any kind of medicine. Subjects determined to have a diagnosed psychotic or personality disorder (based on DSM-IV criteria), or a seizure disorder, were excluded. They did not differ according to the variables such as age, sex, IQ scores and severity of learning disorders. Subjects were solicited to participate in a study of two approaches to employing brainwave biofeedback to evaluate algorithmic thinking abilities. Training sessions were conducted at the Bioinformatics and Human Electrophysiology Laboratory, Department of Informatics, Ionian University, Corfu, Greece. Participants of this case study are adult's undergraduate students of the Department of Informatics of the Ionian University in Corfu. The range of participant's age was 18–25 years old with average 19.802 and 20.307 for the neurofeedback and the control group respectively.

3 Material

The Att-Oncee (Algorithmic Thinking Training—Online Neuro-Cognitive Educational Evaluation) test incorporates with ten algorithmic tasks which are mainly derived from the computer science, graph, and game theory. In particular, "Divide and conquer" algorithm is involved for the case of "Hanoi Tower". Backtracking algorithmic based tasks like "The Eight Queens Problem" and "The Knight Tour" are also used. Additional algorithmic tasks used like the "Minimum Spanning Tree", the "Traveling Salesman Problem" and the "Chinese Postman Problem", are derived from the graph theory (the optimal path is required). Finally, crossing river puzzles and tasks based on "Four Color theorem" are also included within the overall algorithmic thinking evaluation.

The virtual environment used in this study, named "Att-Oncee", was developed with the intention of generally stimulating reasoning by generating positive attitudes. In addition, it was also developed to confront users with tasks within the frame of pedagogical aspects. The environment is accessed with an Internet Explorer-compatible web browser [4]. The algorithmic interactive tasks were materialized with the use of "Scratch" software. Scratch is a free educational programming language that was developed by the Lifelong Kindergarten Group at the Massachusetts Institute of Technology (MIT). Scratch is a visual programming language used for a range of educational projects, in order to create quizzes and games that stimulate the mind and interact with students. During the game, a visual guidance is provided in order to keep stress effect and anxiety at low levels. Recurrent audiovisual reminders of encouragement are provided to the participants in order to remain calm and stay focused. In addition, they were frequently repeatedly to check their breathing and breathe tranquility with the use of audiovisual guidance which is related to performance expectancy. The neurofeedback trained group participants were asked to control their brain activity within the theta rhythm range in order to maintain the maximum concentration while dealing with algorithmic tasks.

4 Methods

The specifications of the Att-Oncee screener are based in human computer interaction theory and focuses on interactive elements and the usability of the platform. Successful interaction design provides consistency, predictability, which improves learnability and usability. Learnability indicates that familiarity and intuition are essential for every interface while usability is a basic element for sufficient interactive in order to build insights into the design [5]. The screener design was based on the proportion, structure, size, and shape and color effect. Sounds effect is used and adjusted in accordance with the three main components i.e. pitch, volume, and timbre or tone quality which can [6]. User's feedback and interaction, based on action and reaction element were enhanced in order the screener environment to be friendly, interesting, and helpful. Each step of the experience included emotions enhancement. Texture can convey emotion as well as the storyline of the virtual environment in order to promote user interaction [7].

5 Neurofeedback Elements Effect

Participants were asked to deal with interactively presented algorithmic procedures in order to assess their ability in algorithmic processing stimuli. The training was conducted within 10 sessions that lasted 60–80 min for approximately nine months. While of their engaging in the given tasks EEG signals of the brain is recorded by means of a sensor system in order to study and analyze their brain activity. This is carried out by protocols based on neuroscience and mainly in neurofeedback methods. In specific two protocols based on neurofeedback were employed namely the Alpha-Theta and the SMR-low Beta protocol in order to reinforce low and high frequencies respectively during the various stages of the neuroeducational approach Neurofeedback treatment was performed based on enhancement of Alpha-Theta ratio in the C4 region and SMR-low Beta ratio enhancement in the P4 region. The reference electrode was placed in the right earlobe [8].

At the initial phase of neurofeedback training, the participants are asked to relax with their eyes closed while listening to relaxing music while Alpha-Theta protocol is applied. In this stage, the low frequencies are reinforced, namely Alpha ratio (8–12 Hz) and Theta ratio (4–8 Hz) respectively while Delta ratio (0.5–4 Hz) and Gamma ratio (30–128 Hz) were suspended [9]. The NFB trained participants to remain for a time in a relaxing state with their eyes closed while they are asked to breathe calmly. This stage is essential in order to feel calm, to minimize the stress effect and enhance the quality of learning during the neurofeedback training. Participants are guided in relaxing situations, to reinforce the intuitive perception and to come into deeper levels of consciousness. The Alpha-Theta ratio state is believed to promote self-awareness and spiritual and intuitive enhancement [10].

After this stage, the participants are free of stress and in the state of mental clarity. They are prepared to deal with the algorithmic tasks in order their performance to be evaluated. During their engagement with the given processes, the SMR-low Beta protocol is applied and signals of their brain are recorded. According to the SMR-low Beta training protocol the Beta ratio (12–15 Hz) is enhanced and this is related to high alertness, concentration and focused Attention. The SMR-low Beta protocol is often used for treating ADD-ADHD, and other disorders [11].

The analysis of the electrical signals obtained from the brain to assess cognitive effects is used in order the learning ability of the brain on the algorithmic thinking to be understood and to provide suggestions for novel and improved learning methods and learning approach. The analysis of the electrical waves is carried out using Acknowledge software and mathematical model analysis.

6 Human–Computer Interaction Effect

Human–computer interaction (HCI) researchers the design and use of computer technology, in order to design efficient and usable experiences between the human and the computer interface. Users interact directly with hardware for the human input and output such as displays, e.g. through a graphical user interface. Computer graphics are effective for gaining attention and can encourage students to create mental images that in turn make it easier for them to learn certain types of information [12]. In addition, authors take into account empirical measurement contacting a pilot study. It is essential to establish quantitative usability specifics such as the time needed to complete the tasks and the number of errors made during the tasks. Thereafter the following iterative design steps were performed. Estimating and evaluating errors and required training time interaction and to interactive interface efficacy was increased. The iterative design process was repeated until a sensible, user-friendly interface is created [13].

7 Color Effect

The color is suggested to be an important visual experience to humans [14]. Color could be significantly effective in learning and education. Color display could produce a higher level of attention and work more effective in working memory. Attention and arousal are important elements in memory performance. Colors can attract attention and produce emotional arousal. Arousal, especially emotional arousal, can play an essential role in keeping the information in the memory system. Colors can enhance the relationship between arousal and memory. Color can produce the emotional arousing effect but the range of arousal vary depending on the emotional element that is being attached with a specific type of color. Especially, the red color is being attached to stronger emotion or feeling compared to the other type colors [15].

The interactive environment was designed using color proposed to enhance concentration. The algorithmic tasks objectives were provided visual and displayed on the platform screen. In addition recorded instructions accompanied instruction visual display. During the tasks, optical and audible encouragement was used in order help participants to continue their effort in the game. This aspect is related to user effort expectancy aspect. Color inspires creativity and helps students to evaluate and solve questions. The benefits of color on learning does not only pertain to infants and children. Eighty percent of the brain receives information visually. Color stimulates the visual sense and encourages the retention of information. Very strong or bright colors should be used sparingly or between dull background tones, whereas soft colors make it easy for us to spend the longer duration of time to read things. For background, the use of soft, dull or neutral colors that allow smaller, bright text or images to stand out vividly were used. Participants that are overly stimulated could benefit from the relaxing colors. The shades of red, orange and yellow colors stimulate and increase brain activity. In our case shades of orange and yellow were mainly used and slightly shades of red. Additionally, a monochromatic color scheme for grouping similar object or facts were used. Finally, a lighter background was used in order to ensure a higher readability level [16].

8 Narration

Narratives are used in the interactive version of the Att-Oncee test. The benefits of using narratives are that narratives instruct user for what they should do while increasing their attention and excitement. It is easier to form memories when something is told as a story, and, therefore, better undergo learning and training and make engagement easier and longer [17]. Narratives provide enjoyment, passionate involvement, structure, motivation, ego gratification, adrenaline, creativity, and interaction. The element of narration is essential and instructions and directions are given during the narration.

9 Statistical Analysis

A descriptive analysis of the study population characteristics was conducted. For all categorical variables absolute and relative frequencies are reported. In the case of continuous variables, the median value and the corresponding interquartile range (IQR, 25th–75th percentile) are reported. Univariate associations were examined in order to evaluate potential differences between the two intervention groups, namely the control and neurofeedback trained group. In specific Fisher's exact test was performed for categorical data and non-parametric tests i.e. Mann-Whitney or Kruskal-Wallis were performed for continuous data as appropriate. Results were computed with the use of the same tests in reference to participants' gender and IQ level. Further, a sensitivity analysis was conducting excluding all the participants that reported learning difficulties. Statistical significance was set at 0.05 level. Analyses were conducted using STATA software, version 13.0 [18].

10 Demographic Data

The case study participants were divided into two equal groups of 91 participants each. The neurofeedback group was parted from 46 male and 35 female participants while the control group by 42 male and 49 female participants respectively (p-value 0.695). The results of the cognition IQ pretest showed that the median score for the neurofeedback trained group was 141 and the median IQ score for the control group was 129 while the difference was not statistically significant (p-value = 0.188), thus the two groups are not considered to be unbalanced in respect with the IQ scale. The mean value of the age of the participants is 19.92 for the control group and 20.3 for the neurofeedback trained group (p-value = 0.776).

11 Algorithmic Tasks Performance Results

The overall correct responses in reference to both NFB and CG responses to the Att-Oncee test is to follow. The correct answers in reference to Hanoi tower task and 5×5 queens for both groups were 96.2% and 100% respectively while referring to the optimal answers the overall percentages was 53.8% and 46.2% respectively. In the case of the knight tour and Chinese postman tasks overall correct answers and optimal solutions, the percentage of success were 96.2% and 0% (correct answers) and 0% (optimal solutions) respectively. The correct answers reached the 30.8% and 26.9% in respect for the traveling salesman and minimum spanning tree and 30.8% and 26.9% for the optimal solution given overall. In the case of the two bridge crossing puzzles, the correct answers given were 80.8% and 11.5% while the overall optimal solutions were 69.2% and 11.5% respectively. In the case of four coloring tasks the correct answers percentage reached the 100% and 96.2% respectively and the optimal solution 38.5% and 69.2% respectively. In Fig. 1, the overall results in terms of the correct and optimal responses and the number of efforts needed for both groups are presented.

12 Overall Results in Algorithmic Thinking

Algorithmic thinking ability is a term which describes the capacity needed in order to complete a task using a series of default actions, aiming to complete a process. This term is somehow a group of skills, concerning the way that an algorithm

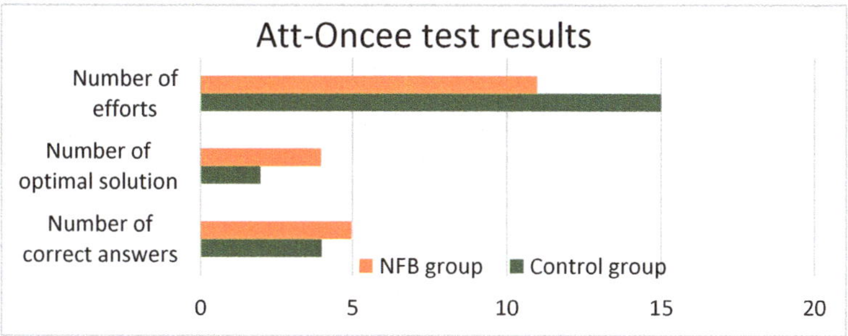

Fig. 1 Att-Oncee test overall results

is conceived, designed, applied and evaluated. The term algorithm describes a finite sequence of actions, which describe how to solve a given problem [19]. Basic algorithmic principles contribute to the development of logical thinking and learning methodologies, which are needed to solve problems [20]. Some of the basic principles needed are the ability to conceive the given problem. Moreover, the design of strategies is needed in order to solve a problem [19]. Moreover, application of strategies, are needed in order to solve a problem [21].

Specifically, concerning problem-solving difficulties, algorithmic thinking difficulties are noticed on three basic aspects, namely perception; procedure and best solution difficulties. Firstly, difficulties in proper perception, evaluation and problem decoding i.e. problem perception difficulty. In addition difficulties in the algorithm, design is noticed i.e. problem-solving design difficulties and difficulties in applying algorithmic steps needed to solve the problem i.e. problem-solving procedure difficulties. On the other hand, lack of perception of the optimized solution is noticed i.e. problem solving the best solution difficulty [3]. Resuming, the basic steps for algorithmic problem solving is the problem perception, the problem solution planning, the problem-solving procedure and the best solution evaluation.

The results presented below are retrieved by the overall analysis of the results obtained within the present study in reference to the participants' tasks performance evaluation. During the evaluation phase, the neurofeedback trained group received a relaxation time period (median value 6.090 s and IQR (5.250, 6.720)).

According to Gerald Futschek in order to solve an algorithmic problem the ability to conceive the given problem is required [19]. In this case, study control group participants needed 3.402 s (median value and IQR (2870, 3.521)) for reading the instructions and trying to perceive instructions of the given algorithmic tasks. The other group participants spend 5.425 s (median value and IQR (5.110, 5.810)) in order to perceive the task problem. These difference is statistically significant (p-value < 0.001) and this result lead to the conclusion that the NFB trained group spent more time in order to read and perceive the tasks specifications. Therefore NFB trained group is considered to present enhanced ability during the problem perception phase.

In addition, the abilities to design strategies and finding the algorithmic step required to solve a problem are a prerequisite in order to solve an algorithmic problem according to Futschek's approach [19]. In this case study, the CG participants used 889 s (median value, IQR (469, 1.260)) for planning their solution while the NFB trained participants needed 966 s (median value and IQR (798, 1.673)) for the solution planning phase. In this case, there was a marginal difference concerning the time spent for planning the problem-solving strategy nevertheless this is not statistically significant at the 5% level (p-value 0.054).

Moreover, in order to solve an algorithmic problem, the algorithmic steps should be efficiently applied according to Schoenfeld's approach [21]. This case study findings suggest that the CG participants spend 5.411 s (median value 5.411 and IQR (4.060, 8.071)) in comparison to the NBF group who spend 2.975 s (median 2.975 and IQR (2.373, 3.472) for the problem-solving phase. This difference is highly significant (p-value $= 0.001$) and this finding implies that the CG spent more time

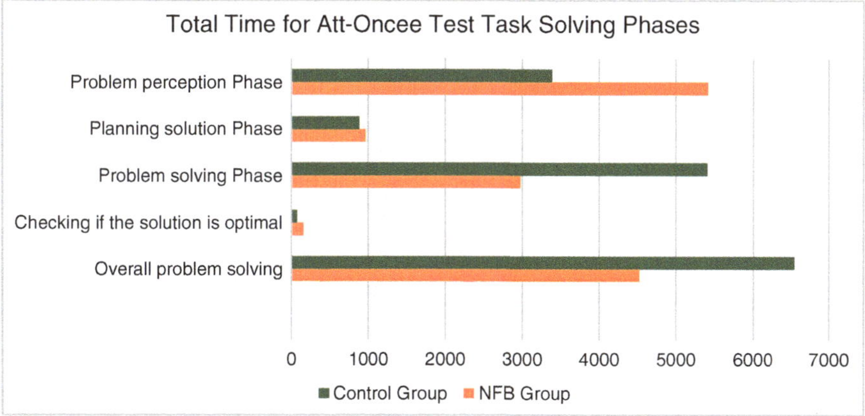

Fig. 2 Att-Oncee algorithmic problem-solving phases and overall time

in the solution phase probably due to the fact that they didn't spend enough time for the problem perception and planning the solution phases, hence these phases were not efficiently completed. It is worth mentioning that their performance was statistically worse comparing to the NFB trained group.

The last step for efficient algorithmic problem solving is the phase where the solution is evaluated whether to be optimal or not i.e., best solution evaluation. The CG participants needed 70 s (median value and (IQR (35, 210)) in order to check whether the solution they provided to the tasks was optimal. On the other hand, 161 s were spent by the NFB trained group (median value and (IQR (133, 210)) while the p-value was computed to be 0.051. This result suggests the neurofeedback trained group spent more time in order to evaluate their solution. The total time needed for the CG to solve all problems was 6.552 (6.552 median value and IQR (4.564, 8.932) in comparison to the NFB group participants who totally needed 4.522 (4.522 medians, IQR (3.500, 6.412)) and p-value = 0.022) for solving the given task. In Fig. 2, the results in respect of the time needed for each phase of the algorithmic task solving for both groups participants are presented.

13 Conclusions and Discussion

Research within the Neuroeducation aims to strengthen its relations between education and neuroscience, namely to examine and comprehend thoroughly the learning ability of the brain in order to provide suggestions for improved learning and for new teaching processes. This case study provides evidence about the neuroeducational efficiency in terms of algorithmic thinking. Results suggest the efficiency of the HCI, color, narration and neurofeedback elements effect in the algorithmic thinking evaluation. Additionally, findings suggest the enhanced performance in the case neurofeedback trained group in terms of the correct answers and solving algorithmic tasks optimally. Furthermore, the skills, concerning the way that an algorithm is conceived, designed, applied and evaluated are improved in the case of the neurofeedback trained group participants. Author's future directions are addressed towards both in a follow-up study to confirm Att-Oncee results stability over time and an overall Att-Oncee test efficiency evaluation in terms of secondary educated participant's implementation.

References

1. Katzir, T., and J. Pare-Blagoev. 2006. Applying Cognitive Neuroscience Research to Education: The Case of Literacy. *Educational Psychologist* 41 (1): 53–74.
2. Hocutt, A.M. 1996. Effectiveness of Special Education: Is Placement the Critical Factor? *The Future of Children, Special Education for Students with Disabilities* 6 (1): 77–102.
3. Plerou, A. 2014. Dealing with Dyscalculia Over Time. International Conference on Information Communication Technologies in Education.
4. Wang, S.-K., and T.C. Reeves. 2006. The Effects of a Web-Based Learning Environment on Student Motivation in a High School Earth Science Course. *Educational Technology Research and Development* 54 (6): 597–621.
5. Green, A., and K.S. Eklundh. 2003. Designing for Learnability in Human-Robot Communication. *IEEE Transactions on Industrial Electronics* 50 (4): 644–650.
6. Hereford, J., and W. Winn. 1994. Non-Speech Sound in Human-Computer Interaction: A Review and Design Guidelines. *Journal of Educational Computing Research* 11 (3): 211–233.
7. Johnson, W.L., J.W. Rickel, and J.C. Lester. 2000. Animated Pedagogical Agents: Face-to-Face Interaction in Interactive Learning Environments. *International Journal of Artificial Intelligence in Education* 11: 47–78.
8. Vernon, D., et al. 2003. The Effect of Training Distinct Neurofeedback Protocols on Aspects of Cognitive Performance. *International Journal of Psychophysiology* 47 (1): 75–85.
9. Basar, E., et al. 1995. Time and Frequency Analysis of the Brain's Distributed Gamma-Band System. *IEEE Engineering in Medicine and Biology Magazine* 14 (4): 400–410.
10. Batty, M.J., et al. 2006. Relaxation Strategies and Enhancement of Hypnotic Susceptibility: EEG Neurofeedback, Progressive Muscle Relaxation and Self-Hypnosis. *Brain Research Bulletin* 71 (1): 83–90.
11. Vernon, D., A. Frick, and J. Gruzelier. 2004. Neurofeedback as a Treatment for ADHD: A Methodological Review with Implications for Future Research. *Journal of Neurotherapy* 8 (2): 53–82.
12. Surjono, H.D. 2015. The Effects of Multimedia and Learning on Student Achievement in Online Electronics Course. *The Turkish Online Journal of Educational Technology* 14 (1): 166–122.
13. Brown, C.M. 1998. *Human-Computer Interface Design Guidelines*.
14. Adams, F.M., and C.E. Osgood. 1973. A Cross-Cultural Study of the Affective Meanings of Color. *Journal of Cross-Cultural Psychology* 4 (2): 135–156.
15. Farley, F.H., and A.P. Grant. 1976. Arousal and Cognition: Memory for Color Versus Black and White Multimedia Presentation. *The Journal of Psychology* 94 (1): 147–150.
16. Dzulkifli, M.A., and M.F. Mustafar. 2013. The Influence of Colour on Memory Performance: A Review. *The Malaysian Journal of Medical Sciences: MJMS* 20 (2): 3–9.

17. Deterding, S., et al. 2011. From Game Design Elements to Gamefulness. Proceedings of the 15th International Academic MindTrek Conference on Envisioning Future Media Environments - MindTrek '11. New York, NY, USA: ACM Press, p. 9.
18. StataCorp L.P. 2007. Stata Data Analysis and Statistical SOFTWARE. In *Special Edition Release 10.*
19. Futschek, G. 2006. Algorithmic Thinking: The Key for Understanding Computer Science. In *Informatics Education – The Bridge between Using and Understanding Computers*, Lecture Notes in Computer Science, edited by R.T. Mittermeir, 159–168. Berlin, Heidelberg: Springer.
20. Voskoglou, M.G., and S. Buckley. 2012. Problem Solving and Computational Thinking in a Learning Environment. Preprint arXiv:1212.0750.
21. Schoenfeld, A. 1992. Learning to Think Mathematically: Problem Solving, Metacognition, and Sense Making in Mathematics. In *Handbook of Research on Mathematics Teaching and Learning*, 334–370.

QM Automata: A New Class of Restricted Quantum Membrane Automata

Konstantinos Giannakis, Alexandros Singh, Kalliopi Kastampolidou, Christos Papalitsas, and Theodore Andronikos

Abstract The term "Unconventional Computing" describes the use of non-standard methods and models in computing. It is a recently established field, with many interesting and promising results. In this work we combine notions from quantum computing with aspects of membrane computing to define what we call QM automata. Specifically, we introduce a variant of quantum membrane automata that operate in accordance with the principles of quantum computing. We explore the functionality and capabilities of the QM automata through indicative examples. Finally we suggest future directions for research on QM automata.

Keywords Unconventional computation • Membrane systems • Quantum computation • QM automata

1 Introduction

Recently, the term "Unconventional Computing" has gained momentum in the literature. It is an umbrella term that describes the implementation of non-standard methods and models in computing. It constitutes an interdisciplinary field, inspired by other scientific areas besides Computer Science, such as Physics, Biology, Material Science etc. (see [1] for a description of the evolution of the Unconventional Computing). Two of the most famous and, at the same time, promising unconventional methods are the bioinspired computation and the quantum computation.

Membrane systems, usually known as P systems, belong to the field of "Natural computing." Natural computing is an area that advocates the use of nature-inspired computing methods and architectures. For example, the functionality of cellular membranes can be seen as a computing device, with specific inputs, rules, and outputs. Combined research efforts from the fields of Computer Science and Biology have achieved several results. P (or membrane) systems were proposed in [2] and since then the field has evolved with many interesting results and ideas.

K. Giannakis (✉) • A. Singh • K. Kastampolidou • C. Papalitsas • T. Andronikos
Department of Informatics, Ionian University, Corfu, Greece
e-mail: kgiann@ionio.gr; p13sing@ionio.gr; p12kast@ionio.gr; c14papa@ionio.gr; andronikos@ionio.gr

P. Vlamos (ed.), *GeNeDis 2016*, Advances in Experimental Medicine and Biology 988, DOI 10.1007/978-3-319-56246-9_15

Membrane computing is inspired by organic cells and their membranes. They can be used in computer science to model phenomena and solve particular problems, especially those exhibiting dynamic behaviour. Well-known variants of P systems that have been extensively studied are: P systems with active membranes, P systems with symport/antiport rules, communicating P systems and tissue P system. P automata are variants of P systems with automata-like accepting behaviour [3, 4].

Quantum theory, which is the theoretical basis of much of modern physics, and which explains the behavior of matter and energy at the atomic and subatomic level, perceives computing as a natural phenomenon that relies on and exploits the laws of quantum mechanics. Quantum computing via the use of quantum mechanical methods strives to expand and evolve the classical computing model. In the quantum paradigm transitions among (quantum) computational states are achieved through the application of unitary operators. It was Feynman in the '80s that pointed out that it is not feasible to simulate efficiently an actual quantum system using a classical computer [5].

Combining the above observation with the fact that Moore's law is reaching its limits, the need to search for novel technologies and computation models becomes obvious. There seem to be particular advantages of quantum computing over classical in terms of complexity [6, 7]. Quantum mechanics and the use of quantum-inspired computing technologies could possibly offer an increase in computational capabilities and efficiency, since the quantum world has an inherently probabilistic nature and non-classical phenomena, such as superposition and entanglement, occur.

Two milestones in this area of research are the algorithms proposed by Shor and Grover. Shor devised a quantum algorithm that solves the problem of integer factorization in polynomial time. The ability to factor integers in polynomial time is a major breakthrough, which affects various fields in computer science, such as cryptography and complexity. The second landmark was Grover's algorithm; a novel quantum algorithm for searching an unsorted array of n elements in $\Theta(\sqrt{n})$.

In this work we investigate the possibility of combining variants of P automata with mechanisms derived from the field of quantum computation. Specifically, we propose a new hybrid type of automaton, the QM automaton (Q for quantum and M for membrane). QM automata are membrane state-machines that operate via unitary transformations. We demonstrate their functionality through indicative examples, which also serve to explain how arbitrary inputs and outputs are transformed into appropriate formats. We explore the computational power of QM automata and suggest potential applications. We then compare the advantages and disadvantages of our constructions with respect to existing similar approaches.

Our paper is organized as follows: After the introduction, we present the analysis of the related literature. In Sect. 3 we provide the necessary mathematical terminology and definitions. Sections 3 and 4 contain the vital part of our approach. Specifically, the definitions of our proposed variant of quantum membrane automata are provided in Sect. 3. Afterwards, in Sect. 4 we sketch the expressive power of the QM automata by giving indicative examples. Finally, Sect. 5 is devoted to the discussion of our results along with directions for future work.

2 Related Work

Current literature is full of elegant and well-presented studies regarding the computation process with membranes. Membrane computing was initiated in the late '90s by G. Păun [2]. Since then, the field has widely spread and many variants have been proposed [8–10]. Ciobanu and Aman in [11] offer an extensive overview of the current trends and updates, such as complexity issues regarding process calculi and natural computation. Recent updates regarding P systems can be found in [12].

P automata are variants of P systems with automata-like accepting behaviour [3, 4]. The first variants of P automata were one-way P automata with only symport rules [3, 13], and analysing P systems, which accept a computation by halting configurations [14]. Besides their computational capabilities, variants of membrane (or P) automata are also used in modeling and describing complex biolocal processes [15, 16]. This observation emphasizes the need for further research in this direction.

Feynman was the first who envisioned quantum computing in [5]. Quantum computation models like automata have already been introduced (see for example [17, 18]). The complexity of quantum systems was described in [19] providing important results about space and time complexity. Below we present to the most relevant literature regarding our approach.

In [20] the authors discuss ways of approaching quantum P systems. Inspired by classical energy-based P systems, the authors describe two models: one based on strictly unitary rules, which implement permutations on the alphabet, and the other based on the use of creation and annihilation operators in a generalised Conditional Quantum Control technique [21], which is used to implement potentially non-unitary operations. In both systems, objects are represented by qudits, with d equal to the cardinality of the alphabet, while multisets are compositions of such individual systems. Energy units, associated with the objects, are incorporated in the system in the form of actual quanta of energy, with values ordered and arranged in an equispaced manner. These are interpreted as the energy levels of a "truncated" quantum harmonic oscillator.

The concept is further refined in [22] and [23] in which to each membrane is associated an infinite-dimensional quantum harmonic oscillator. The state of the oscillator represents the energy assigned to the membrane. Again the lowering and raising of energy levels is achieved through creation and annihilation operators. In this model objects can change their state but can never cross membranes to move to another region. Instead, interactions happen through the modification of energy of the oscillators in each membrane.

Such systems can compute any partial recursive function $f : N^a \to N^b$. These results apparently depend on the use of post-selection operators constructed from the creation/annihilation ones, whose use is known to give efficient results not only for NP-complete problems but also for PP-complete ones [24].

2.1 Our Approach

In QM automata we eschew the use of energy-based rules, oscillators, and non-unitary rules in favour of more conventional quantum computing techniques. In particular, our rules are strictly unitary, permuting input and output registers associated with each membrane. We also avoid the problems associated with the notion of "transferring" systems/objects, which is inherent in the aforementioned works. We tackle this obstacle by providing registers with fixed "depths" that can easily be manipulated with standard unitary operators.

3 QM Automata

We begin with a review of the mathematical background, nomenclature and definitions of quantum computation and then proceed to define of our QM automata.

3.1 Mathematical Background and Terminology

Quantum computation can be described by the successive application of unitary operators in order to evolve an initial state, followed finally by a measurement to obtain results. A Hilbert space $\mathcal{H}_n$ is a n-dimensional vector space equipped with an inner product, such that it is also a complete metric space with respect to the metric induced by the inner product. We focus on finite-dimensional Hilbert spaces over the complex numbers, where our quantum systems reside. A state of our quantum system is a unit vector $|\psi\rangle \in \mathcal{H}_n$. We consider states of the form $|\psi\rangle = a|0\rangle + b|1\rangle$ which represent qubits. Composition of systems of qubits is given in terms of the tensor product $H \otimes V$ of their respective spaces. In the finite dimensional case, such a product can be computed via the Kronecker product $\otimes$ of matrices and vectors.

A unitary operator U is an operator that preserves the norms of vectors and (if operating on a finite-dimensional space) is represented by a unitary matrix. Equivalently, a complex-valued matrix U is unitary if it has an inverse and if $\|U\psi\| = \|\psi\|$ holds for every vector ψ. For unitary matrices it also holds that $U^{-1} = U^\dagger$ (which is the same as $U^\dagger U = UU^\dagger = I$) and also $\|U\| = 1$, where $U^\dagger$ is the Hermitian conjugate of U. A square complex matrix that is equal to its conjugate transpose, i.e. $H = \bar{H}^T$, is called a self-adjoint or Hermitian matrix and describes the observables of our system.

3.2 P Systems in Short

In P systems objects are described by strings evolving under string processing rules like rewriting rules. There are also evolution rules acting upon these objects, acting in specified compartments or on the membranes themselves. A membrane system is represented either by a Venn diagram or by a tree. The root represents the skin membrane. The outer membrane is called *skin* and the space which surrounds the membrane system is called *environment*.

Membranes can contain other membranes creating an hierarchical structure. Each membrane contains a series of objects and rules. P systems evolve via rules that are executed in a non-deterministic, parallel way, and the whole system behaves as being synchronized by a global clock. There are three fundamental characteristics that membrane systems possess: the membrane structure, multisets of objects, and rules. Through the application of rules, transitions among configurations take place. A sequence of transitions is interpreted as computation.

3.3 Introducing the QM Automata

Let us now present the definition for our proposed variant of membrane systems.

Definition 1 A QM automaton is a tuple

$$\Pi = (\Gamma, \mu, W, R, F),$$

where

1. Γ is an alphabet, whose elements we will refer to as *objects*.
2. μ is a membrane structure, in which membranes are nested in hierarchically arranged layers, in a way such that inputs and outputs form a pipeline through the layers. Each membrane consists of two Hilbert spaces, an input and an output one, any of which is potentially shared with other membranes, as described bellow. We consider the outermost membrane to contain the result of a computation.
3. W contains the initial configurations of the membranes. To each membrane $m \in \mu$ corresponds a $w_m \in W$ that describes the initial configuration of m. Each w_m is composed of $|\Gamma|$ qubits, each signifying the amplitude of the corresponding object, and can be thought of as a unit ket in a $|\Gamma|$-dimensional Hilbert space.
4. R contains the rules of the system. To each membrane $m \in \mu$ we assign a rule R_m which acts as some unitary operator on the kets of $m \in \mu$ and as the identity everywhere else.
5. F is the set of accepting configurations. After completing the computation, if the state of the outermost membrane matches any of the accepting configurations, then we say that we accept the initial configuration defined by W. Otherwise it is rejected.

We can recursively define the state spaces for each layer, starting from the inner-most one, which is considered to be layer 0, and reaching the outer-most one:

- Input and output state spaces:
  Each membrane in layer 0 has its own input space and a shared output space. For each layer $k > 0$, the membranes of layer k have as inputs the output space of layer $k-1$, and share an output space, which in turn is the input of layer $k+1$, except for the outermost membrane which does not share its output space with other membranes. The global state of our system is the tensor product of all the constituent input and output systems.
  Let us also describe the rules in more detail, as well as how to chain them together to perform computations:
- The rules:
  Each rule R_m acts as some unitary operator $|\psi\rangle_{m_{in}} \otimes |\phi\rangle_{m_{out}} \mapsto |\psi'\rangle_{m_{in}} \otimes |\phi'\rangle_{m_{out}}$ and as the identity everywhere else in the global state—that is for any other membrane $k \in \mu \neq m$, R_m is the map $|\psi\rangle_{k_{in}} \otimes |\phi\rangle_{k_{out}} \mapsto |\psi\rangle_{k_{in}} \otimes |\phi\rangle_{k_{out}}$
- Computation:
  The input region of each membrane m is initialised with instances of the objects defined by w_m. Computation starts from the innermost layer (layer 0) by applying the of rules R_m to each membrane m that belongs to layer 0. The computation proceeds successively to layer 1, layer 2, etc., until the outermost layer is processed. The output space of the outermost layer contains the result of the computation.

The example depicted in Fig. 1 shows a membrane structure that consists of 3 membranes, namely M1, M2, and M3. Inside them, there are some "objects", represented as letters from the alphabet $\Gamma = \{a, b\}$. For each membrane m, the input and output state kets $|ab\rangle = |a\rangle \otimes |b\rangle$ are composed of two qubits, whose values represent the "degrees of existence" for each letter, and ultimately correspond to the probabilities of finding "a" and "b" in membrane m after measurement. For example, M1's initial state is $|10\rangle = |1\rangle \otimes |0\rangle$, as we have an occurrence of the letter "a" but not of "b". The whole initial configuration of the system can be described as the tensor product of its constituent systems, i.e. each membrane:

$$|10\rangle_{M1_{in}} \otimes |11\rangle_{M2_{in}} \otimes |00\rangle_{M1_{out}/M2_{out}/M3_{in}} \otimes |00\rangle_{M3_{out}}$$

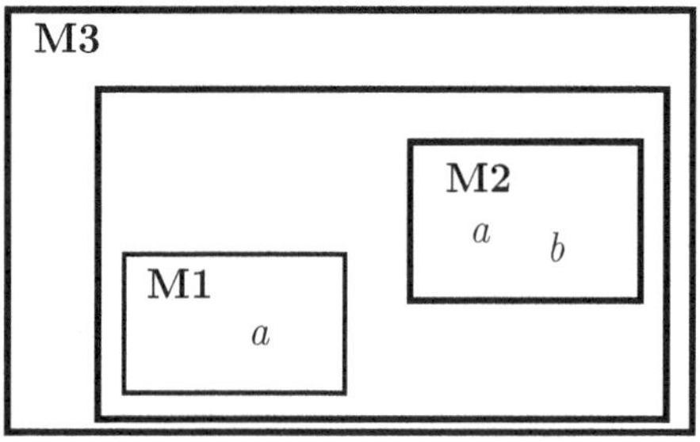

Fig. 1 Our example of a system with its initial state

Here K_{in}/K_{out} are the input and output spaces of a membrane K, respectively; for instance, $M1_{in}$ is the input space of M1. We have that $M1_{out} = M2_{out} = M3_{in}$ and so we use the symbol "/" in a membrane's indexing to indicate the chains among the membranes, where one output is identified with an input.

We also define an arbitrary state to be our accepting state, say $F = \{|11\rangle\}$.

Below we present the membrane rules that govern the evolution of each membrane's state.

- Membrane 1 rule: $R1 = |10\rangle_{M1_{in}} \otimes |00\rangle_{M1_{out}} \rightarrow |00\rangle_{M1_{in}} \otimes |10\rangle_{M1_{out}}$
- Membrane 2 rule: $R2 = |11\rangle_{M2in} \otimes |10\rangle_{M2_{out}} \rightarrow |00\rangle_{M2_{in}} \otimes |11\rangle_{M2_{out}}$
- Membrane 3 rule: $R3 = |11\rangle_{M3in} \otimes |00\rangle_{M3_{out}} \rightarrow |00\rangle_{M3_{in}} \otimes |11\rangle_{M3_{out}}$

Again, K_{in}/K_{out} are the input and output spaces of a membrane K, respectively. Of course $M1_{out} = M2_{out} = M3_{in}$, but we choose to use $M1_{out}$ when talking about the rule $R1$, $M2_{out}$ when talking about $R2$ etc, for the sake of clarity.

The rules would actually work on the whole space:

$$M1_{in} \otimes M2_{in} \otimes M1_{out}/M2_{out}/M3_{in} \otimes M3_{out}$$

with R1 acting as the identity on $M2_{in}$, $M3_{out}$, R2 acting as the identity in $M1_{in}$, $M3_{out}$ and R3 acting as the identity in $M1_{in}$, $M2_{in}$. We omit the identical parts, again for the sake of clarity.

We apply our rules in sequence, starting from the innermost membranes and moving outwards, i.e. $R1, R2, R3$—which can be done by multiplying the respective matrices $R3 \cdot R2 \cdot R1$. Starting with an initial state of:

$$|10\rangle_{M1_{in}} \otimes |11\rangle_{M2_{in}} \otimes |00\rangle_{M1_{out}/M2_{out}/M3_{in}} \otimes |00\rangle_{M3_{out}}$$

after the application of the rules we get the final state:

$$|00\rangle_{M1_{in}} \otimes |00\rangle_{M2_{in}} \otimes |00\rangle_{M1_{out}/M2_{out}/M3_{in}} \otimes |11\rangle_{M3_{out}}$$

The results of the computations are present in the last membrane's output, which in this instance is $M3_{out}$. Because $M3_{out}$ matches our sole accepting configuration, we say that we accept the initial configuration described above. In the above example, we showed how the quantum-inspired rules are being applied on the proposed membrane system. By applying the rules $R3 \cdot R2 \cdot R1$ (starting from $R1$ up to $R3$), the structure switches from one configuration to another. Since we refer to automata, these configurations are actually states of the automaton.

4 Simulating Classical Automata

In this section we sketch a constructive proof for the simulation of classical finite automata by QM automata. Given a fixed $k \in \mathbb{N}$, we are able to build a QM automaton that simulates classical automata running on words of length k.

Construction:

We build a QM automaton whose alphabet consists of the alphabet of the automaton we are simulating, plus all its states (represented as tokens/letters).

Consider k nested membranes, with input/output spaces coupled as before. Each space consists of two components: a letter and a state so that it looks something like this:

$$|letter\rangle \otimes |state\rangle$$

Starting from the inner membrane, we initialize the letter kets to the value of the corresponding letter of the input word such that the k-th membrane contains the k-th letter (counting from the innermost membrane towards the skin).

All state kets are initialised to $|q_0\rangle$, where q_0 is the initial state of the automaton we simulate. Without loss of generality, we can assume that this will be $|0\rangle$. Then to each membrane is assigned the sum of $n = |\Sigma|$ rules of the form:

$$|letter\rangle \langle letter| \otimes U,$$

where $|\Sigma|$ is the length of the automaton's alphabet and U changes the output state's ket to $|newState\rangle$ based on the automaton's transition function:

$$\delta(letter, currentState) = newState$$

The accepting configurations of our simulator matches that of the classical automaton. If for example the accepting states of the classical automaton to be simulated are:

$$F = \{s0, s3\}.$$

Then the corresponding accepting configurations of our simulator are $|s0\rangle$ and $|s3\rangle$.

4.1 Simulation Example

Below we present and describe the example we use as a simulation of the run on our proposed automata variant (see the automaton in Fig. 2). Consider the following classical automaton:

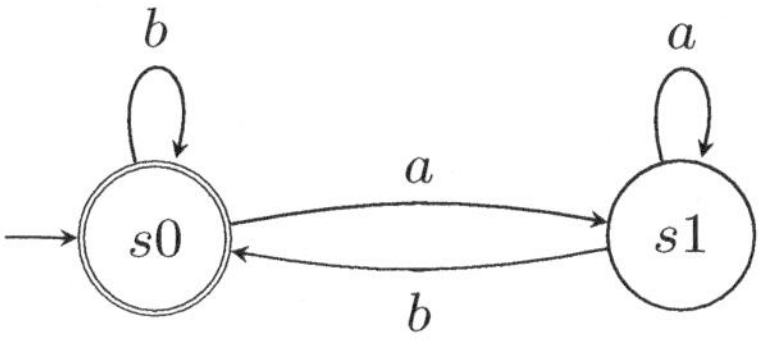

Fig. 2 This figure depicts the automaton of our example

- $\Sigma = \{a, b\}$
- $Q = \{s0, s1\}$
- $F = \{s0\}$
- $\delta(a, s0) = s_1$,
  $\delta(b, s0) = s_0$,
  $\delta(a, s1) = s_1$,
  $\delta(b, s1) = s_0$

The first membrane's rule is:

$$|a\rangle \langle a| \otimes |s0\rangle \langle s0| \otimes I \otimes \mathit{flip} \otimes I+$$
$$|b\rangle \langle b| \otimes |s0\rangle \langle s0| \otimes I \otimes I \otimes I+$$
$$|a\rangle \langle a| \otimes |s1\rangle \langle s1| \otimes I \otimes \mathit{flip} \otimes I+$$
$$|b\rangle \langle b| \otimes |s1\rangle \langle s1| \otimes I \otimes I \otimes I$$

While the second one's is:

$$I \otimes I \otimes |a\rangle \langle a| \otimes |s0\rangle \langle s0| \otimes \mathit{flip}+$$
$$I \otimes I \otimes |b\rangle \langle b| \otimes |s0\rangle \langle s0| \otimes I+$$
$$I \otimes I \otimes |a\rangle \langle a| \otimes |s1\rangle \langle s1| \otimes \mathit{flip}+$$
$$I \otimes I \otimes |b\rangle \langle b| \otimes |s1\rangle \langle s1| \otimes I$$

In the above expression, I denotes the identity operator and *flip* is the operator that "flips" a qubit's value.

Let us simulate a run at depth $k = 2$ for the word "ab".

Our membrane system's initial state is:

$$|a\rangle_{m1_{in}} \otimes |s0\rangle_{m1_{in}} \otimes |b\rangle_{m1_{out}/m2_{in}} \otimes |s0\rangle_{m1_{out}/m2_{in}} \otimes |s0\rangle_{m2_{out}}$$

We can see how each membrane corresponds to one "phase" of our simulation. The first membrane's input corresponds to reading the first letter of our word, namely "*a*" and the second one's to the second letter "*b*". We also initialise all state kets to $|s0\rangle$, since we start our simulation from the initial state "*s0*". Note that we omitted the letter ket of $m2_{out}$, as it wouldn't be used in the computations (as we also did for the rules above).

Applying the first rule to our initial state we get:

$$|a\rangle_{m1_{in}} \otimes |s0\rangle_{m1_{in}} \otimes |b\rangle_{m1_{out}/m2_{in}} \otimes |s1\rangle_{m1_{out}/m2_{in}} \otimes |s0\rangle_{m2_{out}}$$

We see that $m2_{in}$'s state ket has changed to $|s1\rangle$, as the automaton would transition to $s1$ after reading "a".

Finally, applying the second rule we get:

$$|a\rangle_{m1_{in}} \otimes |s0\rangle_{m1_{in}} \otimes |b\rangle_{m1_{out}/m2_{in}} \otimes |s1\rangle_{m1_{out}/m2_{in}} \otimes |s0\rangle_{m2_{out}}$$

The result of our simulation is in the output of the last membrane, i.e. it is $m2_{out}$'s state ket, which is $|s0\rangle$. This corresponds to the fact that after reading "b", our automaton would transition back to $s0$. Because $|s0\rangle_{m2_{out}}$ matches our accepting state, we accept the initial configuration that corresponds to "ab", as described above.

If instead we started with:

$$|a\rangle_{m1_{in}} \otimes |s0\rangle_{m1_{in}} \otimes |a\rangle_{m1_{out}/m2_{in}} \otimes |s0\rangle_{m1_{out}/m2_{in}} \otimes |s0\rangle_{m2_{out}}$$

which corresponds to simulating a run with the word "aa", the final space after the application of the two rules would be:

$$|a\rangle_{m1_{in}} \otimes |s0\rangle_{m1_{in}} \otimes |a\rangle_{m1_{out}/m2_{in}} \otimes |s1\rangle_{m1_{out}/m2_{in}} \otimes |s1\rangle_{m2_{out}}$$

We see that after reading the first letter, "a", our simulated automaton switched to state $s1$ (which is $m1_{out}/m2_{in}$'s state ket). Finally, after reading the second letter, "a", the simulated automaton remains in $s1$ and so $m2_{out}$'s state ket is $|s1\rangle$, so our automaton rejects the initial configuration that corresponds to "aa".

5 Conclusions

Computational models inspired by concepts of Unconventional Computing (and, specifically, quantum computing) have drawn the attention of the research community for quite some time now. There are a lot of works that try to combine a classical approach with a modern, unconventional one. In this study, we followed this line of research and we proposed a variant of P automata that make use of quantum computing techniques, which we called QM automata.

In particular, we proposed membrane state-machines that operate under unitary transformations. We demonstrated the functionality of our model through examples, and we explained our methodology regarding the construction of the quantum rules. Finally, we compared our approach with similar ones, like those in [20, 22]. The novelty of QM automata lies mainly in their use of strictly unitary rules.

As far as the future directions are concerned, of primary interest would be investigating and fully describing languages accepted by QM automata. Variations on the QM automata such as infinitary QM automata would also be of interest. Another possibility is the exploration of game-theoretic connections. Quantum games are widely researched and the association of strategies of these games with inputs in appropriate automata types has already been proposed [25]. Finally, it would be interesting to examine whether our automata may be used to describe complex bioinspired computational models efficiently.

References

1. Calude, C. 2015. Unconventional computing: A brief subjective history. Technical Report, Department of Computer Science, The University of Auckland.
2. Păun, G. 2000. Computing with membranes. *Journal of Computer and System Sciences* 61(1): 108–143.
3. Csuhaj-Varjú, E., and G. Vaszil. 2003. P automata or purely communicating accepting P systems. In *Membrane Computing*, 219–233. New York: Springer.
4. Csuhaj-Varjú, E., and G. Vaszil. 2013. On the power of P automata. In *Unconventional Computation and Natural Computation*, 55–66. Berlin: Springer.
5. Feynman, R.P. 1982. Simulating physics with computers. *International Journal of Theoretical Physics* 21(6): 467–488.
6. Bernstein, E., and U. Vazirani. 1997. Quantum complexity theory. *SIAM Journal on Computing* 26(5): 1411–1473.
7. Fortnow, L. 2003. One complexity theorist's view of quantum computing. *Theoretical Computer Science* 292(3): 597–610.
8. Barbuti, R., A. Maggiolo-Schettini, P. Milazzo, G. Pardini, and L. Tesei. 2011. Spatial P systems. *Natural Computing* 10(1): 3–16.
9. Martın-Vide, C., G. Păun, J. Pazos, and A. Rodrıguez-Patón. 2003. Tissue P systems. *Theoretical Computer Science* 296(2): 295–326.
10. Păun, G., G. Rozenberg, and A. Salomaa. 2010. *The Oxford Handbook of Membrane Computing*. Oxford: Oxford University Press.
11. Aman, B., and G. Ciobanu. 2011. *Mobility in Process Calculi and Natural Computing*. Berlin: Springer Science & Business Media.
12. Păun, G., M.J. Pérez-Jiménez, et al. 2012. Languages and P systems: Recent developments. *Computer Science* 20(2): 59.
13. Csuhaj-Varjú, E. 2010. P automata: Concepts, results, and new aspects. In *Membrane Computing*, 1–15. Berlin: Springer.
14. Freund, R., and M. Oswald. 2002. A short note on analysing P systems with antiport rules. *Bulletin of the EATCS* 78: 231–236.
15. Giannakis, K., and T. Andronikos. 2017. Membrane automata for modeling biomolecular processes. *Natural Computing* 16: 151–163. doi:10.1007/s11047-015-9518-1. http://dx.doi.org/10.1007/s11047-015-9518-1.
16. Giannakis, K., and T. Andronikos. 2015. Mitochondrial fusion through membrane automata. In *Advances in Experimental Medicine and Biology*, GeNeDis 2014, eds. P. Vlamos, and A. Alexiou vol. 820, 163–172. Berlin: Springer International Publishing.
17. Giannakis, K., C. Papalitsas, and T. Andronikos. 2015. Quantum automata for infinite periodic words. In *2015 6th International Conference on Information, Intelligence, Systems and Application (IISA)*, 1–6. New York: IEEE.

18. Hirvensalo, M. 2011. Quantum automata theory–a review. In *Algebraic Foundations in Computer Science*, 146–167. Berlin: Springer.
19. Bernstein, E., and U. Vazirani. 1993. Quantum complexity theory. In *Proceedings of the Twenty-Fifth Annual ACM Symposium on Theory of Computing*, 11–20. New York: ACM.
20. Leporati, A., D. Pescini, and C. Zandron. 2004. Quantum energy–based P systems. In: *Proceedings of the First Brainstorming Workshop on Uncertainty in Membrane Computing*, 145–168.
21. Barenco, A., D. Deutsch, A. Ekert, and R. Jozsa. 1995. Conditional quantum dynamics and logic gates. *Physical Review Letters* 74(20): 4083
22. Leporati, A., G. Mauri, and C. Zandron. 2005. Quantum sequential P systems with unit rules and energy assigned to membranes. In *International Workshop on Membrane Computing*, 310–325. Berlin: Springer.
23. Leporati, A. 2007. (UREM) P systems with a quantum-like behavior: background, definition, and computational power. In *International Workshop on Membrane Computing*, 32–53. Berlin: Springer.
24. Aaronson, S. 2005. Quantum computing, postselection, and probabilistic polynomial-time. In *Proceedings of the Royal Society of London A: Mathematical, Physical and Engineering Sciences*, vol. 461, 3473–3482. London: The Royal Society.
25. Giannakis, K., C. Papalitsas, K. Kastampolidou, A. Singh, and T. Andronikos. 2015. Dominant strategies of quantum games on quantum periodic automata. *Computation* 3(4): 586–599.

On the Detection of Overlapped Network Communities via Weight Redistributions

Stavros I. Souravlas and Angelo Sifaleras

Abstract A community is an important attribute of networking, since people who join networks tend to join communities. Community detection is used to identify and understand the structure and organization of real-world networks, thus, it has become a problem of considerable interest. The study of communities is highly related to network partitioning, which is defined as the division of a network into a set of groups of approximately equal sizes with minimum number of edges. Since this is an NP-hard problem, unconventional computation methods have been widely applied.

This work addresses the problem of detecting overlapped communities (communities with common nodes) in weighted networks with irregular topologies. These communities are particularly interesting, firstly because they are more realistic, i.e., researchers may belong to more than one research community, and secondly, because they reveal hierarchies of communities: i.e., a medical community is subdivided into groups of certain specialties. Our strategy is based on weight redistribution: each node is examined against all communities and weights are redistributed between the edges. At the end of this process, these weights are compared to the total connectivity of each community, to determine if overlapping exists.

Keywords Networks • Community detection

1 Introduction

Several systems of high interest to the scientific community can be successfully represented as networks. Network examples are the Internet, the World-Wide Web, and the social networks [1]. Perhaps the latest are the type of networks in which researchers are most interested nowadays. Examples of social networks are Facebook (1.6 billion users), Instagram (400 million users), or Twitter

S.I. Souravlas • A. Sifaleras (✉)
Department of Applied Informatics, University of Macedonia, 156 Egnatias Street, 54636 Thessaloniki, Greece
e-mail: sourstav@uom.gr; sifalera@uom.gr

P. Vlamos (ed.), *GeNeDis 2016*, Advances in Experimental Medicine and Biology 988, DOI 10.1007/978-3-319-56246-9_16

(320 million users) [2]. Typically, social networks are organized in groups of users [3]. These users join a network, create their own profile, publish information and find other users with the same interests. In this way, small or larger groups of users are formed within networks. Such groups are referred to as *communities*. Although there is no universally acceptable definition of a community, one can define it as a set of nodes and links in a network such that, its internal connections are more than its external connections [4]. The nodes of a community are considered similar to each other, dissimilar to the other nodes of the network [5], and represent its users. The edges represent the connections between the users of one community or between users of different communities.

A lot of effort has focused on detecting communities in social and data networks. Communities give a more realistic approach of the networks and they can also be used to help the relay of messages within a network [6–9]. The majority of the research papers focused on this subject state that common experience shows that communities really exist in social and data networks [3, 10]. The study of community structures is highly related to the problem of network partitioning. Typically, the network partitioning problem is defined as the partitioning of a network into a set of groups of approximately equal sizes with minimum number of edges [5, 11]. This problem is NP-hard and thus, heuristics and approximation algorithms are usually employed [12]. Much of this work involves parallel computations to enhance performance. Specifically, the nodes of the system represent computations while the edges represent communications and the general idea is to equally assign nodes among the processors, so that the communications (edges) between them are minimized. However, network partitioning is not the ideal method for the analysis of networks and for community detection. This is firstly due to the fact that in real networks, the communities formed rarely have approximately the same size, secondly because the connections are not minimized and finally because network partitioning does not consider the similarities between nodes (or users), which are inherited in a social network.

Most of the papers found in the literature agree that the members of a community are more strongly connected between them than they are connected to members of other communities. These papers introduced a number of algorithms to identify communities and measures to test and compare these algorithms [11, 13]. The most popular metric in the literature, which is used to quantify the strength of a community structure is *modularity*. This metric was defined by Newman and Girvan [11] and it measures the difference between the density of edges inside the communities and the density that would have occurred if the edges were randomly distributed over the entire network. A number of papers have been influenced by this metric [10, 14–18], although it has some flaws like the resolution limit, which can be partially overcomed by techniques like [19].

Members in the same community share the same interests, viewpoints, preferences, hobbies, professions, newsgroups, and, as far as social networks are concerned, they are more likely to comment or "like" topics of the same interest. In most of the papers, this whole activity is modeled via analyzing the link weights [5, 9, 15, 20–23] of the network. In order for a node to be considered as member of a community, its internal/external links to this community should follow some well-defined properties. Some algorithms also detect *overlapping* communities [9, 20, 21, 24, 25]

The main problem in these techniques is that, the aforementioned analysis usually suffers high complexity times, although there are also linear-time approaches [5, 26]. Lu et al. [9] proposed a strategy which can detect both overlapping and non-overlapping communities by adding one node to a community at each *expanding step*. In each expanding step, the authors initially compute the *belonging degree* of a number of nodes to a community C and they pick the one with the highest belonging degree. This node is temporarily attached to C, forming $C^{'}$. Then, if the *conductance* of $C^{'}$ is lower compared to the conductance of C, the selected node is finally attached to C. The worst time complexity can grow as large as n^2, where n is the number of nodes in the network.

Another interesting strategy proposed by Newman and Girvan [11] repeatedly calculates the *betweenness* scores between all the edges in the network and remove the edge with the highest score. Again, this strategy operates in worst-case time that can be as large as $O(n^3)$. More strategies for modeling the relationships between the nodes of a network have quadratic or worse complexities [15, 22, 23]. An interesting linear-complexity approach was presented by Raghavan et al. [5], which uses a simple label propagation algorithm. Unlike most of the strategies found in the literature this strategy does not use a predefined objective function (like the conductance mentioned in the previous paragraph) to identify the communities. Every node is initialized with a unique label and at every step each node adopts the label that most of its neighbors currently have. There are two concerns regarding this technique. First, it is assumed that each node joins the community in which most of its neighbors belong and second there is no clear updating strategy that shows what would happen when new nodes entering the network have equal number of neighbors between many communities or when some nodes leave the network temporarily or permanently.

This paper presents a weight-redistribution technique used to find overlapping communities in networks with irregular topologies. Our strategy performs a depth-first based search over the network and redistributes the data weights across the links. Finally, the values computed by the redistribution function are compared to the connectivity values of each community to determine if there are overlaps. The remaining of this paper, is organized as follows: Section 2 briefly sets up the problem studied, presents the weight-redistribution strategy, and performs a theoretical analysis. Section 3 concludes the paper and offers aspects for future research.

2 Overlapped Communities Detection Strategy

In this section, we typically describe the proposed strategy and briefly describe some aspects of the problem studied.

2.1 Problem Aspects

Consider a small network of irregular topology, like the one shown in Fig. 1, where there are three communities C_1, C_2, and C_3. The problem we address is to find whether there is overlapping between the communities. Figure 1 reveals some aspects that need to be considered before trying to uncover these overlaps:

`Topic 1:` User A is directly connected to users E and F in C_2. Thus, there is a chance that A belongs to C_2 as well.

`Topic 2:` User A is connected to C_3 via E or via the rout B, C. Thus, there is a chance that A belongs to C_3 as well. In this case, the relays E or B, C may also belong to C_3.

`Topic 3:` Continuing Topic 2, if there is a high percentage of overlapping between two communities, chances are that they are actually one community.

`Topic 4:` Not all users of a community are necessarily related (in Facebook for example, two users may have common friends while they are not related). In Fig. 1, A is not directly connected H, it is connected to E. In its turn, E is connected to H. To continue with the Facebook analogy, this is a case that a user (for example A) is unaware of a community C_3 until he/she visits the page of another user that has some relationship to members of this community (like E). A way to quantify the *similarities* (see next section) between unrelated users, like A and H is required.

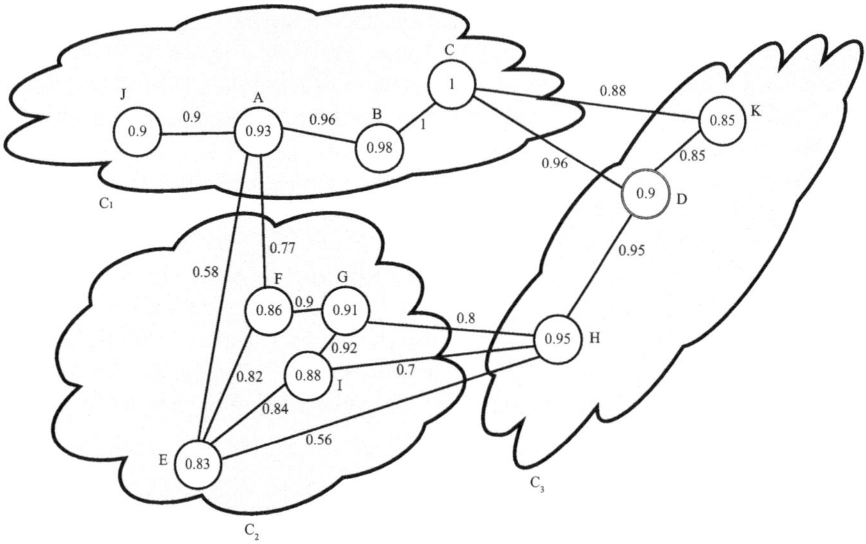

Fig. 1 A network with three communities, C_1, C_2, and C_3, depicting nodes similarities (numbers next to each edge) and network connectivity degrees (numbers inside each node)

The four topics referred in this section will be considered in our proposed strategy, which is typically presented in the next section.

2.2 Finding Community Overlaps

Let $G = (V, E)$ be a weighted, undirected graph, where V is the set of nodes and E is the set of edges. Nodes represent users and edges represent the connections between two users (for example friendship in Facebook, etc.). The *similarity* between users i and j is given by $w_{i,j}$, that is, the weight of the edge that connects i and j. This value lies in the interval $[0 \ldots 1]$. For example, a value of 0.78 shows that the views of i and j can be considered quite converging, at a percentage of 78%. Bu et al. [3] present a table of user phrases that indicate supportive or opposing attitude towards a comment or opinion, etc. These phrases are accompanied by a value that shows the degree of support.

The network nodes are also weighted: the weight of a node i indicates its *network connectivity degree*, that is, how well the preferences, likes, views of a user are fitted to a community. The network connectivity degree is also between $[0 \ldots 1]$. The letters are the node names, the edge values indicate view similarities and the node values indicate similarity degrees. The *Network Connectivity Degree* of a user i, NCD_i, is computed as follows:

$$NCD_i = \frac{\sum w_{i,j}}{m} \tag{1}$$

where $w_{i,j}$ is the weight of any edge that connects user i with any user j that lies in the same community and m is the number of such edges. For example, consider community C_2. User G has two internal links, namely with users F and I and one external, with user H. To compute NCD_G, we only consider the internal links and we have $NCD_G = \frac{(0.9+0.92)}{2} = 0.91$. This value is stored in node G. The values of the external links will be used to determine overlapping, as will be seen later.

Topics 1 and 2 of Sect. 2.1 indicate that a user can be connected to a community, either *directly*, through a relationship with a user or many users of the community or *indirectly*, through a relationship with a user that is related to a member of the community. Thus, we need to quantify the similarities between unrelated nodes (Topic 4), in order to uncover possible overlaps. Since there is no relationship (thus we can't use data like the ones proposed by Bu et al. [3]), we have to intuitively describe similarity. Our approach suggests that the similarity of two unconnected users is generally higher when the number of links between them is small (for example a friend of a friend in Facebook probably has more similarities compared to a friend of a friend ... of a friend ...), and when the weights on the links that connect them are as large as possible. Thus, the largest possible similarities are likely to be found in the shortest paths from each node to each community. The depth-first based procedure described next, finds the highest similarities between unrelated nodes. During this procedure, we also redistribute the weights:

Procedure 1 (Finding the Similarities Between Unrelated Users)

`Step 1:` Start from a user node, which is examined for possible overlaps with a community C. This user will be denoted as the root. Set $a = 0$ (a variable to keep the current highest similarity).

`Step 2:` Find all the users connected to the root, such that the weight of their corresponding connecting links is $> a$.
`If` there are direct connections to the desired community

- `Step 2.1:` Select the link with the highest weight value, assign it as the *working link* and move to `Step 3`

`else`

- `Step 2.2:` Select the link from the root to an unprocessed node j that has the highest $w_{root,j}$ and mark j as processed. Set j as the root. The chosen link is the *working link*. Move to `Step 3`.

`Step 3:` Update the weight of the working link using the computed product of the previous link (or 1 for the first iteration) and the weight of the working link.

`Step 4:` **`If`** C is reached **`then`**,

- `Step 4.1:` **`If`** the working link weight is $> a$, **`then`**
  - `Step 4.1.a:` Update a
  - `Step 4.1.b:` Exclude from further examination all the links with weight less than a (they will never produce a similarity $> a$).
- `Step 4.2:` Return to the root and set the link leading to the root as the working link.
- `Step 4.3:` **`If`** there is a link from the root to a node j (out of the examined community) such that $w_{root,j} > a$, **`then go to`** `Step 2`, **`else go to`** `Step 5`.

`else return` to `Step 2`.

`Step 5:` Work backwards **`until`** there is no unprocessed node and link.

Let us use the example of Fig. 1 to see how this procedure will compute the similarity between node A and the nodes in community C_3. From Step 1, user A is set as the root. From `Step 2`, there are 4 users connected to the root, namely B, J, F, E (in decreasing order of link weights). The weights of all links are greater than a ($a = 0$ from `Step 1`) and there is no direct connection to a member of C_3. So, according to `Step 2.2`, user B is selected, and $w_{A,B} = 0.96$. User B is marked as processed and becomes the root. The link from A to B is the working link. In `Step 3`, a weight equal to 1 is assigned to the working link (first iteration, there is no previous link), and its newly assigned weight will be $0.96 \times 1 = 0.96$. From `Step 4`, C_3 is not reached, so we have to return to `Step 2`. From the root, user B, there is only one connection to user C. There is no direct connection to C_3, so we move to `Step 2.2`. Since $w_{B,C} = 1 > a$, C is marked as processed and becomes the new root The link from B to C becomes the working link. From `Step 3`, the weight of the previous node is redistributed to the working link, and the product is $0.96 \times 1 = 0.96$. So $w_{B,C} = 0.96$. From `Step 4`, C_3 is not reached, so we repeat `Step 2`. The users connected to the root are D and K. The link weights are greater than a and there is direct connection to C_3. So (`Step 2.1`), we choose the link to user D, since its weight is 0.96, while the weight of link to user K is 0.88. The chosen link is the working link. From `Step 3`, the weight from the previous link is redistributed, so a weight of 0.96 is multiplied with $w_{C,D}$. Thus, the new $w_{C,D} = 0.96 \times 0.96 \approx 0.92$. Now, the condition of `Step 4` is satisfied. The working link's weight is greater than a, so a becomes 0.92. Now, all the links with weights less than 0.92 have to be excluded from further searching, since multiplications with values less than 1 will always be producing similarities less than a. Thus, the algorithm must terminate and the largest similarity of a to a node of C_3 is 0.92.

We now show the importance of eliminating some links in `Step 4.1.b`. Assume that, we let the algorithm continue, thus we ignore `Step 4.1.b`. The algorithm will continue, subsequently assigning as roots nodes F and G. Then, it reaches node H inside C_3 and computes $W_{GH} = 0.77 \times 0.9 \times 0.8 \approx 0.55$. Since C is reached, we move to `Step 4.1` but the condition is false. So we return to the root, G, assign the link connecting F and G as the working link and move to user I. From I, there is direct connection to a member of C_3, namely H. Thus, $w_{I,H} = 0.77 \times 0.9 \times 0.92 \times 0.7 \approx 0.44$. Similarly, the algorithm will compute $w_{E,H} = 0.77 \times 0.9 \times 0.92 \times 0.56 \approx 0.35$. Assuming that the condition $w_{root,j} > a$ of `Step 4.3` is also not checked, the algorithm will take us to `Step 5` and from E we will move to the initial root A, since this is the only unprocessed node, when H is the root. When we reconsider A as root, there are two possible options: to compute the similarity for E to H, $w_{E,H} = 0.58 \times 0.56 \approx 0.32$ and to move to user J that has no further connections. Apparently, `Step 4.1.b` saves us from many unnecessary computations.

Having found the highest similarity between a node and a community, we have to define the conditions under which the node can be considered as member of the community. We define the *Average Community Connectivity* (*ACC*) for a community C as the average of the *NCD*s of all the nodes in the C, that is:

$$ACC_C = \frac{\sum_{i=1}^{n} NCD_i}{n} \tag{2}$$

where n is the number of nodes. For example, the *ACC* of C_3 is $ACC_{C_3} = \frac{0.85+0.9+0.95}{3} = 0.9$.

To allow a user to become a member of the community, we must test if its inclusion will improve the community's *ACC*. For example, we have computed the similarity of A and D, which is 0.92. Since A's similarity to C is greater than 0.9, A can be considered as member of C_3. The checking procedure is described into two steps:

Procedure 2 (Checking a Node's Inclusion to a Community)

`Step 1:` From Procedure 1, find the highest similarity of a node i to a member j of a community C.

`Step 2:` **If** $w_{i,j} > ACC_C$, **then**

`Step 2.1:` Include the node to the community.

`Step 2.2:` Run Procedure 1 for all the nodes on the path from i to j to check for more overlaps.

else check another node.

`Step 2.2` indicates that there is a high probability that more nodes on the path from the examined node i to a community member j can also be added to the community. Generally, if two communities have several common nodes, they can be considered as a larger community.

2.3 *Performance of the Communities Detection Strategy*

To evaluate performance, we have to analyze Procedure 1. Proposition 1 gives us an upper bound for the complexity of this procedure.

Proposition 1 *The time required for Procedure 1 to complete its computations is at most $O(m)$, where m is the number of edges.*

Proof Assume that the procedure starts from a node i. Each node is marked as unprocessed once and each link is processed two times, one during the redistribution step and one in case we return back to the root, and there are still unprocessed nodes. Thus, the total time is dictated by the number of edges and it is limited to this value, since the procedure involves a number of edges (and consequently nodes), which are not examined at all (see `Step 4.1.b` of the procedure). Thus, we can conclude that the proposed method is completed at linear time, which is a great advantage if compared to the times of other well-known strategies [15, 22, 23] among others.

3 Conclusions: Future Work

In this paper, we presented a computational method to find overlaps between communities in a network. Communities are important, since they give a realistic view of a network. Moreover, they can be used to develop forwarding (routing) algorithms (for example see [15]). Our technique is based on a depth-first based search through the network, in combination with a redistribution of weights on the links, to compute similarities between nodes that are not connected via a link.

This work gives rise to several issues that need to be further investigated. First, the adaptability to dynamic networks, where nodes arbitrarily enter and leave. This is a very complex problem, since any newly inserted node may be included in more than one communities, while a node leaving the network will necessarily force some reorganization over the network. Another point of interest is to test the strategy on a real network with real users and data and compare our results with other well-known strategies.

References

1. Clauset, A., M.E. Newman, and C. Moore, 2004. Finding community structure in very large networks. *Physical Review E* 70(6): 066111.
2. Statista 2017. http://www.statista.com/statistics/272014/global-social-networks-ranked-by-number-of-users. Accessed 14 May 2017.
3. Bu, Z., C. Zhang, Z. Xia, and J. Wang. 2013. A fast parallel modularity optimization algorithm (FPMQA) for community detection in online social network. *Knowledge-Based Systems* 50: 246–259.
4. Fortunato, S. 2010. Community detection in graphs. *Physics Reports* 486(3–5): 75–174.
5. Raghavan, U.N., R. Albert, and S. Kumara. 2007. Near linear time algorithm to detect community structures in large-scale networks. *Physical Review E* 76(3): 036106.
6. Gao, W., G. Cao, T.L. Porta, and J. Han. 2013. On exploiting transient social contact patterns for data forwarding in delay-tolerant networks. *IEEE Transactions on Mobile Computing* 12(1): 151–165.
7. Gao, W., Q. Li, B. Zhao, and G. Cao. 2012. Social-aware multicast in disruption-tolerant networks. *IEEE/ACM Transactions on Networking* 20(5): 1553–1566.
8. Hui, P., J. Crowcroft, and E. Yoneki, 2011. Bubble rap: Social-based forwarding in delay-tolerant networks. *IEEE Transactions on Mobile Computing* 10(11): 1576–1589.
9. Lu, Z., X. Sun, Y. Wen, G. Cao, and T.L. Porta. 2015. Algorithms and applications for community detection in weighted networks. *IEEE Transactions on Parallel and Distributed Systems* 26(11): 2916–2926.
10. Bu, Z., Z. Xia, and J. Wang. 2013. A sock puppet detection algorithm on virtual spaces. *Knowledge-Based Systems* 37: 366–377.
11. Newman, M.E. and M. Girvan, 2004. Finding and evaluating community structure in networks. *Physical Review E* 69(2): 026113.
12. Leskovec, J., K.J. Lang, and M. Mahoney. 2010. Empirical comparison of algorithms for network community detection. In *Proceedings of the 19th International Conference on World Wide Web, WWW '10*, 631–640. New York, NY: ACM
13. Lancichinetti, A., and S. Fortunato, 2009. Benchmarks for testing community detection algorithms on directed and weighted graphs with overlapping communities. *Physical Review E* 80(1): 016118.
14. Hui, P., E. Yoneki, S.Y. Chan, and J. Crowcroft. 2007. Distributed community detection in delay tolerant networks. In *Proceedings of 2nd ACM/IEEE International Workshop on Mobility in the Evolving Internet Architecture, MobiArch '07*, vol. 7, 1–7:8. New York, NY: ACM.
15. Lu, Z., Y. Wen, and G. Cao. 2013. Community detection in weighted networks: Algorithms and applications. In *IEEE International Conference on Pervasive Computing and Communications (PerCom)*, 179–184.
16. Newman, M.E. 2004. Analysis of weighted networks. *Physical Review E* 70(5): 056131.
17. Xia, Z., and Z. Bu. 2012. Community detection based on a semantic network. *Knowledge-Based Systems* 26: 30–39
18. Zhao, Z., S. Feng, Q. Wang, J.Z. Huang, G.J. Williams, and J. Fan. 2012. Topic oriented community detection through social objects and link analysis in social networks. *Knowledge-Based Systems* 26: 164–173.
19. Berry, J.W., B. Hendrickson, R.A. LaViolette, and C.A. Phillips. 2011. Tolerating the community detection resolution limit with edge weighting. *Physical Review E* 83(5): 056119.

20. Chen, D., M. Shang, Z. Lv, and Y. Fu. 2010. Detecting overlapping communities of weighted networks via a local algorithm. *Physica A: Statistical Mechanics and its Applications* 389(19): 4177–4187.
21. Gregory, S. 2010. Finding overlapping communities in networks by label propagation. *New Journal of Physics* 12(10): 103018.
22. Nguyen, N.P., T.N. Dinh, S. Tokala, and M.T. Thai. 2011. Overlapping communities in dynamic networks: Their detection and mobile applications. In *Proceedings of the 17th Annual International Conference on Mobile Computing and Networking, MobiCom '11*, 85–96. New York, NY: ACM.
23. Nguyen, N.P., T.N. Dinh, Y. Xuan, and M.T. Thai. 2011. Adaptive algorithms for detecting community structure in dynamic social networks. In *IEEE International Conference on Computer Communications (INFOCOM)*, 2282–2290.
24. Lancichinetti, A., S. Fortunato, and J. Kertész, 2009. Detecting the overlapping and hierarchical community structure in complex networks. *New Journal of Physics* 11(3): 033015.
25. Xie, J., S. Kelley, and B.K. Szymanski. 2013. Overlapping community detection in networks: The state-of-the-art and comparative study. *ACM Computing Surveys* 45(4): 43.
26. Wu, F., and B.A. Huberman. 2004. Finding communities in linear time: A physics approach. *European Physical Journal B* 38(2): 331–338.

PerSubs: A Graph-Based Algorithm for the Identification of Perturbed Subpathways Caused by Complex Diseases

Aristidis G. Vrahatis, Angeliki Rapti, Spyros Sioutas, and Athanasios Tsakalidis

Abstract In the era of Systems Biology and growing flow of omics experimental data from high throughput techniques, experimentalists are in need of more precise pathway-based tools to unravel the inherent complexity of diseases and biological processes. Subpathway-based approaches are the emerging generation of pathway-based analysis elucidating the biological mechanisms under the perspective of local topologies onto a complex pathway network. Towards this orientation, we developed PerSub, a graph-based algorithm which detects subpathways perturbed by a complex disease. The perturbations are imprinted through differentially expressed and co-expressed subpathways as recorded by RNA-seq experiments. Our novel algorithm is applied on data obtained from a real experimental study and the identified subpathways provide biological evidence for the brain aging.

Keywords Systems biology • Subpathway • RNA-seq data • Graph topology

1 Introduction

The recent advances in high-throughput technologies such as microarray, next generation sequencing, mass spectrometry, metabolomics and large scale mutagenesis have produced a massive influx of data regarding genes and their products. Differential expression analysis, i.e., the comparison of expression across conditions, has been a commonly used approach for designating biomarkers, drug targets, and candidates for further research [1]. At this point systems-level methodologies pushed forward the transition of gene-by-gene analysis to signaling pathways and

A.G. Vrahatis (✉) • A. Rapti • A. Tsakalidis
Department of Computer Engineering and Informatics, University of Patras, Patras, Greece
e-mail: agvrahatis@upatras.gr; arapti@ceid.upatras.gr; tsak@ceid.upatras.gr

S. Sioutas
Department of Informatics, Ionian University Corfu, Corfu, Greece
e-mail: sioutas@ionio.gr

P. Vlamos (ed.), *GeNeDis 2016*, Advances in Experimental Medicine and Biology 988, DOI 10.1007/978-3-319-56246-9_17

complex interaction networks, thus these approaches have gained ground in the research field of systems biology [2]. Pathway-based approaches, a flourishing research area of Systems Biology, have become the first choice in complex disease analysis for gaining more delicate biological insights [3–5].

Towards this orientation, initial efforts to relate expression data to pathways were based on 'gene set' analysis. One of the most widely applied pathway analysis methods is the over representation approach (ORA), the first generation in pathway analysis, which compares the number of interesting genes that hit a given pathway with the number of genes expected to hit the given pathway by chance [6]. A statistical model, such as the hypergeometric test, is used to calculate the enrichment significance (P-values). Another prominent method is the Gene Set Enrichment Analysis (GSEA), where the identified genes are ranked based on expression values [7, 8]. Significance of enriched gene sets is determined from a maximum running sum, which is estimated for each gene set by simultaneously parsing the ranked gene list and increasing or decreasing the score based on set membership.

Other methods calculate set-based scores via different metrics and distributions [9, 10]. These methods are categorized into those that compare gene sets relative to others (known as enrichment analysis or competitive approaches) and those that compare individual gene sets across conditions without considering other sets (known as self-contained approaches) [11]. Later approaches incorporated pathway topology into analysis with the scope to test an entire pathway for differential expression and on second level to identify the path that contributes significantly to that differential expression. Studies showed that pathway structure information can provide more delicate biological insights and enables the comprehension of the higher-order functions of the biological system [12]. One widely implemented method is signaling pathway impact analysis (SPIA), which combines a set analysis score with a cumulative pathway score [13]. Activation and inhibitory relationships are considered based on a multiplier on the expression values. The main attribute of SPIA is that it takes into consideration the complete pathway structure and not just the differentially expressed genes.

The next generation in pathway analysis shifted the focus towards subpathways (local area of the entire biological pathway), which represent the underlying biological phenomena more accurately, and have emerged as even more targeted and context-specific molecular candidate communities for the treatment of complex diseases [14]. This emerging generation based on the assumption that the mechanism of complex diseases and biological processes can be optimally described through local topologies within each pathway. Hence, several subpathway-based tools have been developed recently focusing though predominantly on differentially expressed (DE) genes [14–23]. Indicatively, an interesting approach is Differential Expression Analysis for Pathways (DEAP) [15]. The main goal of DEAP is to detect the most differentially subpathway of an overall differentially expressed pathway. DEAP calculates the path within each pathway with the maximum absolute running sum score where catalytic/inhibitory edges are taken as positive/negative summands, while statistical significance is evaluated with the use of a random rotation approach. Another related method is Topology Enrichment Analysis frameworK (TEAK)

[20]. TEAK employs an adapted Clique Percolation Method to extract linear and nonlinear subpathways and scores them using the Bayes Net Toolbox to fit a context specific Gaussian Bayesian network for each subpathway. For an extensive review on (sub)pathway topology-based methods the reader should refer to the review of Mitrea et al. [24].

We developed a novel in-house algorithm for the detection of "perturbed subpathways" (PerSub) in the form of local topologies in pathway networks with differentially expressed linked genes. The algorithm operates in a directed pathway network using KEGG pathway maps and the "perturbed subpathways" are formed starting from a user-defined genes in order to record the perturbation caused the corresponding genes. PerSub extracts subpathways with simultaneously differentially expressed and correlated gene expressions. Specifically, the differential expression status of each gene is evaluated using edgeR, a well-established RNA-seq differential expression analysis tool. Also, a measure based on two multivariate logistic functions is used for the co-expression status since it captures pairs of nodes with similar high positive, negative, or strongly opposed fold change values. A unique feature of PerSubs is that builds a "perturbed subpathway" starting from a Node of Interest (NoI), a user-defined gene. Thus, the user can scan the topology around a specific gene and capture the exact perturbation caused in the network by this gene.

2 Methods

In this work, we present a method to extract perturbed subpathways from pathways taking into account graph topology and differential expression. The Fold Change Interactivity (*FCI*) score is used as weight of edges of the pathway graph. We aim to extract subpathways in the form of a densely connected subgraph around a node of interest based on topological criteria. For this, we follow a "seed growing" approach similarly to [25], where we start from an initial Node of Interest (NoI) and we identify the perturbation caused by this node in the entire pathway network. Users can provide a list of genes of interest, but here we selected as nodes of interest the most differentially expressed genes.

The change in the expression level of a gene is adequately controlled by measuring its fold change activity. Under the perspective of pathway networks, genes are considered as unified interconnected groups. Hence, it is very crucial to examine fold change (relative to a control condition) in terms of interacting genes belonging to a specific pathway. To accomplish this, CHRONOS uses the *FCI* score (Eq. 1) based on two multivariate logistic functions, which captures pairs of nodes with similar high positive, negative, or strongly opposed fold change values [26]. For two connected nodes i and j the *FCI* score is:

$$FCI_{i,j} = \left(1 + c\sum_{k=i,j} e^{-K(f_k - T)}\right)^{-1} - \left(1 + c\sum_{k=i,j} e^{-K(-f_k - T)}\right)^{-1} \quad (1)$$

where f_i and f_j are the log2-fold change values of nodes i and j respectively. C, K denote the parameters controlling the shape of the multivariate logistic distribution, and T is a shifting parameter.

Differential expression analysis of RNA-seq expression profiles was performed by Bioconductor software package edgeR [27] in order to detect the most differentially expressed genes. More specifically, the genewise dispersions are estimated by conditional maximum likelihood, conditioning on the total count for that gene. Also, an empirical Bayes procedure is used in order to shrink the dispersions towards a consensus value, effectively borrowing information between genes. Differential expression is assessed for each gene using an exact test analogous to Fisher's exact test, but adapted for overdispersed data.

In order to extract perturbed subpathways from pathways, we use some graph theoretical properties to determine a densely connected neighborhood of a node. Let G = (V, E) a weighted directed graph, where V is the node set and E the edge set, with w_{vu} denoting the edge weight from node v to node u. With N(v) we represent the neighbors of node v. For a subgraph $S \subseteq G$, the internal degree $N_{INT}(v, S)$ of a node $v \in S$ is defined as the number of edges connecting v with nodes within S and the external degree as the number of edges connecting v with nodes not belonging to S. The weighted internal degree is defined as the sum of weights of internal edges divided by internal degree:

$$NW_{INT}(v, S) = \frac{1}{N_{INT}(v, s)} \sum_{u \in N(v) \cap S} w_{vu} \quad (2)$$

Similarly, we define external weighted degree. The density of a graph is defined as the number of edges divided by the number of all possible edges. The weighted density of a (sub)graph is defined as the sum of all edge weights over the number of all possible edges:

$$DW(G) = \frac{1}{|V|\,(|V| - 1)} \sum_{(v,u) \in E} w_{vu} \quad (3)$$

The algorithm operates on two phases, first the node set is expanded by selecting some of the external neighbors and second the selected node set is pruned. Initially, we start with a set S including only the NoI node s. Then, for each NoI's neighbor $v \in N(s)$ we compute internal and external unweighted and weighted degree. In order to select a highly connected subset, a node v is included in the set S, if it satisfies the following two criteria:

$$\text{Criterion 1}: \frac{N_{INT}(v,S)}{N_{INT}(v,S)+N_{EXT}(v,S)} > a \tag{4}$$

$$\text{Criterion 2}: NW_{INT}(v,S) > NW_{EXT}(v,S) \tag{5}$$

where α is a parameter set to 0.45 for direct neighbors of NoI and 0.75 for other nodes.

In the second phase we aim to obtain a more compact set by maximizing weighted density. For this, we remove one by one nodes until we reach to a maximum value. The order of nodes is determined by magnitude of first criterion, with less significant nodes examined first for removal. The algorithm is iterated for external neighbors of the selected nodes, until no more nodes are added to the set S.

Pseudocode of PerSub Algorithm

Input: NoI, G, $\alpha 1$, $\alpha 2$
Output: final subpathway S
I. S = {NoI} // initialize
II. For each v in S // inclusion step
  a. Find neighbors N(v)
  b. Keep not included neighbors: N(v) = N(v) − S
  c. For every u in N(v)
    i. Calculate N_{INT}, N_{EXT}, NW_{INT}, NW_{EXT}
    ii. If $u \in N(NoI)$
      1. Evaluate if Criterion1 > $\alpha 1$
    iii. Else
      1. Evaluate if Criterion1 > $\alpha 2$
    iv. Evaluate Criterion2
    v. if Criterion1 = true AND Criterion2 = true
      1. Include u: $S = S \cup u$
III. For each v in S ordered by increasing Criterion1 // pruning step
  a. if DW(S-v) > DW(S)
    i. Remove v: S = S - v
IV. Repeat steps II and III until no new nodes added

3 Application in Real Data

We applied our algorithm approach on high throughput sequencing expression profiles data based on a recent study which explores the transcriptional changes in the aging mouse hippocampus [28]. In particular, the authors use specific pathogen free (SPF) C57Bl6/J wild type mice where we kept the sequenced mRNA from 3 months (3M) and 29 months (29M) which considered as the young and old state of the mice respectively. The complete RNA-seq dataset is deposited in NCBI's Gene Expression Omnibus (GEO, http://www.ncbi.nlm.nih.gov/geo/) and

Table 1 Annotation in pathway terms of enriched subpathways after applying PerSub on the real expression data

KEGG ID	KEGG pathway term	P-value (log10)	References
mmu04610	Complement and coagulation cascades	−6	[28]
mmu04514	Cell adhesion molecules (CAMs)	−5	[28]
mmu05416	Viral myocarditis	−5	[28]
mmu05168	Herpes simplex infection	−12	[31]
mmu05010	Alzheimer's disease pathway	−14	[32]
mmu04915	Estrogen signaling pathway	−3	[33]
mmu04210	Apoptosis	−4	[34]
mmu04350	TGF-beta signaling pathway	−4	[34]

is accessible through the accession number GSE61915. Also, the expressions not corresponding to the organism under study were discarded. By applying PerSub, we detected subpathways which contain both differentially expressed and co-expressed associated genes, as to their expression change between their young and old state. All non-metabolic pathway maps of Mus musculus (mmu) were downloaded from KEGG [29] by keeping the genes that we have expression profiles form the corresponding dataset [28]. 287 KEGG pathway maps were downloaded and converted to gene-gene networks based on the CHRONOS package [14].

As is well known, hippocampus which is a major component of the brains of humans and other vertebrates, have a crucial role in the consolidation of information from short-term memory to long-term memory and spatial navigation. The authors of this study [28] succeed to decipher in depth the hippocampal transcriptional program linked to aging and to propose the links between the aging-dependent changes in gene expression and the late onset of Alzheimer Disease (AD). Via PerSub, we achieved to evolve the corresponding results and elucidate the exact mechanisms which cause aging hippocampus, through local topologies in the form of differentially expresses subpathways. More specifically, PerSub was applied using as 'staring' genes, those with the highest FDR adjusted p-value related to their differential expression from the two states (young and old) based on the edgeR package.

PerSub exports differentially expressed subpathways for 47 pathway maps since in the rest maps the algorithm stopped in the starting gene without built a DE subpathway. We use strict thresholds in our criteria in order to extract significant and robust results. Outcomes were tested for enrichment (Benjamini-corrected Fisher exact P-value < 0.05) by means of pathway terms using DAVID tool, a web-based database for annotation, visualization, and integrated discovery [30]. Table 1 illustrates the KEGG pathway terms of the enrichment analysis, with the highest p-value, providing that our results corroborated by the original study and the recent literature.

Focusing on specific subpathways, we observed that our analysis results corroborate to established knowledge about pathways known to be triggered during the aging hippocampus, nevertheless elucidate the exact subpathways to be responsible for this process. An indicative example is the pathway with KEGG terms 'Complement and coagulation cascades' (mmu04610), 'Cell adhesion molecules (CAMs)' (mmu04514) and 'Viral myocarditis' (Viral myocarditis) which are strongly related with the scope of this study [28]. With our approach, we provide robust evidence about the exact area of the pathway topology that affects the aging of hippocampus. More specifically, concerning the first term there are three pathways of complement activation: the classical pathway, the lectin pathway, and the alternative pathway and both generate a crucial enzymatic activity which generates the effector molecules of complement. The CAMs consists of proteins that are located on the cell surface involved in binding with other cells or with the extracellular matrix (ECM) in the process called cell adhesion. Also, myocarditis is a cardiac disease associated with inflammation and injury of the myocardium and may be caused by direct cytopathic effects of virus, a pathologic immune response to persistent virus, or autoimmunity triggered by the viral infection. The analysis of the original study showed that the activation of the above pathways are key features of the aging hippocampus, however we extracted the exactly subpathways that played the major role in the aging hippocampus.

In Alzheimer's disease (AD), the hippocampus is one of the first regions of the brain to suffer damage, thus the hippocampus aging is strongly related with AD. In our analysis, a subpathway with high value was obtained from the pathway with term 'Herpes simplex infection' (mmu05168), which is related to the well-known Herpes simplex virus (HSV), the main cause of herpes infections that lead to the formation of characteristic blistering lesion. Our result is in accordance with the authors [31] since the activation of herpes simplex infection increases the risk of Alzheimer's disease. Two subpathways with high score were exported from our analysis, both of them belonging to the pathway with term 'Alzheimer's disease pathway' (mmu05010). It is evident that this term has direct relation with hippocampus aging [32]. Similarly, a subpathway enriched in pathway term 'Estrogen signaling pathway' was detected, a related with aging term [33]. The authors of another one study deciphered the regional and cell type changes in AD recently [34]. Their analysis exported a list with differentially expressed genes which in their turn were overrepresented on some ontology categories such as 'TGF signaling pathway' (TGF-beta signaling pathway) and 'Apoptosis' (mmu04210). Our analysis elucidated the exact subpathway from these pathway terms which are very relevant with the AD.

We illustrate two significant subpathways in Fig. 1, both including in the KEGG pathway with term 'Alzheimer Disease'. The first includes four genes, Fas (TNF receptor superfamily member 6), Fadd (Fas (TNFRSF6)-associated via death domain), Casp8 (caspase 8) and Bid (BH3 interacting domain death agonist). Study [35] reports Fas gene as a brain-derived nerve growth factor contributing in brain age where there is a continuum increase of stress-related gene expression with increased pro-inflammatory and decreased anti-inflammatory gene signaling in hippocampus. Our analysis deciphered the entire gene cascade that triggered from this crucial gene. The second subpathway starts from the Psen1 (presenilin 1) gene and triggers

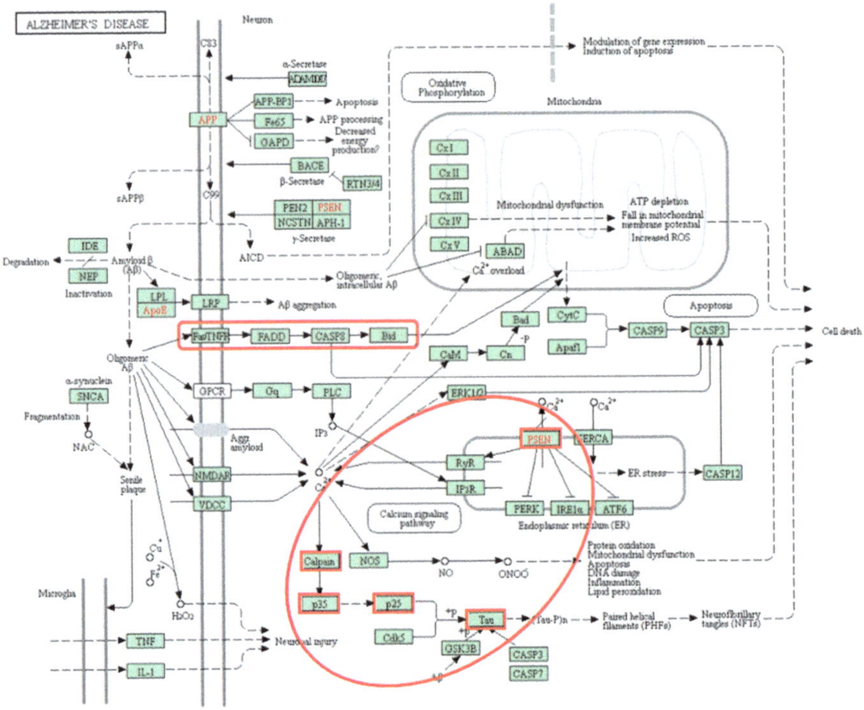

Fig. 1 Two differentially expressed subpathways onto the Alzheimer's disease KEGG pathway map, after applying PerSub in real expression data

the following genes: RyR (ryanodine receptor 3), Capn1 (calpain 1), Cdk5r1 or p35 (cyclin-dependent kinase 5, regulatory subunit 1), Mapt (microtubule-associated protein tau). Capn1 is known for its action since the inhibition of calpains improves memory and synaptic transmission in a mouse model of AD [36] and Psen is related with brain aging [37].

4 Conclusion

We introduced the PerSub, a graph-based algorithm for the identification of subpathways perturbed by a complex disease under study. Starting from a user-defined gene, a subpathway is build recording the perturbation in the pathway network caused by the corresponding gene. The application of PerSubs on transcriptomics expressions data of a real experimental study, provides evidence that our novel algorithm is a robust subpathway-based tool. Its capabilities render PerSub as an essential algorithm for identifying perturbed subpathways, which can be considered as potential network biomarkers for the disease under study.

Acknowledgement Our thanks to C.Caratheodory Research Program from University of Patras, Greece to support this research.

References

1. Walker, M.G. 2001. Drug Target Discovery by Gene Expression Analysis Cell Cycle Genes. *Current Cancer Drug Targets* 1 (1): 73–83.
2. Leung, E.L., Z.W. Cao, Z.H. Jiang, H. Zhou, and L. Liu. 2013. Network-Based Drug Discovery by Integrating Systems Biology and Computational Technologies. *Briefings in Bioinformatics* 14 (4): 491–505.
3. Jin, L., et al. 2014. Pathway-Based Analysis Tools for Complex Diseases: A Review. *Genomics, Proteomics & Bioinformatics* 12 (5): 210–220.
4. Khatri, P., M. Sirota, and A.J. Butte. 2012. Ten Years of Pathway Analysis: Current Approaches and Outstanding Challenges. *PLoS Computational Biology* 8 (2): e1002375.
5. Wang, K., M. Li, and M. Bucan. 2007. Pathway-Based Approaches for Analysis of Genomewide Association Studies. *The American Journal of Human Genetics* 81 (6): 1278–1283.
6. Zhang, S., J. Cao, Y.M. Kong, and R.H. Scheuermann. 2010. GO-Bayes: Gene Ontology-Based Overrepresentation Analysis Using a Bayesian Approach. *Bioinformatics* 26 (7): 905–911.
7. Huang, D.W., B.T. Sherman, and R.A. Lempicki. 2009. Bioinformatics Enrichment Tools: Paths Toward the Comprehensive Functional Analysis of Large Gene Lists. *Nucleic Acids Research* 37 (1): 1–13.
8. Subramanian, A., et al. 2005. Gene Set Enrichment Analysis: A Knowledge-Based Approach for Interpreting Genome-Wide Expression Profiles. *Proceedings of the National Academy of Sciences* 102 (43): 15545–15550.
9. Jiang, Z., and R. Gentleman. 2007. Extensions to Gene Set Enrichment. *Bioinformatics* 23 (3): 306–313.
10. Tian, L., S.A. Greenberg, S.W. Kong, J. Altschuler, I.S. Kohane, and P.J. Park. 2005. Discovering Statistically Significant Pathways in Expression Profiling Studies. *Proceedings of the National Academy of Sciences of the United States of America* 102 (38): 13544–13549.
11. Goeman, J.J., and P. Bühlmann. 2007. Analyzing Gene Expression Data in Terms of Gene Sets: Methodological Issues. *Bioinformatics* 23 (8): 980–987.
12. Emmert-Streib, F., S. Tripathi, and R. de Matos Simoes. 2012. Harnessing the Complexity of Gene Expression Data from Cancer: From Single Gene to Structural Pathway Methods. *Biology Direct* 7 (1): 1.
13. Tarca, A.L., S. Draghici, P. Khatri, S.S. Hassan, P. Mittal, J.S. Kim, C.J. Kim, J.P. Kusanovic, and R. Romero. 2009. A Novel Signaling Pathway Impact Analysis. *Bioinformatics* 25 (1): 75–82.
14. Vrahatis, A.G., K. Dimitrakopoulou, P. Balomenos, A.K. Tsakalidis, and A. Bezerianos. 2016. CHRONOS: A Time-Varying Method for microRNA-Mediated Subpathway Enrichment Analysis. *Bioinformatics* 32 (6): 884–892.
15. Haynes, W.A., R. Higdon, L. Stanberry, D. Collins, and E. Kolker. 2013. Differential Expression Analysis for Pathways. *PLoS Computational Biology* 9 (3): e1002967.
16. Martini, P., G. Sales, E. Calura, S. Cagnin, M. Chiogna, and C. Romualdi. 2014. timeClip: Pathway Analysis for Time Course Data Without Replicates. *BMC Bioinformatics* 15 (5): 1.
17. Nam, S., H.R. Chang, K.T. Kim, M.C. Kook, D. Hong, C. Kwon, H.R. Jung, H.S. Park, G. Powis, H. Liang, and T. Park. 2014. PATHOME: An Algorithm for Accurately Detecting Differentially Expressed Subpathways. *Oncogene* 33 (41): 4941–4951.

18. Li, C., J. Han, Q. Yao, C. Zou, Y. Xu, C. Zhang, D. Shang, L. Zhou, C. Zou, Z. Sun, and J. Li. 2013. Subpathway-GM: Identification of Metabolic Subpathways Via Joint Power of Interesting Genes and Metabolites and Their Topologies Within Pathways. *Nucleic Acids Research* 41 (9): e101–e101.
19. Li, C., et al. 2012. Identifying Disease Related Sub-pathways for Analysis of Genome-Wide Association Studies. *Gene* 503 (1): 101–109.
20. Judeh, T., C. Johnson, A. Kumar, and D. Zhu. 2013. TEAK: Topology Enrichment Analysis Framework for Detecting Activated Biological Subpathways. *Nucleic Acids Research* 41 (3): 1425–1437.
21. Vrahatis, A.G., P. Balomenos, A.K. Tsakalidis, and A. Bezerianos. 2016. DEsubs: An R Package for Flexible Identification of Differentially Expressed Subpathways Using RNA-seq Experiments. *Bioinformatics* 32: btw544.
22. Vrahatis, A.G., G.N. Dimitrakopoulos, A.K. Tsakalidis, and A. Bezerianos. 2015, August. Identifying miRNA-Mediated Signaling Subpathways by Integrating Paired miRNA/mRNA Expression Data with Pathway Topology. 37th Annual International Conference of the IEEE Engineering in Medicine and Biology Society (EMBC), 3997–4000. IEEE.
23. Dimitrakopoulos, G.N., A.G. Vrahatis, P. Balomenos, K. Sgarbas, and A. Bezerianos. 2015, July. Age-Related Subpathway Detection Through Meta-Analysis of Multiple Gene Expression Datasets. IEEE International Conference on Digital Signal Processing (DSP), 539–542. IEEE.
24. Mitrea, C., et al. 2013. Methods and Approaches in the Topology-Based Analysis of Biological Pathways. *Frontiers in Physiology* 4: 278.
25. Maraziotis, I.A., K. Dimitrakopoulou, and A. Bezerianos. 2007. Growing Functional Modules from a Seed Protein Via Integration of Protein Interaction and Gene Expression Data. *BMC Bioinformatics* 8 (1): 1.
26. Kim, Y., et al. 2011. Principal Network Analysis: Identification of Subnetworks Representing Major Dynamics Using Gene Expression Data. *Bioinformatics* 27 (3): 391–398.
27. Robinson, M.D., D.J. McCarthy, and G.K. Smyth. 2010. edgeR: A Bioconductor Package for Differential Expression Analysis of Digital Gene Expression Data. *Bioinformatics* 26 (1): 139–140.
28. Stilling, R.M., E. Benito, M. Gertig, J. Barth, V. Capece, S. Burkhardt, S. Bonn, and A. Fischer. 2014. De-regulation of Gene Expression and Alternative Splicing Affects Distinct Cellular Pathways in the Aging Hippocampus. *Frontiers in Cellular Neuroscience* 8: 373.
29. Kanehisa, M., and S. Goto. 2000. KEGG: Kyoto Encyclopedia of Genes and Genomes. *Nucleic Acids Research* 28 (1): 27–30.
30. Dennis, G., et al. 2003. DAVID: Database for Annotation, Visualization, and Integrated Discovery. *Genome Biology* 4 (9): 1.
31. Lövheim, H., J. Gilthorpe, R. Adolfsson, L.G. Nilsson, and F. Elgh. 2015. Reactivated Herpes Simplex Infection Increases the Risk of Alzheimer's Disease. *Alzheimer's & Dementia* 11 (6): 593–599.
32. Heppner, Frank L., Richard M. Ransohoff, and Burkhard Becher. 2015. Immune Attack: The Role of Inflammation in Alzheimer Disease. *Nature Reviews Neuroscience* 16 (6): 358–372.
33. Cui, J., Y. Shen, and R. Li. 2013. Estrogen Synthesis and Signaling Pathways During Aging: From Periphery to Brain. *Trends in Molecular Medicine* 19 (3): 197–209.
34. Miller, J.A., R.L. Woltjer, J.M. Goodenbour, S. Horvath, and D.H. Geschwind. 2013. Genes and Pathways Underlying Regional and Cell Type Changes in Alzheimer's Disease. *Genome Medicine* 5 (5): 1.
35. Von Bernhardi, R., J.E. Tichauer, and J. Eugenín. 2010. Aging-Dependent Changes of Microglial Cells and Their Relevance for Neurodegenerative Disorders. *Journal of Neurochemistry* 112 (5): 1099–1114.
36. Trinchese, F., et al. 2008. Inhibition of Calpains Improves Memory and Synaptic Transmission in a Mouse Model of Alzheimer Disease. *The Journal of Clinical Investigation* 118 (8): 2796–2807.
37. Yankner, B.A., T. Lu, and P. Loerch. 2008. The Aging Brain. *Annual Review of Pathology: Mechanisms of Disease* 3: 41–66.

MiR-140-3p Downregulation in Association with PDL-1 Overexpression in Many Cancers: A Review from the Literature Using Predictive Bioinformatics Tools

Nikolaos Kapodistrias, Catherine Bobori, and Georgia Theocharopoulou

Abstract Programmed death-ligand 1 (PD-L1) has been speculated to play a critical role in suppression of the immune system and it can be upregulated in cancer cells, which may allow cancers to evade the host immune system. MicroRNAs (miRNAs) are small non-coding RNA molecules (containing about 22 nucleotides), that function in RNA silencing and post-transcriptional regulation of gene expression. MiRNAs were found deregulated (upregulated or downregulated) and implicated in cancer development with various roles which depend on their gene target. Using targetscan web server prediction algorithm, we concluded that miR-140-3p is a targeting mirRNA with conserved consequential pairing of target region for PD-L1. Moreover, by reviewing all the available cancer studies in Pub/Medline about miR-140-3p, was found permanently down regulated. Furthermore, in recent immunotherapy related clinical trials in most cancers, evaluated PD-L1, it is found overexpressed. In the near future, in vitro or in vivo studies need to validate whether there is direct correlation between PD-L1 overexpression and miR-140-3p downregulation as targetscan performed algorithm predicted.

Keywords Programmed cell death • Human cancer • miNA • miRNA • PD-L1

1 Introduction

PD-1 (Programmed cell death protein 1) is a key immune-checkpoint receptor expressed by activated T cells, which mediates immunosuppression. PD-1 functions primarily in peripheral tissues, where T cells may encounter the immunosuppressive PD-1 ligands, namely PD-L1 (B7-H1) and PD-L2 (B7-DC), which are expressed

N. Kapodistrias
Cancer Biobank Center, University of Ioannina, Ioannina, Greece
e-mail: kapi_nik@yahoo.gr

C. Bobori (✉) • G. Theocharopoulou
Department of Informatics, Ionian University, Corfu, Greece
e-mail: p12bobo@ionio.gr; zeta.theo@ionio.gr

P. Vlamos (ed.), *GeNeDis 2016*, Advances in Experimental Medicine and Biology 988, DOI 10.1007/978-3-319-56246-9_18

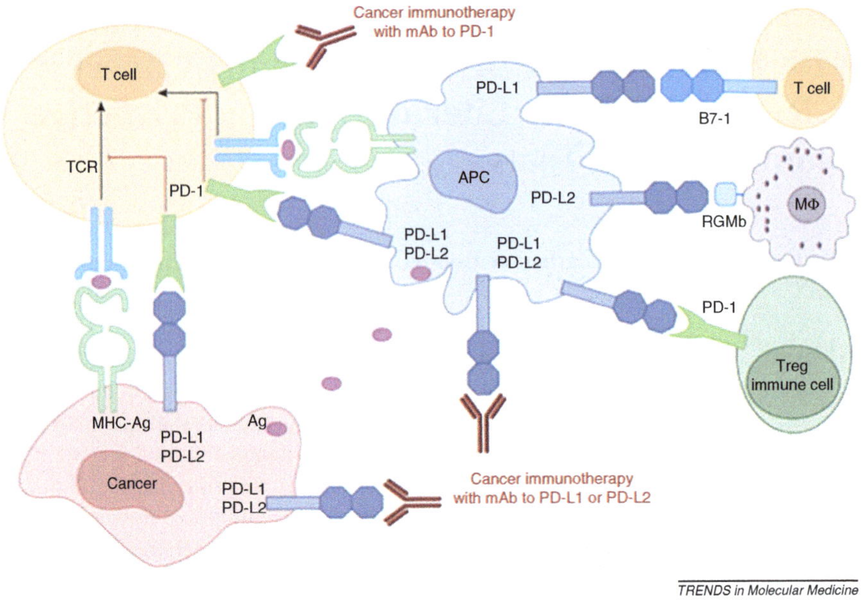

Fig. 1 Human cancer immunotherapy with antibodies to the PD-1 and PD-L1 pathway [5]

by tumor cells, stromal cells, or both [1]. It appears that upregulation of PD-L1 may allow cancers to evade the host's immune system through the transmission of an immunosuppressing signal from tumor cells. Many recent studies show that PD-L1 expression in tumor is associated with impaired survival [2]. An analysis of 196 tumor specimens from patients with renal cell carcinoma found that high tumor expression of PD-L1 was associated with increased tumor aggressiveness and a 4.5-fold increased risk of death [3]. Therefore, many PD-L1 inhibitors are in development as immuno-oncology therapies, showing good results in clinical trials [4] and consist a new treatment approach against cancer treatment (Fig. 1).

An effective strategy of preventing immunosuppression by antibodies blocking the PD-1, or its ligand (PD-L1), was shown in phase I trials by inducing a 30–50% response in several cancer types [6]. More recent phase II and III studies took accelerated approval of anti-PD-1 antibodies for metastatic melanoma (MM) [7, 8], non-small cell lung cancer (NSCLC) [9] and renal cell cancer (RCC) [10]. In another very recent phase II study atezolizumab, an anti-PD-L1 antibody was approved for metastatic bladder cancer treatment [11]. The importance of PD-L1 expression in prognosis and in prediction of treatment outcome, promote identifying other molecules modulators or regulators of it's expression. MicroRNAs (miRNAs) are a class of small, non-coding, endogenous single RNA molecules that play important roles in gene expression through binding to the 3 UTR of the targeted gene mRNA, leading to mRNA cleavage or translational repression [12]. miRNAs are transcribed by RNA polymerase II as part of capped and polyadenylated primary

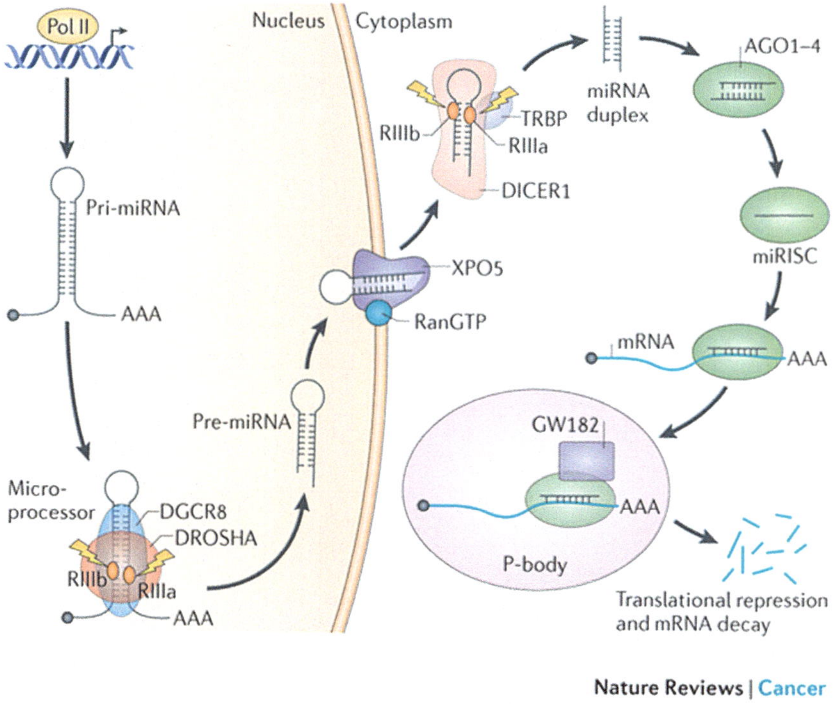

Fig. 2 Overview of miRNA biogenesis pathway [13]

transcripts (pri-miRNAs) that can be either protein-coding or non-coding. The primary transcript is cleaved by the Drosha ribonuclease III enzyme to produce an approximately 70-nt stem-loop precursor miRNA (pre-miRNA), which is further cleaved by the cytoplasmic Dicer ribonuclease to generate the mature miRNA and antisense miRNA star (miRNA*) products. The mature miRNA is incorporated into a RNA-induced silencing complex (RISC), which recognizes targeted mRNAs through imperfect base pairing with the miRNA and most commonly results in translational inhibition or destabilization of the target mRNA (Fig. 2).

Numerous studies show that miRNAs participate in various biological processes, such as cell differentiation, cell growth, apoptosis, death and timing development [14]. In the present study we aim to investigate if PD-L1 is a predicted target by microRnas, using the targetscan web server prediction algorithm. The results will be analyzed by reviewing the available publication studies of the target miRnas in various tumors and PD-L1 expression as well.

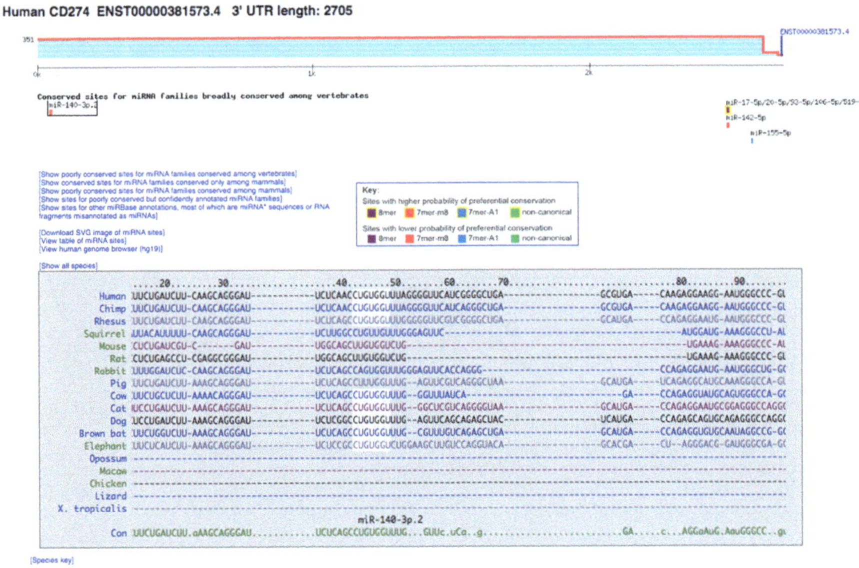

Fig. 3 Results evince PD-L1 as predictive target for mir-140-3p

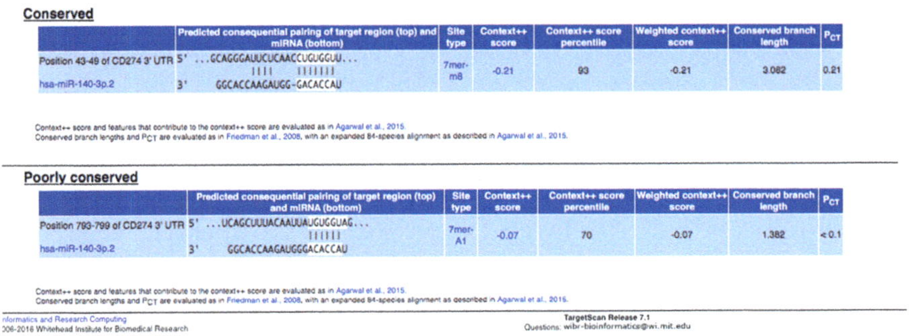

Conserved

	Predicted consequential pairing of target region (top) and miRNA (bottom)	Site type	Context++ score	Context++ score percentile	Weighted context++ score	Conserved branch length	P_{CT}
Position 43-49 of CD274 3' UTR hsa-miR-140-3p.2	5' ...GCAGGGAUUCUCAACCUGUGGUU... 3' GGCACCAAGAUGG-GACACCAU	7mer-m8	-0.21	93	-0.21	3.082	0.21

Context++ score and features that contribute to the context++ score are evaluated as in Agarwal et al., 2015.
Conserved branch lengths and P_{CT} are evaluated as in Friedman et al., 2008, with an expanded 84-species alignment as described in Agarwal et al., 2015.

Poorly conserved

	Predicted consequential pairing of target region (top) and miRNA (bottom)	Site type	Context++ score	Context++ score percentile	Weighted context++ score	Conserved branch length	P_{CT}
Position 793-799 of CD274 3' UTR hsa-miR-140-3p.2	5' ...UCAGCUUUACAAUUAUGUGGUAG... 3' GGCACCAAGAUGGGACACCAU	7mer-A1	-0.07	70	-0.07	1.382	<0.1

Context++ score and features that contribute to the context++ score are evaluated as in Agarwal et al., 2015.
Conserved branch lengths and P_{CT} are evaluated as in Friedman et al., 2008, with an expanded 84-species alignment as described in Agarwal et al., 2015.

nformatics and Research Computing
)06-2016 Whitehead Institute for Biomedical Research
TargetScan Release 7.1
Questions: wibr-bioinformatics@wi.mit.edu

Fig. 4 mir-140-3p predicted conserved consequential pairing of 3-UTR mRNA for PD-L1

2 Method

In bioinformatics, TargetScan is a web server that predicts biological targets of microRNAs by searching for the presence of sites that match the seed region of each miRNA. Performing targetscan, results showed PD-L1 as predictive target for miR-140-3p (Fig. 3).

The prediction targetscan method indicates that miR 140-3p.2 predicted target PD-L1(CD274) with conserved consequential pairing of target region and miRna, in position 43–49, near the stop codon end of the 3-UTR, a region with high efficacy for repression [15] (Fig. 4).

The prediction targetscan method, indicate that miR 140-3p.2 is the only miRna that predicted target PD-L1(CD274) with conserved consequential pairing of target region and miRna, in this position. Searching in Pubmed/Medline database, all available publications about: 'miR-140-3p and cancer', in almost all studies' results that referred to cells or tissues samples analysed, miR140-3p was found down regulated and higher levels of miR 140-3p expression, favored tumor's deprivation. In a study analyzing miR-140-3p in breast cancer cell lines, were found evidence that suggest there is functional synergy between the canonical hsa-miR-140-3p and the newly identified 5'isomiR-140-3p in suppressing growth and progression of breast cancer by simultaneously targeting genes related to differentiation, proliferation and migration. In patients with higher levels there was a trend for better survival [16]. In others studies, miR-140-3p expression level was lower in lung cancer tissues compared to adjacent normal lung cancer tissue and after miR-140-3p was upregulated in A549 or H1299 cells, cell proliferation, invasion and migration was notably attenuated [17, 18]. Moreover, analyzing miRNA expression signatures in angioimmunoblastic T-cell lymphomas (AITLs) of 760 microRNAs in 30 nodal AITL, miR-140-3p, was found to be downregulated to a significant level [19]. The same results found in multiple myeloma cells study, demonstrating miR-140-3p being repressed [20].

In contrast, plasma levels of miR-140-3p were significantly higher in the papillary thyroid carcinoma (PTC) patients than in those with benign nodules or the healthy controls [21]. This is not supported by others studies referred to circulated miRnas in tumors. In a study assessing circulating microRna in blood in five different tumors, miR-140-3p was a potential biomarker, among those selected to distinguish patients with cancer or other diseases from healthy controls. Significantly miR-140-3p was downregulated in all five tumors samples (lung, ovarian, colon, nervous system, hematologic) [22]. Furthermore, from the rest of available studies evaluating miR-140-3p in cancer, published in pubmed database, the conclusion is the same. There was a clear miR-140-3p underexpression in all different tumor samples detected in these studies (cutaneous squamous cell carcinoma, basal cell carcinoma, colon adenocarcinoma, oral squamous cell carcinoma, ovarian carcinoma, lung squamous cell carcinoma, multiple myeloma) [23–29].

Moreover, the expression of PD-l1 in various tumor types has been widely explored in context to predict the most effective immunotherapy for the disease. In studies evaluating PD-L1 expression, as a predictive biomarker in immunotherapy, solid tumors with most increased expression were in concordance with tumor types in which miR-140-3p was found downregulated before [30] (Fig. 5).

Tumor histology	Percent tumor samples expressing PD-L1	Tumor surface expression cutoff for positivity
Melanoma	38%-100%	>10%-5%
RCC	44% (primary tumor), 54% (metastasis)	10%-1%
NSCLC	49%-95%	11%-1%
Bladder	28%-21% (37% squamous, 22% transitional)	5%-1%
Head and neck	66%-31%	10%-5%
Ovarian	87%-87%	10%-5%
Cervical	29%-19%	εως 5%
Colorectal	56% [(MSI-H) vs. 21% (MSS)]-53%	>10%
Glioblastoma multiforme	45%-25%	50%-5%
Breast cancer	18% TNBC (0% ER/PR/HER2+)-50%	5%-1%
Esophageal	44%(squamos)-20%	10%-5%
Hepatocellular carcinoma	25%-15%	5%
Pancreatic	39%	10%
Gastric	42%	NR
Sarcoma	12% (27% GIST)	1%
Unknown primary	28%	5%
Acute myeloid leukemia	37%	5%
Leukemias (various)	57%	NR
B-cell lymphomas	58%	5%
Multiple myeloma	93%	NR

Fig. 5 Table 1

3 Discussion

The association of miR-140-3p with his target gene PD-L1, as was predicted by targetscan prediction algorithm, is more complex and unclear than it seems. In a study at chordomas, miR-140-3p was found upregulated and correlated with recurrence, tumor invasion and worse recurrence-free survival [31]. However, at a recent study analyzing PD-L1 overexpression in tumor infiltrating lymphocytes, in spinal. In spinal chordoma samples, there was no association between PD-L1 with local recurrence-free survival (LRFS) or overall survival (OS), but instead with favorable prognosis [32]. Certainly, its apoptotic role is more elucidated in preclinical or clinical studies that used miR-140-3p, in fields other than in oncology. Preclinically, accessing miR-140-3p with others apoptotic miRnas in myocardial infarction post transplantation with skeletal myoblast in rats, there was apparently increased expression of this miRna, starting apoptosis [33]. Clinically, targeting synovial fibroblast by intraarticular delivery of miRNAs 140-3p and 140-5p can ameliorate autoimmune arthritis [34]. These findings indicate that miR-140-3p has an apparent role in apoptosis, which might implicate additionally PD-L1 regulation. All the above studies suggest that miR-140-3p functions through apoptosis network but the only way to assure that it is inversely correlated or even downregulate PD-L1 expression, are in vivo studies. In a previous study exploring miRNA profiling of normal mammary stem cells and cancer stem-like cells (CSC) from DCIS (ductal carcinoma in situ) tumors, revealed that miR-140 is significantly downregulated in cancer stem-like cells compared with normal stem cells [35]. Moreover, in ER-negative/basal-like DCIS, a cancer stem cell fenotype, restoration of miR-140-3p reduced tumor growth in vivo. Considering that PD-L1 in basal like breast cancer cells is significantly upregulated compare to others breast cancer cells [36] there is an obvious indirect correlation with miR-140-3p downregulation.

4 Conclusions

MiR140-3p, evaluated in many tissue cancers, has been found downregulated. Moreover, his target gene PD-L1, with high clinical importance in most cancer, is overexpressed. In view of the fact that, miR-140-3p is corroborated as apoptotic related miRna, and its ability in pairing conserved with his target PD-L1 in a unique way (near the stop codon end of the 3-UTR) as targetscan performed, makes warrant the confirmation with further in vivo or in vitro experiments as a direct PD-L1 regulator.

References

1. Pardoll, D.M. 2012. The blockade of immune checkpoints in cancer immunotherapy. *Nature Reviews Cancer* 12(4): 252–264.
2. Pyo, J., G. Kang, and J. Kim. 2016. Prognostic role of pd-l1 in malignant solid tumors: A meta-analysis. *The International Journal of Biological Markers* 2:32(1): e68–e74.
3. Thompson, R.H., M.D. Gillett, J.C. Cheville, C.M. Lohse, H. Dong, W.S. Webster, K.G. Krejci, J.R. Lobo, S. Sengupta, L. Chen et al. 2004. Costimulatory b7-h1 in renal cell carcinoma patients: Indicator of tumor aggressiveness and potential therapeutic target. *Proceedings of the National Academy of Sciences of the United States of America* 101(49): 17174–17179.

4. Velcheti, V., K.A. Schalper, D.E. Carvajal, V.K. Anagnostou, K.N. Syrigos, M. Sznol, R.S. Herbst, S.N. Gettinger, L. Chen, and D.L. Rimm. 2014. Programmed death ligand-1 expression in non-small cell lung cancer. *Laboratory Investigation* 94(1): 107–116.
5. Ohaegbulam, K.C., A. Assal, E. Lazar-Molnar, Y. Yao, and X. Zang. 2015. Human cancer immunotherapy with antibodies to the pd-1 and pd-l1 pathway. *Trends in Molecular Medicine* 21(1): 24–33.
6. Topalian, S.L., F.S. Hodi, J.R. Brahmer, S.N. Gettinger, D.C. Smith, D.F. McDermott, J.D. Powderly, R.D. Carvajal, J.A. Sosman, M.B. Atkins et al. 2012. Safety, activity, and immune correlates of anti–pd-1 antibody in cancer. *New England Journal of Medicine* 366(26), 2443–2454.
7. Robert, C., G.V. Long, B. Brady, C. Dutriaux, M. Maio, L. Mortier, J.C. Hassel, P. Rutkowski, C. McNeil, E. Kalinka-Warzocha et al. 2015. Nivolumab in previously untreated melanoma without braf mutation. *New England Journal of Medicine* 372(4): 320–330.
8. Robert, C., J. Schachter, G.V. Long, A. Arance, J.J. Grob, L. Mortier, A. Daud, M.S. Carlino, C. McNeil, M. Lotem et al. 2015. Pembrolizumab versus ipilimumab in advanced melanoma. *New England Journal of Medicine* 372(26): 2521–2532.
9. Garon, E.B., N.A. Rizvi, R. Hui, N. Leighl, A.S. Balmanoukian, J.P. Eder, A. Patnaik, C. Aggarwal, M. Gubens, L. Horn et al. 2015. Pembrolizumab for the treatment of non–small-cell lung cancer. *New England Journal of Medicine* 372(21): 2018–2028.
10. Motzer, R.J., B. Escudier, D.F. McDermott, S. George, H.J. Hammers, S. Srinivas, S.S. Tykodi, J.A. Sosman, G. Procopio, E.R. Plimack et al.: Nivolumab versus everolimus in advanced renal-cell carcinoma. *New England Journal of Medicine* 373(19): 1803–1813.
11. Hoffman-Censits, J.H., P. Grivas, M.S. Van Der Heijden, R. Dreicer, Y. Loriot, M. Retz, N.J. Vogelzang, J.L. Perez-Gracia, A. Rezazadeh, S. Bracarda et al. 2016. Imvigor 210, a phase ii trial of atezolizumab (mpdl3280a) in platinum-treated locally advanced or metastatic urothelial carcinoma (muc). In *ASCO Annual Meeting Proceedings*, vol. 34, 355.
12. Bartel, D.P. 2004. Micrornas: Genomics, biogenesis, mechanism, and function. *Cell* 116(2): 281–297.
13. Lin, S., and R.I. Gregory. 2015. Microrna biogenesis pathways in cancer. *Nature Reviews Cancer* 15(6): 321–333.
14. Ambros, V. 2004. The functions of animal micrornas. *Nature* 431(7006): 350–355.
15. Grimson, A., K.K.H. Farh, W.K. Johnston, P. Garrett-Engele, L.P., Lim, and D.P. Bartel. 2007. Microrna targeting specificity in mammals: Determinants beyond seed pairing. *Molecular Cell* 27(1): 91–105.
16. Salem, O., N. Erdem, J. Jung, E. Münstermann, A. Wörner, H. Wilhelm, S. Wiemann, and C. Körner. 2016. The highly expressed 5-isomir of hsa-mir-140-3p contributes to the tumor-suppressive effects of mir-140 by reducing breast cancer proliferation and migration. *BMC Genomics* 17(1): 566.
17. Kong, X.M., G.H. Zhang, Y.K. Huo, X.H. Zhao, D.W. Cao, S.F. Guo, A.M. Li, and X.R. Zhang. 2015. Microrna-140-3p inhibits proliferation, migration and invasion of lung cancer cells by targeting atp6ap2. *International Journal of Clinical and Experimental Pathology* 8(10): 12845.
18. Dong, W., C. Yao, X. Teng, J. Chai, X. Yang, and B. Li. 2016. Mir-140-3p suppressed cell growth and invasion by downregulating the expression of atp8a1 in non-small cell lung cancer. *Tumor Biology* 37(3): 2973–2985.
19. Reddemann, K., D. Gola, A. Schillert, J. Knief, C. Kuempers, J. Ribbat-Idel, S. Ber, J. Schemme, V. Bernard, N. Gebauer et al. 2015. Dysregulation of micrornas in angioimmunoblastic t-cell lymphoma. *Anticancer Research* 35(4): 2055–2061.
20. Yuan, L., G.C.F. Chan, K.L. Fung, and C.S. Chim, Rankl expression in myeloma cells is regulated by a network involving rankl promoter methylation, dnmt1, microrna and tnfα in the microenvironment. *Biochimica et Biophysica Acta (BBA)-Molecular Cell Research* 1843(9): 1834–1838.
21. Li, M., Q. Song, H. Li, Y. Lou, and L. Wang. 2015. Circulating mir-25-3p and mir-451a may be potential biomarkers for the diagnosis of papillary thyroid carcinoma. *PloS One* 10(7): e0132403.

22. Taguchi, Y., and Y. Murakami. 2013. Principal component analysis based feature extraction approach to identify circulating microrna biomarkers. *PloS One* 8(6): e66714.
23. Sand, M., M. Skrygan, D. Georgas, D. Sand, S.A., Hahn, T. Gambichler, P. Altmeyer, and F.G. Bechara. 2012. Microarray analysis of microrna expression in cutaneous squamous cell carcinoma. *Journal of Dermatological Science* 68(3): 119–126.
24. Sand, M., M. Skrygan, D. Sand, D. Georgas, S. Hahn, T. Gambichler, P. Altmeyer, and F. Bechara. 2012. Expression of micrornas in basal cell carcinoma. *British Journal of Dermatology* 167(4): 847–855.
25. Piepoli, A., F. Tavano, M. Copetti, T. Mazza, O. Palumbo, A. Panza, F.F. Di Mola, V. Pazienza, G. Mazzoccoli, G. Biscaglia et al. 2012. Mirna expression profiles identify drivers in colorectal and pancreatic cancers. *PloS One* 7(3): e33663.
26. Serrano, N.A., C. Xu, Y. Liu, P. Wang, W. Fan, M.P. Upton, J.R. Houck, P. Lohavanichbutr, M. Kao, L.P. Zhao et al. 2012. Integrative analysis in oral squamous cell carcinoma reveals dna copy number–associated mirnas dysregulating target genes. *Otolaryngology–Head and Neck Surgery* 0194599812442490.
27. Miles, G.D., M. Seiler, L. Rodriguez, G. Rajagopal, and B. Bhanot. 2012. Identifying microrna/mrna dysregulations in ovarian cancer. *BMC Research Notes* 5(1): 1.
28. Tan, X., W. Qin, L. Zhang, J. Hang, B. Li, C. Zhang, J. Wan, F. Zhou, K. Shao, Y. Sun et al. 2011. A 5-microrna signature for lung squamous cell carcinoma diagnosis and hsa-mir-31 for prognosis. *Clinical Cancer Research* 17(21): 6802–6811.
29. Lionetti, M., M. Biasiolo, L. Agnelli, K. Todoerti, L. Mosca, S. Fabris, G. Sales, G.L., Deliliers, S., Bicciato, L. Lombardi et al. 2009. Identification of microrna expression patterns and definition of a microrna/mrna regulatory network in distinct molecular groups of multiple myeloma. *Blood* 114(25): e20–e26.
30. Patel, S.P., and R. Kurzrock. 2015. Pd-l1 expression as a predictive biomarker in cancer immunotherapy. *Molecular Cancer Therapeutics* 14(4): 847–856.
31. Zou, M.X., W. Huang, X.B. Wang, G.H., Lv, J. Li, and Y.W. Deng. Identification of mir-140-3p as a marker associated with poor prognosis in spinal chordoma. *International Journal of Clinical and Experimental Pathology* 7(8): 4877.
32. Zou, M.X., A.B. Peng, G.H., Lv, X.B. Wang, J. Li, X.L. She, and Y. Jiang. 2016. Expression of programmed death-1 ligand (pd-l1) in tumor-infiltrating lymphocytes is associated with favorable spinal chordoma prognosis. *American Journal of Translational Research* 8(7): 3274.
33. Liu, Q., G.Q. Du, Z.T. Zhu, C. Zhang, X.W. Sun, J.J. Liu, X. Li, Y.S. Wang, and W.J. Du. 2015. Identification of apoptosis-related micrornas and their target genes in myocardial infarction post-transplantation with skeletal myoblasts. *Journal of Translational Medicine* 13(1): 1.
34. Peng, J.S., S.Y. Chen, C.L. Wu, H.E. Chong, Y.C. Ding, A.L. Shiau, and C.R. Wang. 2016. Amelioration of experimental autoimmune arthritis through targeting of synovial fibroblasts by intraarticular delivery of micrornas 140-3p and 140-5p. *Arthritis & Rheumatology* 68(2): 370–381.
35. Li, Q., Y. Yao, G. Eades, Z. Liu, Y. Zhang, and Q. Zhou. 2014. Downregulation of mir-140 promotes cancer stem cell formation in basal-like early stage breast cancer. *Oncogene* 33(20): 2589–2600.
36. Ali, H., S.E. Glont, F. Blows, E. Provenzano, S.J. Dawson, B. Liu, L. Hiller, J. Dunn, C. Poole, S. Bowden et al. 2015. Pd-l1 protein expression in breast cancer is rare, enriched in basal-like tumours and associated with infiltrating lymphocytes. *Annals of Oncology* 26(7): 1488–1493.

Spontaneous Neuronal Network Persistent Activity in the Neocortex: A(n) (Endo)phenotype of Brain (Patho)physiology

Pavlos Rigas, Leonidas J. Leontiadis, Panagiotis Tsakanikas, and Irini Skaliora

Abstract Abnormal synaptic homeostasis in the cerebral cortex represents a risk factor for both psychiatric and neurodegenerative disorders, from autism and schizophrenia to Alzheimer's disease. Neurons via synapses form recurrent networks that are intrinsically active in the form of oscillating activity, visible at increasingly macroscopic neurophysiological levels: from single cell recordings to the local field potentials (LFPs) to the clinically relevant electroencephalography (EEG). Understanding in animal models the defects at the level of neural circuits is important in order to link molecular and cellular phenotypes with behavioral phenotypes of neurodevelopmental and/or neurodegenerative brain disorders. In this study we introduce the novel idea that recurring persistent network activity (Up states) in the neocortex at the reduced level of the brain slice may be used as an endophenotype of brain disorders that will help us understand not only how local microcircuits of the cortex may be affected in brain diseases, but also when, since an important issue for the design of successful treatment strategies concerns the time window available for intervention.

Keywords Cerebral cortex • Persistent activity • Up states • Endophenotype • Neurodevelopment • Neurodegeneration

Funding sources: Marie Curie Actions—International Re-integration Grants (IRG) (INTRICA-256592), Hellenic Ministry of Education—General Secretariat of Research and Technology, Greece (ESCORT-LS5(1130)).

P. Rigas (✉) • L.J. Leontiadis • P. Tsakanikas • I. Skaliora
Neurophysiology Laboratory, Center for Basic Research, Biomedical Research Foundation of the Academy of Athens (BRFAA), Soranou Efessiou 4, Athens 11527, Greece
e-mail: prigas@bioacademy.gr; pavlosrigas@gmail.com

P. Vlamos (ed.), *GeNeDis 2016*, Advances in Experimental Medicine and Biology 988, DOI 10.1007/978-3-319-56246-9_19

1 Introduction

Abnormal excitatory and inhibitory synapses of the cerebral cortex have emerged as key cellular components in the pathogenesis of several psychiatric and neurodegenerative disorders, including autism spectrum disorders (ASDs), schizophrenia and Alzheimer's disease, all characterized by marked disruptions in information processing and cognition both dependent upon normal cortical function. As accumulating evidence supports altered synaptic morphology, connectivity and/or dynamics in the brains of affected individuals [1–4], it has been suggested that the unique deficits in cognition and behavior associated with these disorders depend on *when* dysregulation of synaptic structure and function occurs across the lifespan [5, 6]. Autism is a neurodevelopmental disorder that affects 1 out of 100–150 children and is characterized by impaired social interaction and communication, absence or delay in language, restrictive and repetitive behavior, with an early diagnosis at around 2–3 years of age [7, 8]. Schizophrenia is a disorder of thought, perceptions of reality, affect and cognition, which affects approximately 0.5–1% of the population and whose symptoms typically emerge in late adolescence or early adulthood [9, 10]. Finally, Alzheimer's disease, the leading form of dementia, which affects approximately 36 million people around the globe, has a typical onset at age 65 and is marked by progressive loss of memory, critical reasoning and other cognitive abilities. Although Alzheimer's disease is a neurodegenerative disease, with amyloid plaques, neurofibrillary tangles and cell death remaining its defining characteristics, much evidence supports synaptic dysfunction as a preceding and contributing insult to eventual neuronal death [11].

The cerebral cortex is organized in local recurrent networks formed by excitatory and inhibitory connectivity that generate stable and self-sustaining periods of persistent activity alternated with periods of no activity, called *Up* and *Down* states, respectively—a prominent feature of the cortical activity during slow wave sleep in vivo. Such activity is maintained in cortical slice preparations, in the absence of sensory inputs or active neuromodulation, indicating that it is chiefly the outcome of intrinsic properties of local networks and hence reflects the 'default' activity of the cortex. The fact that cortical Up states can be sustained in the absence of subcortical or long range inputs, has fuelled studies of Up state activity, in vitro, in brain slices as a model of the basic operation of the cortex whose mechanisms may form the substrate for cognitive functions during attention [12]. In recent work of ours we showed that spontaneous Up states recorded in cortical slices by means of local field potentials can be used to draw the lifetime trajectory of network dynamics of the mouse neocortical microcircuit [13]. This work revealed early postnatal development and ageing as the two time periods with the most intense changes in the function of local circuits, closely following changes in synaptic density in the cortex over the lifespan [14]. Since Up states are synaptically mediated network events that reflect the balance of excitation and inhibition in the neocortex [15–17] we propose that spontaneous Up states at the reduced level of the brain slice may serve as an endophenotype to pinpoint and study the onset of age-dependent, synapse-based, brain disorders.

In order to test this hypothesis we chose autism as an example. As a neurodevelopmental disorder we decided to compare Up states-based developmental trajectories of the primary somatosensory cortex of the whiskers (i.e. barrel cortex, S1BF) in the normal mouse and in a mouse model of autism. We took advantage of the stereotyped and well-characterized development of the murine primary somatosensory cortex of the whiskers to examine cortical maturation. The Fmr1KO mouse, an excellent animal model of the fragile X syndrome (FXS), the most common inherited form of intellectual disability and an identified genetic cause of autism in humans, has been widely employed as an animal model for autism research [18–20]. Importantly, FXS has a well describable line of neurobiology traceable from genes and molecules to cells, synapses and circuits to behavior a fact that renders the Fmr1KO mouse a highly promising animal model for the understanding of neurodevelopmental disorders from genes to behavior, and their treatment [21]. It has been hypothesized that the cognitive and behavioral deficits in autism and intellectual disability stem from altered cortical function with synaptic dysfunction lying at the heart of the pathophysiology [6, 7, 22–25]. In support of this theory respective research has revealed molecular, cellular, synaptic and circuit alterations in the cortex [23, 25–30], however, the bulk of this work has focused on describing *how* the Fmr1KO differs from the normal mouse while the question of *when* cortical development goes awry has remained largely neglected.

Mental retardation and autism spectrum syndromes are often associated with developmental delays of motor and speech skills, as well as excess responses to sensory stimuli, which all rely on intact cortical processing [31–34]. Since normal cortical function requires the functional balance between excitation and inhibition, it has been suggested that impaired excitatory-inhibitory balance may cause a shifted critical period for cortical development in ASDs [35] which could be manifested as a shift/delay in developmental trajectories between normal and autistic cortices [5, 6, 21]. Although a delay in cortical development as the core neurobiology underlying these neurodevelopmental disorders has been theoretically suggested, it has yet to be demonstrated.

2 Materials and Methods

2.1 Animals

C57Bl/6J mice were bred in the animal facility of the Center for Experimental Surgery of the Biomedical Research Foundation of the Academy of Athens. The facility is registered as a breeding and experimental facility according to the Presidential Decree of the Greek Democracy 160/91, which harmonizes the Greek national legislation with the European Council Directive 86/609/EEC on the protection of animals used for experimental and other scientific purposes. Mice were weaned at 21 days old (do), housed in groups of 5–10, in 267 × 483 × 203 mm cages supplied with bedding material and kept at a 12–12 dark-light schedule. Food was provided ad libitum.

2.2 Brain Slice Preparation

Coronal brain slices (400 μm) from primary somatosensory cortex of the whiskers (i.e. barrel cortex, S1BF; Anterior-Posterior from Bregma (A/P): 0.58–1.58 mm, Medial-Lateral (M/L): 2.5–4 mm) were prepared from male mice of ages ranging from 7 to 37do (considering P0 the day of birth). After the mouse was sacrificed (cervical dislocation for mice older than 10 days or decapitation for mice younger than 10 days), the brain was removed and placed in an oxygenated (95% O_2–5% CO_2) ice-cold dissection buffer containing, in mM: KCl 2.14; $NaH_2PO_4{\cdot}H_2O$ 1.47; $NaHCO_3$ 27; $MgSO_4$ 2.2; D-Glucose 10; Sucrose 200; and $CaCl_2{\cdot}2H_2O$ 2; osmolarity (mean ± sd): 298 ± 5 mOsm, pH: 7.4. Slices were cut using a vibratome (VT 1000S, Leica) and placed in a holding chamber with artificial cerebrospinal fluid (ACSF). The ACSF contained, in mM: NaCl 126; KCl 3.53; $NaH_2PO_4{\cdot}H_2O$ 1.25; $NaHCO_3$ 26; $MgSO_4{\cdot}1$; D-Glucose 10 and $CaCl_2{\cdot}2H_2O$ 2 (osmolarity (mean ± sd): 317 ± 4 mOsm, pH: 7.4) and were left to recover at room temperature (RT: 24–26 °C) for at least 1 h before use.

2.3 In Vitro Electrophysiology

Following recovery, slices were transferred to a submerged chamber (Luigs and Neumann), where they were gravity-perfused at high flow rates (10–15 ml/min) to ensure optimal oxygenation of the cortical tissue [36, 37]. Recordings were performed in "in-vivo like" ACSF (composition as above but with 1 mM $CaCl_2$,), since this ionic buffer is thought to better mimic cerebrospinal fluid in vivo [38, 39] and we and others have previously shown that under these conditions cortical slices are spontaneously active [16, 40–43]. Recordings were performed at RT after at least 30 min of incubation in 1 mM [$CaCl_2$] ACSF buffer. To stabilize slices we used a modified submerged type of chamber that included a surface of transparent silicone onto which up to 4 slices could be pinned. The advantage of this modification was that we could perform simultaneous recordings from different ages and/or brain regions thus maximizing the yield and permitting a direct comparison of the different experimental groups under identical conditions.

Spontaneous network activity was assessed by means of local field potential (LFP) recordings (sampled at 10 kHz, band-passed filtered at 1 Hz–3 kHz) which were obtained from cortical layers II/III using low impedance (∼0.5 MΩ) glass pipettes filled with ACSF. Signals were acquired and amplified (MultiClamp 700B, Axon Instruments), digitized (Instrutech, ITC-18) and viewed on-line with appropriate software (Axograph). All reagents and drugs were purchased from Sigma except for KCl and K-gluconate, which were purchased from Carlo Erba Reagents and Fluka, respectively.

2.4 Data Analysis

For visualization and analysis of spontaneous LFP Up states, traces were exported to MatLab format and analyzed with custom-made MatLab scripts (LFPAnalyzer) that automatically detected the LFP events and marked their onsets and offsets. Preprocessing of the recordings included low-pass filtering at 200 Hz with a 3rd order Butterworth filter, and DC offset subtraction (Fig. 1a). Detection of individual Up states (Fig. 1b, upper trace) was performed with the following automated method: (a) the signal is transformed using the Hilbert Transform [44] and simultaneously but independently, using the Short-Time Energy Transform [45], (b) a dynamic and data-driven threshold is then automatically estimated via Gaussian Mixture Modelling [46] and finally (c) the detected signal segments from each transformed signal are combined via an OR logical operation, resulting to the final LFP event.

For each Up state we could *measure* a number of different parameters, such as for example duration, based on the automatically detected onset and offset, or maximal negative peak as a measure of amplitude (Fig. 1b). In the present analysis we measured the rectified area of each Up state as an index of its size which combines duration and amplitude (Fig. 1c). In addition, we *calculated* the occurrence of spontaneous Up states as the number of events divided by the duration of the recording session. Finally, we *calculated* an overall Up state activity index as the product of occurrence * mean rectified area of Up states within each recording.

The comparison of Up state activity in primary somatosensory cortex of the WT and Fmr1KO mouse was performed using the Up state index measure. For both cortices the developmental profile had the form of an inverted U with a peak around which the changes were more pronounced. In order to describe and compare the timing of this developmental progression we used the DataFit Curve Fitting and Data Plotting Software available online by Oakdale Engineering (http://www.oakdaleengr.com) as well as the Matlab Curve Fitting Software, and fitted a peak function $y = a\exp(-0.5((\ln(x/b)/c)^2)$ where x is age in postnatal days. We identified the peak as the time at which the first derivative of this function equals to zero, i.e. $y' = \frac{-a\, e^{-\frac{(\ln(x)-\ln(b))^2}{2c^2}} (\ln(x)-\ln(b))}{c^2 x} = 0 \Rightarrow x = b$ (using the online available derivative calculator (http://www.derivative-calculator.net/). Finally, to determine if the timing of development for the two cortices was significantly different we performed a two-sample t-test using the peak parameter b and its respective standard deviation, on the assumption that Up state index values are normally distributed within each age group. Similarly, a z-test was also performed to test whether Up state index values of the two cortices at specific ages were significantly different. Up state index means and standard deviations could be predicted in Matlab according to the wellness of fit of the peak function model applied.

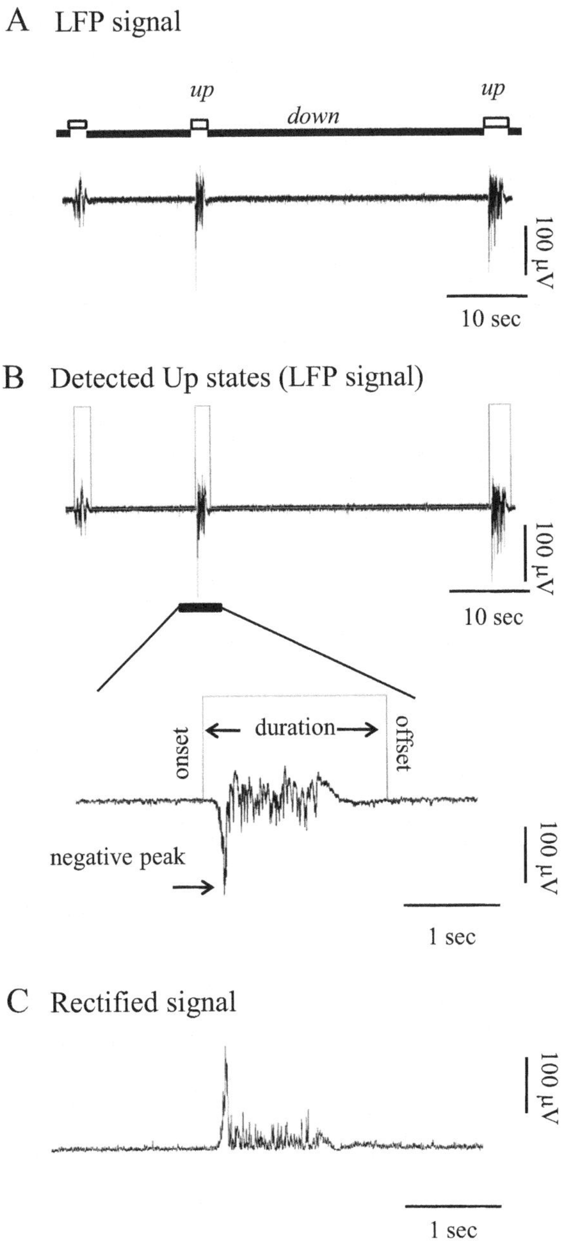

Fig. 1 Detection and quantification of the local field potential (LFP) Up state. (**a**) Continuous LFP recording (1–200 Hz) of spontaneous Up/Down state activity from a cortical slice. (**b**) *Top panel*: Automatically detected LFP Up states of the signal in (**a**) are outlined by *gray rectangles. Bottom panel*: Signal at high magnification provides view of individual Up state. *Gray line* is the automatically detected onset and offset of the event, based on which duration is calculated. (**c**) Rectified signal (absolute valued signal) of the Up state, from which the rectified area is calculated

3 Results

Local field potential (LFP) recordings from layer 2/3 of S1BF cortex revealed periodic bursts of persistent activity containing fast oscillations, or "Up states," interspersed by quiescent periods, or "Down states" (Fig. 1a) as previously described [13, 47]. Recordings were sampled from WT and Fmr1KO mice at various ages over the first five postnatal weeks distributed as follows: WT: 7–37do, total $n_{recordings} = 12$ ($n_{slices} = 12$, $n_{animals} = 10$): *7do*; $n_{slices} = 1$, $n_{animals} = 1$, *9do*; $n_{slices} = 1$, $n_{animals} = 1$, *11do*; $n_{slices} = 1$, $n_{animals} = 1$, *14do*; $n_{slices} = 2$, $n_{animals} = 1$, *15do*; $n_{slices} = 1$, $n_{animals} = 1$, *16do*; $n_{slices} = 1$, $n_{animals} = 1$, *21do*; $n_{slices} = 2$, $n_{animals} = 2$, *30do*; $n_{slices} = 2$, $n_{animals} = 1$, *37do*; $n_{slices} = 1$, $n_{animals} = 1$) and FXS: 9–35do, total $n_{recordings} = 10$ ($n_{slices} = 10$, $n_{animals} = 8$): *9do*; $n_{slices} = 2$, $n_{animals} = 1$, *12do*; $n_{slices} = 1$, $n_{animals} = 1$; *14do*; $n_{slices} = 1$, $n_{animals} = 1$, *15do*; $n_{slices} = 1$, $n_{animals} = 1$, *16do*; $n_{slices} = 1$, $n_{animals} = 1$, *22do*; $n_{slices} = 1$, $n_{animals} = 1$, *31do*; $n_{slices} = 2$, $n_{animals} = 1$, *35do*; $n_{slices} = 1$, $n_{animals} = 1$. For each genotype (WT, FXS) we calculated the Up state index of each recording and plotted it against respective age in order to draw developmental trajectories (Fig. 2). In both cortices, the developmental progression of Up state index has the form of an inverted U with an ascending and a descending phase which was well-fitted by peak functions (see Sect. 2): WT cortex: Up State Index = $0.38 \times \exp(-0.5 \times (\ln(\text{Age}/19.21)/0.27)^2$, curve-fit $R^2 = 0.61$, $p < 0.05$, FXS cortex: Up State Index = $0.63 \times \exp(-0.5 \times (\ln(\text{Age}/22.22)/0.29)^2$, curve fit $R^2 = 0.89$; $p < 0.05$. This allowed us to identify and compare the peaks of the two trajectories. We found that, in line with previous work of ours [13], the WT developmental trajectory peaked around postnatal day 19 (19.21 ± 2.23 do), however the respective peak for the FXS cortex was significantly delayed by 3 days (22.22 ± 1.36 do) ($z_{(22)} = -3.89$, $p < 0.001$, z-test) in support of a developmental shift/delay in FXS.

Besides comparing the age at which Up state activity peaks, developmental trajectories also allow us to predict the mean and standard deviation of the Up state index at any age of interest which we could then compare between the WT and FXS mouse. Interestingly, according to Fig. 2 the development of cortical Up states in the two mice begin to deviate at around P15, in other words after the onset of whisking and somatosensory experience [48]. Interestingly, in two-weeks old rodents, which corresponds to the human age of a toddler [49, 50], the intracortical circuitry undergoes a rapid phase of experience/activity dependent development [51–54]. In order to pinpoint at which specific age the two mice become phenotypically distinct, based on Up state development, we calculated and compared mean values and standard deviations of the Up state index at P15, P16, P17, P18, P19 and P20. We found that while the two mice begin to deviate at around P15 they become significantly different at the age of P18 (*P15*: 0.25 ± 0.11 (WT) vs. 0.26 ± 0.10 (FXS), $z_{(22)} = -0.04$, $p = 0.97$; *P16*: 0.31 ± 0.11 (WT) vs. 0.34 ± 0.11 (FXS), $z_{(22)} = -0.65$, $p = 0.52$; *P17*: 0.35 ± 0.13 (WT) vs. 0.42 ± 0.11 (FXS), $z_{(22)} = -1.35$, $p = 0.18$; *P18*: 0.37 ± 0.14 (WT) vs. 0.49 ± 0.13 (FXS),

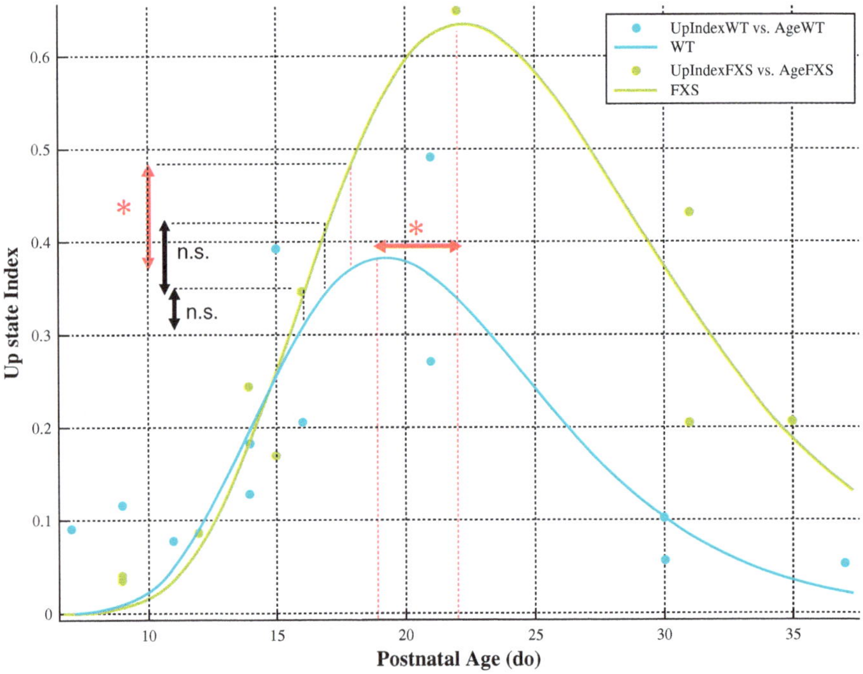

Fig. 2 Comparing developmental trajectories of the WT and FXS cortices. Cortical Up state index values of the normal (WT) and FXS mouse were plotted as a continuum from early postnatal age till the end of the fifth postnatal week of life. Data points, each representing the mean Up state index at a given age, were fitted by peak functions (WT cortex: Up State Index = 0.38 × exp(−0.5 × (ln(Age/19.21)/0.27)^2, curve-fit $R^2 = 0.61$, $p < 0.05$, FXS cortex: Up State Index = 0.63 × exp(−0.5 × (ln(Age/22.22)/0.29)^2, curve fit $R^2 = 0.89$; $p < 0.05$.). *Red asterisks* indicate statistically significant differences, while n.s. stands for non-significant

$z_{(22)} = -2.04$, p = 0.042 <0.05; *P19*: 0.38 ± 0.15 (WT) vs. 0.55 ± 0.14(FXS), $z_{(22)} = -2.68$, p = 0.007 <0.01; *P20*: 0.38 ± 0.15 (WT) vs. 0.59 ± 0.16(FXS), $z_{(22)} = -3.28$, p = 0.001). In addition we compared Up state indices between WT and FXS mice at the postnatal ages of P30 and P40 and we found that respective values in the FXS remained significantly higher than the normal mouse (*P30*: 0.24 ± 0.15 (WT) vs. 0.58 ± 0.15 (FXS), $z_{(22)} = -5.37$, p < 0.001; *P40*: 0.10 ± 0.14 (WT) vs. 0.49 ± 0.13 (FXS), $z_{(22)} = -5.11$, p < 0.001) which is in good agreement with previous work showing enhanced network excitability in FXS [55, 56]. However, as opposed to the few previous attempts made in the past to draw and compare developmental trajectories in the FXS and WT mouse based on molecular, cellular and synaptic parameters, which showed transient developmental differences between the two mice [27, 57], in our study differences in network dynamics once established are sustained at least through the range of ages that we tested. Sustained, non-transient, differences in the development of FXS is in line with both previous work using Up states [56] as well as the reality that autism is a life-long psychiatric disorder with an early onset. These results stress the necessity for research that will further our understanding at the integrated level of the local microcircuit since the latter cannot be necessarily predicted by changes of its individual constituents.

4 Discussion

Although there is an abundance of published information currently available on the genetic, molecular, cellular and synaptic changes that take place in neurodevelopmental and neurodegenerative disorders, much less is known on whether and to what extent these changes actually contribute to higher levels of organization such as the neuronal network, as a common *final functional pathway* that contributes to the pathophysiology of the disease and ultimately respective behavior. This limiting gap of knowledge is due to (1) the fact that research of mental disorders is often complicated by behavior, thus stressing for the need to study simpler (behavior independent) biological markers or "endophenotypes" of the disorder and (2) the lack of such appropriate physiological markers to study. By definition, an endophenotype can be the biological manifestation of a disease at a *reduced* level of biological organization as opposed to the macro-level of behaviour [58–62]. Thus, in order for biological research of mental disorders to proceed, it is essential to 'decompose' the disorder into simpler parameters that can serve as its *endophenotypes*. In this perspective, studying the activity of the cortical microcircuitry may serve to fill the cell and molecules to behavior gap in our understanding of the pathology underlying sever psychiatric disorders. In addition, since it's likely that entirely new properties emerge at higher levels of neuronal organization (networks) compare to those of the constituent parts (molecules, cells and synapses) [63, 64], not only can cortical microcircuits not be understood in terms of a mere extrapolation of the properties of their particles, but rather the understanding of their function will lead to new directions of integrative knowledge.

Understanding the defects at the level of neural circuits is important in order to link molecular and cellular phenotypes with behavioral phenotypes of neurodevelopmental and/or neurodegenerative brain disorders. In this perspective, we propose the idea that spontaneous persistent network activity in the neocortex (Up states) may serve as a model to study network dynamics in both development and maturation and/or ageing since Up states change systematically with cortical development and ageing [13, 47]. This spontaneous activity of the cortex is of particular neurobiological and clinical interest. Up/Down states consist the 'default' activity of the cortex reflecting its hardwiring as shaped by genes and experience, the intrinsic properties of its cells and the dynamics of their synapses; and they form the background upon which incoming sensory stimuli interact determining cortical responses and behavior. Moreover, Up/Down states are cellular correlates of the slow-oscillation the electroencephalographic (EEG) hallmark of quiescent states of the brain such as deep, non-REM sleep, anesthesia and quiet wakefulness (rest)

[65, 66]. In humans, recordings of spontaneous network cortical activity during rest are employed for the discovery of biomarkers of autism, schizophrenia and Alzheimer's disease [67–70].

In this study we introduced the novel idea that spontaneous neuronal network persistent activity (Up states) may help us understand not only *how* local micro-circuits of the cortex may be affected in brain disorders, but also *when* and we animated this idea using an example of a neurodevelopmental disorder. Pinpointing when development (and, respectively, ageing) deviates from normal is important for several reasons: (1) There is evidence that although dysregulation of synaptic structure and function may underlie a number of diverse psychiatric disorders manifested at distinct stages of life, however it may be when these disruptions occur across the lifespan that correlates them to unique cognitive and behavioral deficiencies, and therefore disorders [5]. (2) It may provide insight into when therapeutic intervention would be most effective in preventing the emergence of defected phenotypes [6, 21]. (3) It will contribute to our understanding of the role of genes, molecules and cells of appropriate and well described animal models in normal cortical development and ageing, and (4) Finally understanding the underlying biology at the time of functional/physiological phenotype onset may better describe brain disorders which often rely on neuropathological studies and descriptions performed at advanced stages of the disease which in turn likely reflect an "endpoint" of the disease making it difficult to distinguish between cause, consequence, compensation or confound.

References

1. Selemon, L.D., and P.S. Goldman-Rakic. 1999. The Reduced Neuropil Hypothesis: A Circuit Based Model of Schizophrenia. *Biological Psychiatry* 45 (1): 17–25.
2. Hutsler, J.J., and H. Zhang. 2010. Increased Dendritic Spine Densities on Cortical Projection Neurons in Autism Spectrum Disorders. *Brain Research* 1309: 83–94.
3. Glantz, L.A., and D.A. Lewis. 2000. Decreased Dendritic Spine Density on Prefrontal Cortical Pyramidal Neurons in Schizophrenia. *Archives of General Psychiatry* 57 (1): 65–73.
4. Tackenberg, C., A. Ghori, and R. Brandt. 2009. Thin, Stubby or Mushroom: Spine Pathology in Alzheimer's Disease. *Current Alzheimer Research* 6 (3): 261–268.
5. Penzes, P., M.E. Cahill, K.A. Jones, J.E. VanLeeuwen, and K.M. Woolfrey. 2011. Dendritic Spine Pathology in Neuropsychiatric Disorders. *Nature Neuroscience* 14 (3): 285–293.
6. Penzes, P., A. Buonanno, M. Passafaro, C. Sala, and R.A. Sweet. 2013. Developmental Vulnerability of Synapses and Circuits Associated with Neuropsychiatric Disorders. *Journal of Neurochemistry* 126 (2): 165–182.
7. Toro, R., M. Konyukh, R. Delorme, C. Leblond, P. Chaste, F. Fauchereau, M. Coleman, M. Leboyer, C. Gillberg, and T. Bourgeron. 2010. Key Role for Gene Dosage and Synaptic Homeostasis in Autism Spectrum Disorders. *Trends in Genetics* 26 (8): 363–372.
8. Fombonne, E. 2005. Epidemiology of Autistic Disorder and Other Pervasive Developmental Disorders. *The Journal of Clinical Psychiatry* 66 (Suppl 10): 3–8.
9. Wilson, T.W., O.O. Hernandez, R.M. Asherin, P.D. Teale, M.L. Reite, and D.C. Rojas. 2008. Cortical Gamma Generators Suggest Abnormal Auditory Circuitry in Early-Onset Psychosis. *Cerebral Cortex* 18: 371–378.

10. McCarley, R.W., M.A. Niznikiewicz, D.F. Salisbury, P.G. Nestor, B.F. O'Donnell, Y. Hirayasu, H. Grunze, R.W. Greene, and M.E. Shenton. 1999. Cognitive Dysfunction in Schizophrenia: Unifying Basic Research and Clinical Aspects. *European Archives of Psychiatry and Clinical Neuroscience* 249 (Suppl. 4): IV/69–IV/82.
11. Selkoe, D.J. 2002. Alzheimer's Disease is a Synaptic Failure. *Science* 298 (5594): 789–791.
12. Castro-Alamancos, M.A. 2009. Cortical Up and Activated States: Implications for Sensory Information Processing. *The Neuroscientist* 15 (6): 625–634.
13. Rigas, P., D.A. Adamos, C. Sigalas, P. Tsakanikas, N.A. Laskaris, and I. Skaliora. 2015. Spontaneous Up States In Vitro: A Single-Metric Index of the Functional Maturation and Regional Differentiation of the Cerebral Cortex. *Frontiers in Neural Circuits* 9: 59.
14. Huttenlocher, P.R. 1979. Synaptic Density in Human Frontal Cortex - Developmental Changes and Effects of Aging. *Brain Research* 163 (2): 195–205.
15. Hasenstaub, A., Y. Shu, B. Haider, U. Kraushaar, A. Duque, and D.A. McCormick. 2005. Inhibitory Postsynaptic Potentials Carry Synchronized Frequency Information in Active Cortical Networks. *Neuron* 47 (3): 423–435.
16. Sanchez-Vives, M.V., and D.A. McCormick. 2000. Cellular and Network Mechanisms of Rhythmic Recurrent Activity in Neocortex. *Nature Neuroscience* 3 (10): 1027–1034.
17. Shu, Y., A. Hasenstaub, and D.A. McCormick. 2003. Turning On and Off Recurrent Balanced Cortical Activity. *Nature* 423 (6937): 288–293.
18. Bakker, C.E., C. Verheij, R. Willemsen, R. van der Helm, F. Oerlemans, M. Vermey, et al. 1994. Fmr1 Knockout Mice: A Model to Study Fragile X Mental Retardation. The Dutch-Belgian Fragile X Consortium. *Cell* 78 (1): 23–33.
19. Kaufmann, W.E., R. Cortell, A.S. Kau, I. Bukelis, E. Tierney, R.M. Gray, C. Cox, G.T. Capone, and P. Stanard. 2004. Autism Spectrum Disorder in Fragile X Syndrome: Communication, Social Interaction, and Specific Behaviors. *American Journal of Medical Genetics. Part A* 129A (3): 225–234.
20. Hagerman, R.J., M.Y. Ono, and P.J. Hagerman. 2005. Recent Advances in Fragile X: A Model for Autism and Neurodegeneration. *Current Opinion in Psychiatry* 18 (5): 490–496.
21. Krueger, D.D., and M.F. Bear. 2011. Toward Fulfilling the Promise of Molecular Medicine in Fragile X Syndrome. *Annual Review of Medicine* 62: 411–429.
22. Geschwind, D.H., and P. Levitt. 2007. Autism Spectrum Disorders: Developmental Disconnection Syndromes. *Current Opinion in Neurobiology* 17 (1): 103–111.
23. Huber, K.M., S.M. Gallagher, S.T. Warren, and M.F. Bear. 2002. Altered Synaptic Plasticity in a Mouse Model of Fragile X Mental Retardation. *Proceedings of the National Academy of Sciences of the United States of America* 99 (11): 7746–7750.
24. Bear, M.F., K.M. Huber, and S.T. Warren. 2004. The mGluR Theory of Fragile X Mental Retardation. *Trends in Neurosciences* 27 (7): 370–377.
25. Irwin, S.A., B. Patel, M. Idupulapati, J.B. Harris, R.A. Crisostomo, B.P. Larsen, F. Kooy, P.J. Willems, P. Cras, P.B. Kozlowski, R.A. Swain, I.J. Weiler, and W.T. Greenough. 2001. Abnormal Dendritic Spine Characteristics in the Temporal and Visual Cortices of Patients with Fragile-X Syndrome: A Quantitative Examination. *American Journal of Medical Genetics* 98 (2): 161–167.
26. Pfeiffer, B.E., and K.M. Huber. 2007. Fragile X Mental Retardation Protein Induces Synapse Loss Through Acute Postsynaptic Translational Regulation. *The Journal of Neuroscience* 27 (12): 3120–3130.
27. Bureau, I., G.M. Shepherd, and K. Svoboda. 2008. Circuit and Plasticity Defects in the Developing Somatosensory Cortex of FMR1 Knock-Out Mice. *The Journal of Neuroscience* 28 (20): 5178–5188.
28. Chattopadhyaya, B., and G.D. Cristo. 2012. GABAergic Circuit Dysfunctions in Neurodevelopmental Disorders. *Frontiers in Psychiatry* 3: 51.
29. Krueger, D.D., E.K. Osterweil, S.P. Chen, L.D. Tye, and M.F. Bear. 2011. Cognitive Dysfunction and Prefrontal Synaptic Abnormalities in a Mouse Model of Fragile X Syndrome. *Proceedings of the National Academy of Sciences of the United States of America* 108 (6): 2587–2592.

30. Bassell, G.J., and S.T. Warren. 2008. Fragile X Syndrome: Loss of Local mRNA Regulation Alters Synaptic Development and Function. *Neuron* 60 (2): 201–214.
31. Ess, K.C. 2006. The Neurobiology of Tuberous Sclerosis Complex. *Seminars in Pediatric Neurology* 13 (1): 37–42.
32. Wiesner, G.L., S.B. Cassidy, S.J. Grimes, A.L. Matthews, and L.S. Acheson. 2004. Clinical Consult: Developmental Delay/Fragile X Syndrome. *Primary Care* 31 (3): 621–625. x.
33. Guerrini, R., R. Carrozzo, R. Rinaldi, and P. Bonanni. 2003. Angelman Syndrome: Etiology, Clinical Features, Diagnosis, and Management of Symptoms. *Paediatric Drugs* 5 (10): 647–661.
34. Hagerman, R.J., E. Berry-Kravis, W.E. Kaufmann, M.Y. Ono, N. Tartaglia, A. Lachiewicz, R. Kronk, C. Delahunty, D. Hessl, J. Visootsak, J. Picker, L. Gane, and M. Tranfaglia. 2009. Advances in the Treatment of Fragile X Syndrome. *Pediatrics* 123 (1): 378–390.
35. Bourgeron, T. 2009. A Synaptic Trek to Autism. *Current Opinion in Neurobiology* 19 (2): 231–234.
36. Hajos, N., T.J. Ellender, R. Zemankovics, E.O. Mann, R. Exley, S.J. Cragg, T.F. Freund, and O. Paulsen. 2009. Maintaining Network Activity in Submerged Hippocampal Slices: Importance of Oxygen Supply. *The European Journal of Neuroscience* 29 (2): 319–327.
37. Bregestovski, P., and C. Bernard. 2012. Excitatory GABA: How a Correct Observation May Turn Out to be an Experimental Artifact. *Frontiers in Pharmacology* 3: 65.
38. Somjen, G.G. 2004. *Ions in the Brain*. Oxford: Oxford University Press.
39. Fishman, R.A. 1992. *Cerebrospinal Fluid in Diseases of the Nervous System*. Philadelphia, PA: Elsevier Health Sciences.
40. Fanselow, E.E., and B.W. Connors. 2010. The Roles of Somatostatin-Expressing (GIN) and Fast-Spiking Inhibitory Interneurons in UP-DOWN States of Mouse Neocortex. *Journal of Neurophysiology* 104 (2): 596–606.
41. MacLean, J.N., B.O. Watson, G.B. Aaron, and R. Yuste. 2005. Internal Dynamics Determine the Cortical Response to Thalamic Stimulation. *Neuron* 48 (5): 811–823.
42. Mann, E.O., M.M. Kohl, and O. Paulsen. 2009. Distinct Roles of GABA(A) and GABA(B) Receptors in Balancing and Terminating Persistent Cortical Activity. *The Journal of Neuroscience* 29 (23): 7513–7518.
43. Rigas, P., and M.A. Castro-Alamancos. 2007. Thalamocortical Up States: Differential Effects of Intrinsic and Extrinsic Cortical Inputs on Persistent Activity. *The Journal of Neuroscience* 27 (16): 4261–4272.
44. Oppenheim, A.V., and R.W. Schafer. 1998. *Discrete-Time Signal Processing*. 2nd ed. Englewood Cliffs, NJ: Prentice-Hall.
45. Jalil, M., F.A. Butt, and A. Malik. 2013. Short-Time Energy, Magnitude, Zero Crossing Rate and Autocorrelation Measurement for Discriminating Voiced and Unvoiced Segments of Speech Signals. In *Technological Advances in Electrical, Electronics and Computer Engineering 2013*, 208–212. IEEE.
46. McLachlan, G.J., and D. Peel. 2000. *Finite Mixture Models*. New York: Wiley.
47. Sigalas, C., P. Rigas, P. Tsakanikas, and I. Skaliora. 2015. High-Affinity Nicotinic Receptors Modulate Spontaneous Cortical Up States In Vitro. *The Journal of Neuroscience* 35 (32): 11196–11208.
48. Erzurumlu, R.S., and P. Gaspar. 2012. Development and Critical Period Plasticity of the Barrel Cortex. *The European Journal of Neuroscience* 35 (10): 1540–1553.
49. Nehlig, A., and P. Vert. 1997. Cerebral Metabolic Consequences of Neonatal Pathologies in the Immature Rat. *Acta Paediatrica Japonica* 39 (Suppl 1): S26–S32.
50. Velisek, L., and S.L. Moshe. 2002. Effects of Brief Seizures During Development. *Progress in Brain Research* 135: 355–364.
51. Stern, E.A., M. Maravall, and K. Svoboda. 2001. Rapid Development and Plasticity of Layer 2/3 Maps in Rat Barrel Cortex In Vivo. *Neuron* 31 (2): 305–315.
52. Bender, K.J., C.B. Allen, V.A. Bender, and D.E. Feldman. 2006. Synaptic Basis for Whisker Deprivation-Induced Synaptic Depression in Rat Somatosensory Cortex. *The Journal of Neuroscience* 26 (16): 4155–4165.

53. Bureau, I., G.M. Shepherd, and K. Svoboda. 2004. Precise Development of Functional and Anatomical Columns in the Neocortex. *Neuron* 42 (5): 789–801.
54. Shepherd, G.M., T.A. Pologruto, and K. Svoboda. 2003. Circuit Analysis of Experience-Dependent Plasticity in the Developing Rat Barrel Cortex. *Neuron* 38 (2): 277–289.
55. Gibson, J.R., A.F. Bartley, S.A. Hays, and K.M. Huber. 2008. Imbalance of Neocortical Excitation and Inhibition and Altered UP States Reflect Network Hyperexcitability in the Mouse Model of Fragile X Syndrome. *Journal of Neurophysiology* 100 (5): 2615–2626.
56. Hays, S.A., K.M. Huber, and J.R. Gibson. 2011. Altered Neocortical Rhythmic Activity States in Fmr1 KO Mice Are Due to Enhanced mGluR5 Signaling and Involve Changes in Excitatory Circuitry. *The Journal of Neuroscience* 31 (40): 14223–14234.
57. Till, S.M., L.S. Wijetunge, V.G. Seidel, E. Harlow, A.K. Wright, C. Bagni, A. Contractor, T.H. Gillingwater, and P.C. Kind. 2012. Altered Maturation of the Primary Somatosensory Cortex in a Mouse Model of Fragile X Syndrome. *Human Molecular Genetics* 21 (10): 2143–2156.
58. Gould, T.D., and I.I. Gottesman. 2006. Psychiatric Endophenotypes and the Development of Valid Animal Models. *Genes, Brain, and Behavior* 5 (2): 113–119.
59. Gottesman, I.I., and T.D. Gould. 2003. The Endophenotype Concept in Psychiatry: Etymology and Strategic Intentions. *The American Journal of Psychiatry* 160 (4): 636–645.
60. Hasler, G., W.C. Drevets, H.K. Manji, and D.S. Charney. 2004. Discovering Endophenotypes for Major Depression. *Neuropsychopharmacology* 29 (10): 1765–1781.
61. Gottesman, I.I., and J. Shields. 1973. Genetic Theorizing and Schizophrenia. *The British Journal of Psychiatry* 122 (566): 15–30.
62. Almasy, L., and J. Blangero. 2001. Endophenotypes as Quantitative Risk Factors for Psychiatric Disease: Rationale and Study Design. *American Journal of Medical Genetics* 105 (1): 42–44.
63. Anderson, P.W. 1972. More is Different. *Science* 177.
64. Alivisatos, A.P., M. Chun, G.M. Church, R.J. Greenspan, M.L. Roukes, and R. Yuste. 2012. The Brain Activity Map Project and the Challenge of Functional Connectomics. *Neuron* 74 (6): 970–974.
65. Steriade, M., A. Nunez, and F. Amzica. 1993. A Novel Slow (<1 Hz) Oscillation of Neocortical Neurons In Vivo: Depolarizing and Hyperpolarizing Components. *The Journal of Neuroscience* 13 (8): 3252–3265.
66. Crochet, S., and C.C. Petersen. 2006. Correlating Whisker Behavior with Membrane Potential in Barrel Cortex of Awake Mice. *Nature Neuroscience* 9 (5): 608–610.
67. Gandal, M.J., J.C. Edgar, R.S. Ehrlichman, M. Mehta, T.P. Roberts, and S.J. Siegel. 2010. Validating Gamma Oscillations and Delayed Auditory Responses as Translational Biomarkers of Autism. *Biological Psychiatry* 68 (12): 1100–1106.
68. Kissler, J., M.M. Muller, T. Fehr, B. Rockstroh, and T. Elbert. 2000. MEG Gamma Band Activity in Schizophrenia Patients and Healthy Subjects in a Mental Arithmetic Task and at Rest. *Clinical Neurophysiology* 111 (11): 2079–2087.
69. Sorg, C., V. Riedl, M. Muhlau, V.D. Calhoun, T. Eichele, L. Laer, A. Drzezga, H. Forstl, A. Kurz, C. Zimmer, and A.M. Wohlschlager. 2007. Selective Changes of Resting-State Networks in Individuals at Risk for Alzheimer's Disease. *Proceedings of the National Academy of Sciences of the United States of America* 104 (47): 18760–18765.
70. Wada, Y., Y. Nanbu, M. Kikuchi, Y. Koshino, and T. Hashimoto. 1998. Aberrant Functional Organization in Schizophrenia: Analysis of EEG Coherence During Rest and Photic Stimulation in Drug-Naive Patients. *Neuropsychobiology* 38: 63–69.

Evaluating the Homeostasis Assessment Model Insulin Resistance and the Cardiac Autonomic System in Bariatric Surgery Patients: A Meta-Analysis

Styliani A. Geronikolou, Konstantinos Albanopoulos, George Chrousos, and Dennis Cokkinos

Abstract Morbid obesity is a severe chronic disease and subject to surgical methods for losing weight. This intervention is expected to drive to better quality of life and health status. Other important aspects which may be influenced are: HOMA-IR (as insulin resistance marker) and heart rate variability (as cardiac function and autonomic nervous system marker), which are independent and valid predictors of future cardiac, neurological, metabolic health. We pooled 4 studies (646 subjects) resulting to HOMA-IR and nine HRV components-grouped in those undergone to gastric bypass (RYGP) and those operated with vertical sleeve gastrectomy (SG) method. We performed a meta-analysis in patients for HOMA-IR and HRV, using Hedge's g correction of Cohen d for small samples. We concluded that RYGP favors insulin resistance decrease, whereas SG increases the vagal tone, improving cardiac function. The severity of cardiovascular diseases history suggests the selection of the surgery method: SG for the most severe cardiovascular cases and RYGP for those with higher HOMA-IR.

S.A. Geronikolou (✉)
First Clinic of Paediatrics, Athens University, Aghia Sophia Children Hospital, Athens, Greece

Clinical, Translational, Experimental Surgery Research Center, Biomedical Research Foundation of Academy of Athens, 4 Soranou Efessiou Street, 11527, Athens, Greece
e-mail: sgeronik@bioacademy.gr

K. Albanopoulos
First Department of Propedeutic Surgery, University of Athens, Hippokration Hospital, Athens, Greece

G. Chrousos
Clinical Translational, Experimental Surgery Research Centre, Biomedical Research Foundation Academy of Athens, 4 Soranou Efessiou Street, 11527, Athens, Greece

First Clinic of Paediatrics, Athens University, Aghia Sophia Children Hospital, Athens, Greece

D. Cokkinos
Clinical Translational, Experimental Surgery Research Centre, Biomedical Research Foundation Academy of Athens, 4 Soranou Efessiou Street, 11527, Athens, Greece

P. Vlamos (ed.), *GeNeDis 2016*, Advances in Experimental Medicine and Biology 988, DOI 10.1007/978-3-319-56246-9_20

Keywords Gastric bypass • Sleeve gastrectomy • HOMA-IR • Heart rate variability

1 Introduction

WHO has identified obesity as a severe chronic disease and ranked morbid obesity as the seventh cause of death. Bariatric surgery advantages can be summed as follows: 90% improvement in related medical conditions as asthma, sleep apnea, type 2 diabetes, hypertension, gastroesophageal reflux disease, decrease of mortality rate up to 0.5%, decrease of the serious complication up to 0.5% [1, 2]. Its complications comprise short and long term neurological disorders ranging from weakness in one extremity or vitamin deficiency up to more severe situations as dysphagia, peripheral neuropathy etc. [3–5]. The latter, was associated with insulin resistance independent of metabolic syndrome (which is a common comorbidity to morbid obesity) [6].

HOMA_IR method for insulin resistance evaluation gained popularity in epidemiological studies, because (1) it adjusts the glucose-insulin feedback system in the fasting state, (2) it corrects for the visceral, peripheral resistance (3) it corrects for renal glucose loss [7].

Heart rate variability (HRV) is a valid marker of autonomic nervous system (ANS) evaluation. ANS activity might be parasympathetic or sympathetic: The sympathetic is considered the "flight or fight" (excitory) system, whilst the parasympathetic is considered the "feed and breed" (inhibiting) system. The parameters of the HRV linked to the parasympathetic tone of the ANS are: RMSSD, HF-and those linked to the sympathetic tone are SDNN, SDANN, SDNNi, LF, LF/HF, VLF. In Table 1, we summarize the definition of each HRV component as described by the European Society of Cardiology [8]. The autonomic nervous system (ANS) and the parasympathetic tone in particular, is proposed to regulate the inflammation

Table 1 Frequency and time domain parameters measured in the included studies

	Parameter	Unit	Description
Time domain	SDNN	ms	Standard deviation of all NN intervals
	SDANN	ms	Standard deviation of the average of NN intervals for each 5 min
	RMSSD	ms	Root mean sum of the squares of successive differences in milliseconds that reflect the vagal influences
Frequency domain	HF	ms^2	High frequency (HF) (0.15–0.4 Hz)
	LF	ms^2	Low frequencies (0.04–0.15 Hz)
	LF/HF	–	Balance
	VLF	ms^2	Very low frequencies (VLF) (0.003–0.04 Hz)

allostatic load. It is also, associated to hypothalamus-pituitary-adrenal axis (HPA) function, glucose regulation [9] and even autoimmune disorders, such as systemic lupus erythematodes (SLE) disease activity [10, 11]. On the other hand, it is well established that the vagus nerve plays a significant role in health and disease [12, 13]. Obesity, in its turn, is associated with cardiac dysfunction and HRV irregularity [14, 15].

In this article we try to retrieve effect of gastric bypass (RYGP) and vertical sleeve gastrectomy (SG) to homeostasis model assessment insulin resistance (HOMA_IR) and heart rate variability (HRV) in bariatric patients, so as to advance the understanding of the neurohormonal–metabolic consequences of such severe interventions (as weight loss surgery) and validate it as a resolution of type II diabetes in bariatric pathology [16], and add to the estimating the balance between benefit and harm of the two operations.

2 Material-Methods

Databases as Scopus and PUBMED (The National Library of Medicine) were searched using the keywords "HOMA-IR and heart rate variability in bariatric patients" or "homeostasis index and heart rate variability in bariatrics." The literature search was conducted in August 2016 and the articles meeting the eligibility criteria were evaluated with the aid of the PRISMA approach [17–19]. The search was conducted with the following inclusion criteria: a. English language, b. Cohort or case control studies were included in this study c. Target to human subjects and not animal models d. Bariatric surgery patients e. Referring both HOMA-IR and HRV changes.

Two investigators (SG and KA) blindly searched and screened the articles and consented to articles' quality: Cohen's kappa for inter-rater agreement was 100% agreement (0.9) for the abstract selection, and 100% (1.0) for the 84 full study inclusion [17].

3 Statistical Analysis

The Comprehensive Meta-Analysis (CMA) software was used to transform results of individual studies into the common effect sizes of Hedges' g. Hedges' g is a measure of the standardized difference between intervention and control condition that corrects for biases associated with small sample sizes and can be interpreted in the same way as Cohen's d, whereby 0.2 represents a small effect, 0.5, a medium effect and 0.8, a large effect [20, 21]. A single meta-analysis for every component separately was conducted to acquire an effect size for HRV or HOMA-IR 6 months after surgery relative to baseline (before surgery) for each study. Then we summarized each component effect according to the surgical method.

In consequence we summarized the components effect for each surgery method according to their ANS activity linking. Positive effect sizes point to higher means 6 months after surgery. In forest plots, null effect falls to zero, decease after operation falls on the left (negative results) and increase after operation falls on the right of 0, making it positive. The sample sizes are small, the number of the studies is also small. Thus, the assumption of a common population effect was tested with the I^2—the percentage of total variation across studies that is due to heterogeneity rather than chance $I^2 = 100\% \times (Q - \mathrm{df})/Q$, *where* x^2—distributed homogeneity test Q, df = degrees of freedom. An estimate of the between study variance in a random effects meta-analysis is known as tau squared (τ2): if τ2 > 1 a substantial statistical between study heterogeneity is present [22].

4 Results

Four studies from two different countries (United States of America and Taiwan) have been identified [23–26] published in 2013 and 2015 [23, 25–27]. All studies except the Casellini study did not separate diabetic and non-diabetic patients. Casellini provided measurement of three groups: those of nondiabetic prediabetic and diabetic type II patients, and have been considered as separate studies in this meta-analysis. The resulting six studies targeted to a total of 646 patients aged 34–52.5 years and BMI > 50. In the Wu et al. [26] and Casellini 3 studies [23], the patients underwent sleeve gastrectomy, whereas, in the Maser [27] and, Perugini [25] studies the patients underwent laparoscopic Roux-en-Y gastric bypass.

Sample sizes ranged from 18 to 32 (28.4 ± 12.77) with a median of 29. The mean age ranged from 34 to 45 (40.82 ± 5.87) with a median 39.7. The Wu publication reported frequency (VLF, LF, HF, LF/HF) and time domain parameters (RMSSD), the Maser study reported frequency domain parameters (LF, HF, LF/HF), while all Casellini and Perugini studies reported only time domain parameters (SDNN, RMSSD, SDANNi, SDNNi). All four studies reported HOMA_IR changes after weight loss surgery.

Initially, we performed separate meta-analysis for HOMA-IR and each heart rate variability component of each study. Consequently, we summarized the components effect sizes for each weight loss surgical method separately. The results are illustrated in Forest Plots Fig. 1. The fixed effects presented in this plot are identical to random effects for variables, due to the small sample sizes.

The summary fixed effect of all components linked to the parasympathetic (summary of RMSSD, HF) or sympathetic tone (summary of SDNN, SDANN, SDNNi, LF, LF/HF, VLF) of the ANS according to RYGP or SG surgery method are presented in Table 2 and illustrated in Figs. 2 and 3. Sleeve gastrectomy has a higher fixed effect on the parasympathetic 0.818 > 0.35 as well as the sympathetic tone compared to the effect of RYGP method: 0.283 > 0.245. The random and the fixed effects in this case (SG) are identical for the parasympathetic tone, but non significant for the sympathetic tone [−0.383, 0.645]. The random effects of this latter meta-analysis are of great interest: the prediction intervals for the

Surgery method effect on each HRV or metabolic component

Studyname	Hedges's g	Variance	Standard error	Lower limit	Upper limit	Z-Value	p-Value
HOMA-IR.gp	-0,491	0,017	0,132	-0,750	-0,232	-3,720	0,000
HOMA-IR.sg	0,156	0,012	0,110	-0,060	0,372	1,418	0,156
SDNN gp	0,364	0,109	0,330	-0,283	1,010	1,102	0,270
SDNN sg	0,285	0,000	0,015	0,256	0,314	19,427	0,000
SDNNi gp	0,364	0,034	0,184	0,003	0,724	1,977	0,048
SDANNi gp	0,364	0,034	0,184	0,003	0,724	1,977	0,048
RMDSSDgp	0,364	0,001	0,034	0,297	0,430	10,753	0,000
RMDSSDsg	0,817	0,000	0,021	0,775	0,859	38,364	0,000
VLF.sg	0,186	0,052	0,227	-0,259	0,632	0,819	0,413
LF gp	0,627	0,036	0,189	0,257	0,997	3,317	0,001
LF sg	0,893	0,073	0,270	0,364	1,422	3,307	0,001
HF gp	0,000	0,030	0,172	-0,337	0,337	0,000	1,000
HF sg	0,893	0,073	0,270	0,364	1,422	3,307	0,001
LF/HF gp	-0,352	0,032	0,178	-0,701	-0,003	-1,978	0,048
LF/HF sg	-0,893	0,073	0,270	-1,422	-0,364	-3,307	0,001
	0,430	0,000	0,011	0,408	0,451	38,639	0,000

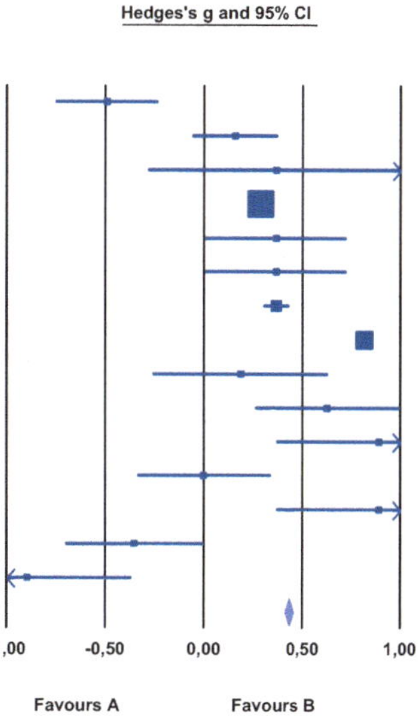

Fig. 1 Surgery method effect on each HRV or metabolic component

measurements of RYGP include 0 for the parasympathetic [−0.127, 0.569] as well as the sympathetic tone [−0.225, 0.557] suggesting that the method has no effect on cardiac ANS, thus no benefit to cardiovascular diseases.

The size of the squares points to a large weight of the observed effect in the Forest plot. The effect size for all measures is quite homogeneous, except from HOMA-IR and SDNN, where a significant variance was observed (see Table 2). No heterogeneity is observed in the rest of the (HRV) components (Table 2). On the contrary, considering the summary ANS activity effect, great heterogeneity is observed because of the variety of parameters included (Table 2). Yet, this was inevitable for a cardiovascular result to be concluded.

In the end, if we summarize separately all frequency domain as well as time domain components for each method, we conclude that SG has greater effect on Frequency domain components $g(sg) = 0.262 > g(rygp) = 0.069$, as well as on time domain components $g(sg) = 0.456 > g(rygp) = 0.364$.

5 Discussion

The PRISMA protocol was applied in this systematic review. We used Hedge's g correction to guard against biased estimates due to small samples [21].

The summary HOMA-IR improvement was more beneficial in RYGP follow up (relative to sleeve gastrectomy) but Casellini et al. [23] have showed that

Table 2 Fixed effect results of the meta-analysis

Method	Variable	n	Hedge's g	SE	Variance	Lower limit	Upper limit	Q	I^2	τ^2
RYGP	LF	32	0.627	0.189	0.036	0.255	0.998	0.00	0.00	0.00
SG		18	0.893	0.270	0.073	0.364	1422	0.00	0.00	0.00
SG	VLF	18	0.186	0.227	5165	−0.26	0.632	0.00	0.00	0.00
RYGP	HF	32	0.000	0.172	0.030	−0.338	0.338	0.00	0.00	0.00
SG		18	0.893	0.27	0.073	0.368	1.422	0.00	0.00	0.00
RYGP	LF/HF	32	−0.352	0.178	0.032	−0.701	−0.003	0.00	0.00	0.00
SG		18	−0.893	0.27	0.073	−1.422	−0.364	0.00	0.00	0.00
RYGP	RMSSD	30	0.364	0.034	0.001	0.297	0.43	0.00	0.00	0.00
SG		88	0.817	0.021	0.000	0.775	0.859	0.26	0.00	0.00
RYGP	SDNN	30	0.364	0.033	0.001	0.298	0.429	0.00	0.00	0.00
SG		70	0.285	0.015	0.000	0.256	0.313	**6.9**	56.58	0.029
RYGP	SDNNi	30	0.364	0.184	0.034	0.003	0.724	0.00	0.00	0.00
RYGP	SDANNi	30	0.364	0.184	0.034	0.003	0.724	0.00	0.00	0.00
RYGP	HOMA-IR	62	−0.491	0.132	0.017	−0.750	−0.233	0.99	0.00	0.00
SG		88	−0.156	0.112	0.013	−0.377	0.064	**26.9**	88.7	0.412
RYGP	Parasympathetic tone	48	0.35	0.033	0.001	0.285	0.415	**4.304**	76.66	0.051
SG		88	0.818	0.021	0.000	0.776	0.859	0.078	0.000	0.000
RYGP	Sympathetic tone	154	0.283	0.088	0.008	0.071	0.418	**16.296**	75.454	0.123
SG		124	0.245	0.015	0.000	0.254	0.311	**24.772**	87.640	0.478

n = number of patients; Bold: p-value of Q pQ < 0.05

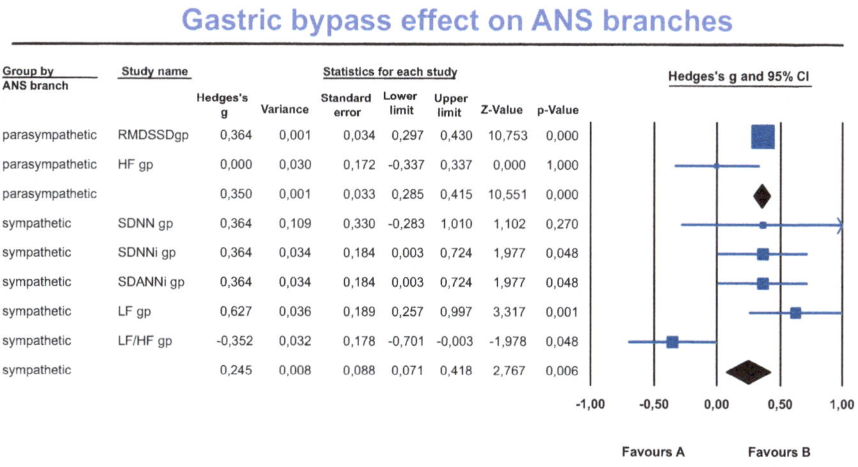

Group by ANS branch	Study name	Hedges's g	Variance	Standard error	Lower limit	Upper limit	Z-Value	p-Value
parasympathetic	RMDSSDgp	0,364	0,001	0,034	0,297	0,430	10,753	0,000
parasympathetic	HF gp	0,000	0,030	0,172	-0,337	0,337	0,000	1,000
parasympathetic		0,350	0,001	0,033	0,285	0,415	10,551	0,000
sympathetic	SDNN gp	0,364	0,109	0,330	-0,283	1,010	1,102	0,270
sympathetic	SDNNi gp	0,364	0,034	0,184	0,003	0,724	1,977	0,048
sympathetic	SDANNi gp	0,364	0,034	0,184	0,003	0,724	1,977	0,048
sympathetic	LF gp	0,627	0,036	0,189	0,257	0,997	3,317	0,001
sympathetic	LF/HF gp	-0,352	0,032	0,178	-0,701	-0,003	-1,978	0,048
sympathetic		0,245	0,008	0,088	0,071	0,418	2,767	0,006

Fig. 2 RYGP effect on ANS branches

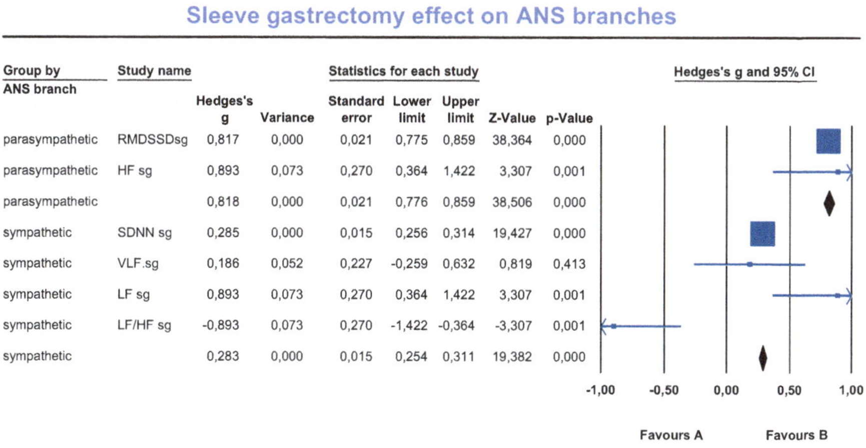

Group by ANS branch	Study name	Hedges's g	Variance	Standard error	Lower limit	Upper limit	Z-Value	p-Value
parasympathetic	RMDSSDsg	0,817	0,000	0,021	0,775	0,859	38,364	0,000
parasympathetic	HF sg	0,893	0,073	0,270	0,364	1,422	3,307	0,001
parasympathetic		0,818	0,000	0,021	0,776	0,859	38,506	0,000
sympathetic	SDNN sg	0,285	0,000	0,015	0,256	0,314	19,427	0,000
sympathetic	VLF.sg	0,186	0,052	0,227	-0,259	0,632	0,819	0,413
sympathetic	LF sg	0,893	0,073	0,270	0,364	1,422	3,307	0,001
sympathetic	LF/HF sg	-0,893	0,073	0,270	-1,422	-0,364	-3,307	0,001
sympathetic		0,283	0,000	0,015	0,254	0,311	19,382	0,000

Fig. 3 SG effect on ANS branches

improvements of cardiac autonomic function are independent of insulin resistance. Decreased HOMA-IR is associated to positive feedback loop between the liver and the β-cell in fasting. Insulin resistance is a valid predictor of type 2 diabetes. In non diabetic as well as diabetic subjects, insulin resistance is related to several cardiovascular risk factors, including hyperglycemia [28], dyslipidemia [28], hypertension [28, 29]. Improved insulin sensitivity after weight loss surgery has already been reported [30, 31]. The case, here, is whether this is related to autonomic nervous system modulation or to diabetic status or inflammation. To resolve this, we could perform meta-regression but our samples are so small that it could not be performed.

Gastric bypass had no effect on HF component, thus, the detrimental effects of the HRV levels (cardiovascular diseases) before surgery persist 6 months after gastric bypass. On the contrary the effect of SG on HF is large g = 0.893. In the literature, increased SDNN, HF, SDANN, HF (observed in SG) is known to have been associated with elevated testosterone levels and intense physical activity (athletic habitat) [8]. Thus, we ascertained an increase in the parasympathetic tone, observed in sleeve gastrectomy: this is beneficial to the cardiovascular system [32–35], although it has not been established yet, how much the markers of the vagal activity have to increase in order to provide protection to the heart as it also result to blood pressure improvement. This observation is, possibly, attributed to the technique: the surgeon preserves the vagal trunk of the lesser curvature. The observed increase in the mixed sympathovagal tone as expressed by VLF (observed in SG again) could have been associated with the renin-angiotensin-aldosterone system [36], that is known to be associated with inflammation as well as predictive of morbidity and mortality [36, 37].

Accordingly, the I^2 values verified that most of the variability across variables (56.58% for SDNN outcome and 88.7% for HOMA-IR) is due to a heterogeneity rather than chance. Indeed, there is variation in diabetic status within the studies including non diabetic, prediabetic and diabetic group.

LF/HF ratio has been criticised for its representing the sympathovagal balance [38]. The ratio is smaller, thus, more beneficial for patients undergoing SG rather than those operated with RYGP (Table 2 and Fig. 1).

Accordingly, an increase in the sympathetic tone (as expressed by LF) is associated to exercise or cardiac ischemia, or heart failure [8]. However, the relation between physiology and HRV needs to be examined in greater depth, in order to untangle the information given by the various techniques of HRV analysis [8, 38]. The RYGP favoured RMSSD decrease which has been attributed to increase plasma rennin activity which in its turn is related to increased vagal modulation-protective against cardiac morbidity [39]. The comparison of the summary effect for each domain (frequency or time), definitely shows that SG has a greater effect size over ANS.

SG method has a greater effect on both branches of the cardiac ANS, making it more beneficial for severe cardiovascular patients. On the other hand, the reduced vagal tone observed in RYGP (relevant to SG) is associated to functional dyspepsia [40] or secretion of glucogen-like peptide [41]. Yet, the effect of RYGP seems to be null, considering the random effects (the effects expected to be observed when the studies/patients become extremely increased).

In animal studies it has been shown that in the gut, the cross talk of the stimulated vagus nerve with immune cells increases the cholinergic tone, reducing mucosal inflammation-driving innate immune cells into a immunologic tolerance [42]: a case that seems to happen after sleeve gastrectomy, where the vagal tone definitely increases Table 2 and Fig. 2). The mechanisms involved have not been groped yet, and literature is lacking data on human subjects.

Bariatric patients included in the studies that we meta-analyzed varied in glycemic profile, treatment, comorbidities. The resistance of central control of heart rate to weight loss surgery is greater in patients that had undergone RYGP rather in sleeve gastrectomy, which seems to be suggested for severe cardiovascular patients. Thus, the link of the gastrointestinal tract to hypothalamus, insulin sensitivity, inflammation autonomic nervous system, cardiac function is a target mechanism that emerges for future research. Additionally, because of limited scale and design heterogeneity of existing studies, future studies might focus on validating above findings in more large-scale and longitudinal studies with elaborate design.

6 Conclusions

This systematic review postulates that the two methods activate different mechanisms for achieving homeostasis. The summary effect of the operational methods on each ANS branch points that the GP technique has no effect on cardiac function (as expressed in HRV parameters). Thus, SG is mostly suggested to severe cardiovascular patients rather than GP. Nevertheless, GP favors insulin resistance decrease, making it beneficial for patients with high glycemic profile and/or hypertension.

Conflict of Interest All authors see no conflict of interest.

References

1. Morino, M., M. Toppino, P. Forestieri, L. Angrisani, M.E. Allaix, and N. Scopinaro. 2007. Mortality After Bariatric Surgery: Analysis of 13,871 Morbidly Obese Patients from a National Registry. *Annals of Surgery* 246: 1002–1007; discussion 1007-9.
2. Wittgrove, A.C., and G.W. Clark. 2000. Laparoscopic Gastric Bypass, Roux-en-Y-500 Patients: Technique and Results, with 3-60 Month Follow-Up. *Obesity Surgery* 10: 233–239.
3. Ba, F., and Z.A. Siddiqi. 2010. Neurologic Complications of Bariatric Surgery. *Reviews in Neurological Diseases* 7: 119–124.
4. Becker, D.A., L.J. Balcer, and S.L. Galetta. 2012. The Neurological Complications of Nutritional Deficiency Following Bariatric Surgery. *Journal of Obesity* 2012: 608534.
5. Pardo-Aranda, F., N. Perez-Romero, J. Osorio, J. Rodriguez-Santiago, E. Munoz, N. Puertolas, and E. Veloso. 2016. Wernicke's Encephalopathy After Sleeve Gastrectomy: Literature Review. *International Journal of Surgery Case Reports* 20: 92–95.
6. Han, L., L. JI, J. Chang, J. Wen, W. Zhao, H. Shi, L. Zhou, Y. Li, R. Hu, J. Hu, and B. Lu. 2015. Peripheral Neuropathy is Associated with Insulin Resistance Independent of Metabolic Syndrome. *Diabetology and Metabolic Syndrome* 7: 14.
7. Wallace, T.M., J.C. Levy, and D.R. Matthews. 2004. Use and Abuse of HOMA Modeling. *Diabetes Care* 27: 1487–1495.
8. Task Force of the European Society of Cardiology and the North American Society of Electrophysiology. 1996. Heart Rate Variability, Standards of Measurement, Physiological Interpretation and Clinical Use. *European Heart Journal* 17: 354–381.
9. Thayer, J.F., and E. Sternberg. 2006. Beyond Heart Rate Variability: Vagal Regulation of Allostatic Systems. *Annals of the New York Academy of Sciences* 1088: 361–372.

10. Aydemir, M., V. Yazisiz, I. Basarici, A.B. Avci, F. Erbasan, A. Belgi, and E. Terzioglu. 2010. Cardiac Autonomic Profile in Rheumatoid Arthritis and Systemic Lupus Erythematosus. *Lupus* 19: 255–261.
11. Thanou, K.E.A. 2014. Heart Rate Variabilty: New Window in Lupus Disease Activity. 14th Annual Meeting of the American College of Rheumatology 2014 Boston.
12. Darwin, C.R. 1872/1999. *The Expression of the Emotions in Man and Animals*. London, UK: Harper Collins.
13. Porges, S.W. 2007. The Polyvagal Perspective. *Biological Psychology* 74: 116–143.
14. Kannel, W.B., J.F. Plehn, and L.A. Cupples. 1988. Cardiac Failure and Sudden Death in the Framingham Study. *American Heart Journal* 115: 869–875.
15. Kim, J.A., Y.G. Park, K.H. Cho, M.H. Hong, H.C. Han, Y.S. Choi, and D. Yoon. 2005. Heart Rate Variability and Obesity Indices: Emphasis on the Response to Noise and Standing. *The Journal of the American Board of Family Practice* 18: 97–103.
16. Singh, A.K., R. Singh, and S.K. Kota. 2015. Bariatric Surgery and Diabetes Remission: Who Would Have Thought It? *Indian Journal of Endocrinology Metabolism* 19: 563–576.
17. Cohen, J. 1960. A Coefficient of Agreement of Nominal Scales. *Educational and Psychological Measurement* XX: 37–46.
18. Moher, D., A. Liberati, J. Tetzlaff, and D.G. Altman. 2009. Preferred Reporting Items for Systematic Reviews and Meta-Analyses: The PRISMA Statement. *Journal of Clinical Epidemiology* 62: 1006–1012.
19. Whitlock, E.P., S.A. Lopez, S. Chang, M. Helfand, M. Eder, and N. Floyd. 2010. AHRQ Series Paper 3: Identifying, Selecting, and Refining Topics for Comparative Effectiveness Systematic Reviews: AHRQ and the Effective Health-Care Program. *Journal of Clinical Epidemiology* 63: 491–501.
20. Cohen, D. 1988. *Statistical Power Analysis for the Behavioral Sciences*. Hillsdale, NJ.
21. Hedges, L. 1981. Distribution Theory for Glass' Estimator of Effect Size and Related Estimators. *Journal of Educational Statistics* 6: 107–128.
22. Borenstein, M., L. Hedges, J. Higgins, and H. Rothstein. 2009. *Introduction to Meta-Analysis*. John Wiley & Sons Ltd: West Sussex, UK.
23. Casellini, C.M., H.K. Parson, K. Hodges, J.F. Edwards, D.C. Lieb, S.D. Wohlgemuth, and A.I. Vinik. 2016. Bariatric Surgery Restores Cardiac and Sudomotor Autonomic C-Fiber Dysfunction Towards Normal in Obese Subjects with Type 2 Diabetes. *PLoS One* 11: e0154211.
24. Maser, R.E., M.J. Lenhard, M.B. Peters, I. Irgau, and G.M. Wynn. 2013. Effects of Surgically Induced Weight Loss by Roux-en-Y Gastric Bypass on Cardiovascular Autonomic Nerve Function. *Surgery for Obesity and Related Diseases* 9: 221–226.
25. Perugini, R.A., Y. Li, L. Rosenthal, K. Gallagher-Dorval, J.J. Kelly, and D.R. Czerniach. 2010. Reduced Heart Rate Variability Correlates with Insulin Resistance but not with Measures of Obesity in Population Undergoing Laparoscopic Roux-en-Y Gastric Bypass. *Surgery for Obesity and Related Diseases* 6: 237–241.
26. Wu, J.M., H.J. Yu, H.S. Lai, P.J. Yang, M.T. Lin, and F. Lai. 2015. Improvement of Heart Rate Variability Decreased Insulin Resistance After Sleeve Gastrectomy for Morbidly Obesity Patients. *Surgery for Obesity and Related Diseases* 11: 557–563.
27. Maser, R.E., M. James Lenhard, P.B. Balagopal, P. Kolm, and M.B. Peters. 2013. Effects of Surgically Induced Weight Loss by Roux-en-Y Gastric Bypass on Osteocalcin. *Surgery for Obesity and Related Diseases* 9: 950–955.
28. Bonora, E., S. Kiechl, J. Willeit, F. Oberhollenzer, G. Egger, G. Targher, M. Alberiche, R.C. Bonadonna, and M. Muggeo. 1998. Prevalence of Insulin Resistance in Metabolic Disorders: The Bruneck Study. *Diabetes* 47: 1643–1649.
29. Bonora, E., G. Targher, M. Alberiche, R.C. Bonadonna, M.B. Zenere, F. Saggiani, and M. Muggeo. 2001. Intracellular Partition of Plasma Glucose Disposal in Hypertensive and Normotensive Subjects with Type 2 Diabetes Mellitus. *The Journal of Clinical Endocrinology and Metabolism* 86: 2073–2079.

30. Bobbioni-Harsch, E., J. Sztajzel, V. Barthassat, V. Makoundou, G. Gastaldi, K. Sievert, G. Chassot, O. Huber, P. Morel, F. Assimacopoulos-Jeannet, and A. Golay. 2009. Independent Evolution of Heart Autonomic Function and Insulin Sensitivity During Weight Loss. *Obesity (Silver Spring)* 17: 247–253.
31. Muscelli, E., G. Mingrone, S. Camastra, M. Manco, J.A. Pereira, J.C. Pareja, and E. Ferrannini. 2005. Differential Effect of Weight Loss on Insulin Resistance in Surgically Treated Obese Patients. *The American Journal of Medicine* 118: 51–57.
32. Dimova, R., T. Tankova, N. Chakarova, G. Groseva, and L. Dakovska. 2015. Cardiovascular Autonomic Tone Relation to Metabolic Parameters and hsCRP in Normoglycemia and Prediabetes. *Diabetes Research and Clinical Practice* 109 (2): 262–270.
33. Guiraud, T., M. Labrunee, K. Gaucher-Cazalis, F. Despas, P. Meyer, L. Bosquet, C. Gales, A. Vaccaro, M. Bousquet, M. Galinier, J.M. Senard, and A. Pathak. 2013. High-Intensity Interval Exercise Improves Vagal Tone and Decreases Arrhythmias in Chronic Heart Failure. *Medicine and Science in Sports and Exercise* 45: 1861–1867.
34. Olshansky, B., H.N. Sabbah, P.J. Hauptman, and W.S. Colucci. 2008. Parasympathetic Nervous System and Heart Failure: Pathophysiology and Potential Implications for Therapy. *Circulation* 276 (118): 863–871.
35. Roy, A., S. Guatimosim, V.F. Prado, R. Gros, and M.A. Prado. 2014. Cholinergic Activity as a New Target in Diseases of the Heart. *Molecular Medicine* 20: 527–537.
36. Taylor, J.A., D.L. Carr, C.W. Myers, and D.L. Eckberg. 1998. Mechanisms Underlying Very-Low-Frequency RR-Interval Oscillations in Humans. *Circulation* 98: 547–555.
37. Bigger, J.T. Jr., J.L. Fleiss, R.C. Steinman, L.M. Rolnitzky, R.E. Kleiger, and J.N. Rottman. 1992. Frequency Domain Measures of Heart Period Variability and Mortality After Myocardial Infarction. *Circulation* 85: 164–171.
38. Billman, G.E. 2013. The LF/HF Ratio Does Not Accurately Measure Cardiac Sympatho-Vagal Balance. *Frontiers in Physiology* 4: 26.
39. Virtanen, R., A. Jula, T. Kuusela, H. Helenius, and L.M. Voipio-Pulkki. 2003. Reduced Heart Rate Variability in Hypertension: Associations with Lifestyle Factors and Plasma Renin Activity. *Journal of Human Hypertension* 17: 171–179.
40. Lorena, S.L., M.J. Figueiredo, J.R. Almeida, and M.A. Mesquita. 2002. Autonomic Function in Patients with Functional Dyspepsia Assessed by 24-Hour Heart Rate Variability. *Digestive Diseases and Sciences* 47: 27–31.
41. Rocca, A.S., and P.L. Brubaker. 1999. Role of the Vagus Nerve in Mediating Proximal Nutrient-Induced Glucagon-Like Peptide-1 Secretion. *Endocrinology* 140: 1687–1694.
42. Matteoli, G., and G.E. Boeckxstaens. 2013. The Vagal Innervation of the Gut and Immune Homeostasis. *Gut* 62: 1214–1222.

Association of Type 1 Diabetes, Social Support, Illness and Treatment Perception with Health Related Quality of Life

Mekonen Yerusalem, Sofia Zyga, and Paraskevi Theofilou

Abstract This study examines the importance of perceived social support, perception of illness and treatment on health related quality of life (HRQOL) in patients with type 1 diabetes. The purpose was to assess how type 1 diabetes, perceived social support, perception of the illness and treatment can be predictors of health related quality of life and how these factors associate. Additional purpose was to establish the reliability of Illness Index and Treatment Index among Greek population. Four questionnaires were used, the Missoula-VITAS Quality of Life Index-15 (MVQoLI-15), the Multidimensional Scale of Perceived Social Support (MSPSS), the Illness Index, and the Treatment Index. Participants were type 1 diabetics. The sample consists of 60 volunteers, 19 males and 41 females. All participants were Greeks. There was a correlation between variables of social support, treatment and illness index and the variables of HRQOL. Several domains of social support, the illness and treatment perception, are statistically significant predictors of the total HRQOL. Treatment and Illness Scales were found highly reliable among Greek population (20 items; $\alpha = 0.93$). Results suggest that social support interacts with relationships, spirituality and the total QOL. Family interacts with spirituality and total QOL. This study indicates friends as the most important and positive factor towards most of the HR-QOL's subsections. Findings support that perception of the condition (II) and perception of the treatment (TI) are predictors of HRQOL. This study aims to establish the reliability of Illness and Treatment Index among Greek population.

M. Yerusalem (✉)
School of Psychology, University of Central Lancashire, Preston, PR1 2HE, UK

Psychology Department, ICPS – Institution for Counselling and Psychological Studies, Peristeri 12131, Greece
e-mail: mekonen.jerry@gmail.com

S. Zyga
Department of Nursing, Faculty of Human Movement and Quality of Life Sciences, University of Peloponnese, Sparta, Greece
e-mail: zygas@uop.gr

P. Theofilou
Department of Nursing, University of Peloponnese, Tripoli 22100, Greece

P. Vlamos (ed.), *GeNeDis 2016*, Advances in Experimental Medicine and Biology 988, DOI 10.1007/978-3-319-56246-9_21

Keywords Type 1 diabetes • HRQOL • Perceived social support • Illness perception • Treatment perception

1 Introduction

Type 1 diabetes is an endocrine and metabolic condition for which there is currently no cure, and high quality care is essential if acute and long-term complications are to be avoided [1]. Diabetes is a largely self-managed disease [2]. Patient education is central to successful self-management but also the factor of perceived social support, appears to have a fundamental impact in controlling diabetes [3]. Strong association is reported between this chronic disease and quality of life. More specifically, individuals with type 1 diabetes have reported low health related quality of life (HRQOL). Successful intensive diabetes management requires significant patient engagement. Patients' beliefs have been shown to influence treatment commitment in 5 chronic illness [4]. Evidence has shown that intensive DM management can increase HRQOL [5].

Research has demonstrated [6] that socio-demographic components are also associated with HRQOL in patients with type 1 diabetes. Socio-demographics context results in differences considering the social support that an individual receives and social support is associated with psychological distress and somatic health problems and socio-economic factors [7].

Studies have resulted to the conclusion that the perception and evaluation that individuals construct on their health condition and illness experience is greatly involved and implicated in the recovery phase from the disease [8], life span [9], commitment to medical care [10] and finally, temper and adjustment [11]. Adequacy of social support is related to HRQOL and the most important diabetes-related psychosocial factors are self-efficacy, and diabetes-related social support [12].

2 Materials and Methods

This was a correlational study in which the MVQOLI test score was a measure of the health related quality of life in five dimensions (symptoms, functionality, interpersonal relationships, wellness, spirituality), and the MSPSS test score was a measure of the patient's evaluation on the amount of social support received in the presence of the type 1 diabetes. Also, the Illness Index (II) test scores was a measure of the outcomes of the disease to the patient, and the Treatment Index (TI) test scores a measure of the outcomes of the therapy. All participants completed all four questionnaires individually. The participants of this study should include specific criteria. Every participant should have type 1 diabetes, be over 18 years old, speak the Greek language, have perceptive ability and fulfill a satisfying level of cooperation. The sample consists of 60 volunteers, 19 males and 41 females, with type 1 diabetes, age range 18–61, mean age 34.2 and $SD = 9.82$, all in perfect state to give written consent form filled along with the questionnaires. All participants are Greeks. Minors and individuals with other health issues and under other medications were not included.

2.1 Data Collection

The procedure in total took place electronically. The four questionnaires, the Vitas Quality of Life Index [13], the Multidimensional Scale of Perceived Social Support [13], the Illness Index and the Treatment Index [15], consent form, the briefing, the debriefing, and demographics, were all formed online. Each participant received invitation to participate in the study individually. The volunteers were referred to a link to have access and complete the questionnaires online. Participation run individually. After fulfilling the questionnaires, participants provided their contact email in order to proceed to the re-test of both Illness and Treatment Index. Participants were assigned to a number according to the turn whereby they answered the questionnaires. Thereby, the scorings of the initial Illness and Treatment Index scales where matched with the scores of the re-test.

2.2 Data Analysis

Pearson's correlation was performed in order to investigate the association between domains of social support (significant others, family, friends), illness and treatment index with the sections of HRQOL (symptoms, functionality, relationships, well-being, spirituality, total quality of life). A multiple linear regression was performed to establish if the items included in the social support variable (significant others, family, friends), illness and treatment index are statistically significant predictors of HRQOL and its domains (symptoms, functionality, relationships, well-being, spirituality, total quality of life).

2.3 Ethical Considerations

The study was conducted with the participation of volunteers. All participants were over 18, capable of fulfilling the questionnaires and proceeding through the project in full consent. The study ensures that the best interest of the participants and of the society in general are met. Participants' rights of confidentiality, voluntary participation, withdrawal, informed consent, deception in research, and debriefing were maintained. This study is justifiable by its potential benefit, including contribution to knowledge and understanding, social welfare and individual well-being. This project is supported by current literature and previous studies.

3 Results

3.1 Participants' Characteristics

The sample consists of 60 volunteers, 19 males and 41 females, with type 1 diabetes, age range 1861, mean age 34.2 and SD = 9.82 (Table 1).

3.2 Correlational Analysis

Significant others and treatment perception were statistically significantly and negatively correlated, $r = -0.3$, $p < 0.05$. Relationships and significant others were statistically significantly correlated, $r = 0.4$, $p < 0.05$. Significant others were also statistically significantly correlated to spirituality and the total quality of life $r = 0.4$, $p < 0.05$, and $r = 0.33$, $p < 0.05$ respectively. Family support was significantly correlated to spirituality $r = 0.25$, $p < 0.05$, and total quality of life $r = 0.27$, $p < 0.05$. Friends support was significantly and negatively correlated to illness index and treatment index, $r = -0.4$ ($p < 0.05$) and $r = -0.37$ ($p < 0.05$) respectively. There was a statistically significant correlation of friends and symptoms $r = 0.34$ ($p < 0.05$), relationships $r = 0.47$ ($p < 0.01$), well-being $r = 0.47$ ($p < 0.01$), spirituality $r = 0.43$ ($p < 0.05$), and total quality of life $r = 0.55$ ($p < 0.01$).

Table 1 Participants' characteristics

	N	%
Gender		
Female	41	68.3
Male	19	31.7
Marital status		
Unmarried	31	51.7
Married	26	43.3
Divorced	3	5.0
Education		
Pre-high School	2	3.3
High School	23	38.3
University	31	51.7
MSc	4	6.7
Professional Status		
Unemployed	23	38.3
Private employee	11	18.3
State employee	8	13.3
Self-employed	9	15.0
Housekeeping	1	1.7
Retired	4	6.7

Table 2 Correlations between illness index, treatment index, items of social support (Significant others, friends, family, total social support) and components of HRQOL (symptoms, functionality, relationships, well-being, spirituality, total quality of life)

	Quality of life components					
	SY	FUN	REL	WB	SP	TQL
Sig. others	–	–	0.43**	–	0.40**	0.33**
Friends	0.34*	–	0.47*	0.47*	0.47*	0.55*
Family	–	–	–	–	0.26*	0.27*
Total social	–	–	0.42*	0.36*	0.45*	0.48*
Illness Perc.	−0.56*	–	–	−0.60*	−0.60*	−0.66*
Treatment Perc.	−0.47*		*−0.27**	−0.56*	*−0.65**	*−0.66**

*p < 0.05; **p = 0.001

Table 3 Summary of multiple regression analysis with significant others, family, friends, illness index and treatment index as predictors of total quality of life

	Quality of life				
	B	*SE (B)*	β	*t*	Sig. (*p*)
Significant others	0.019	0.037	0.056	0.51	0.61
Family	0.012	0.018	0.069	0.69	0.49
Friends	0.101	0.035	0.36	2.91	0.005
Illness Index	−0.007	0.006	−0.25	–1.24	0.22
Treatment Index	−0.008	0.005	−0.29	−1.47	0.15

Total social support was significantly and negatively correlated to illness index and treatment index, $r = -0.33$ ($p < 0.05$), $r = -0.37$ ($p < 0.05$) respectively. Total social support was significantly correlated to relationships $r = 0.42$ ($p < 0.05$), well-being $r = 0.36$ ($p < 0.05$), spirituality $r = 0.45$ ($p < 0.01$) and total quality of life $r = 0.48$ ($p < 0.01$) (Table 2).

3.2.1 1st Reg.

A multiple linear regression was calculated to predict quality of life based on significant others, family, friends, illness index and treatment index. A significant regression equation was found F (5, 54) $= 13.99$, $p < 0.001$, $R^2 = 0.56$, $R^2_{Adjusted} = 0.52$. The only significant predictor as reported in Table 3 was Friends ($\beta = 0.306$, $p < 0.05$).

3.2.2 2nd Reg.

A multiple linear regression was calculated to predict symptoms based on significant others, family, friends, illness index and treatment index. A significant regression equation was found F (5, 54) $= 5.76$, $p < 0.001$, $R^2 = 0.59$, $R^2_{Adjusted} = 0.35$.

Table 4 Summary of multiple regression analysis with significant others, family, friends, illness index and treatment index as predictors of symptoms

	Symptoms				
	B	*SE* (*B*)	β	*t*	Sig. (*p*)
Significant others	−0.14	0.11	−0.17	–1.28	0.21
Family	0.04	0.05	0.08	0.69	0.49
Friends	0.15	0.10	0.19	1.49	0.14
Illness Index	−0.04	0.02	−0.51	–2.07	0.04
Treatment Index	0.001	0.02	0.02	0.09	0.93

Table 5 Summary of multiple regression analysis with significant others, family, friends, illness index and treatment index as predictors of relationships

	Relationships				
	B	*SE* (*B*)	β	*t*	Sig. (*p*)
Significant others	0.32	0.17	0.26	1.89	0.06
Family	−0.006	0.08	−0.009	−0.07	0.94
Friends	0.42	0.16	0.36	2.66	0.01
Illness Index	0.009	0.03	0.08	0.33	0.74
Treatment Index	−0.01	0.02	−0.14	−0.53	0.59

The only significant predictor as reported in Table 4 was Illness Index ($\beta = -0.51$, $p < 0.05$).

3.2.3 4th Reg.

A multiple linear regression was calculated to predict relationships based on significant others, family, friends, illness index and treatment index. A significant regression equation was found F (5, 54) = 4.54, $p < 0.05$, $R^2 = 0.30$, $R^2_{Adjusted} = 0.23$. The only significant predictor as reported in Table 5 was Friends ($\beta = 0.36$, $p < 0.05$).

3.2.4 5th Reg.

A multiple linear regression was calculated to predict well-being based on significant others, family, friends, illness index and treatment index. A significant regression equation was found F (5, 54) = 8.67, $p < 0.001$, $R^2 = 0.44$, $R^2_{Adjusted} = 0.39$. The only significant predictor as reported in Table 6 was Friends ($\beta = 0.30$, $p < 0.05$).

Table 6 Summary of multiple regression analysis with significant others, family, friends, illness index and treatment index as predictors of well-being

	Well—Being				
	B	*SE* (*B*)	β	*t*	Sig. (*p*)
Significant others	−0.18	0.17	−0.13	-1.02	0.31
Family	0.09	0.08	0.12	1.07	0.29
Friends	0.41	0.16	0.30	2.49	0.02
Illness Index	−0.04	0.03	−0.32	-1.43	0.16
Treatment Index	−0.02	0.02	−0.18	−0.79	0.43

Table 7 Summary of multiple regression analysis with significant others, family, friends, illness index and treatment index as predictors of spirituality

	Spirituality				
	B	*SE* (*B*)	β	*t*	Sig. (*p*)
Significant others	0.23	0.15	0.19	1.58	0.12
Family	0.02	0.07	0.03	0.30	0.76
Friends	0.17	0.14	0.14	1.23	0.22
Illness Index	−0.02	0.02	−0.16	−0.73	0.47
Treatment Index	−0.04	0.02	−0.40	−1.84	0.07

3.2.5 6th Reg.

A multiple linear regression was calculated to predict spirituality based on significant others, family, friends, illness index and treatment index). A significant regression equation was found $F\ (5, 54) = 10.66$, $p < 0.001$, $R^2 = 0.50$, $R^2{}_{Adjusted} = 0.45$. There was not a significant predictor as reported in Table 7.

4 Discussion

The aim of the current study was to understand how the chronic disease of type 1 diabetes, along with the variables of perceived social support, perception on illness and treatment, can predict the HR-QOL of a patient. The results indicated in a sufficient extent what has been established from the literature review. Suffering from type 1 diabetes, good monitoring [5], perceived social support, illness and treatment perception of the patient are correlated variables, and subscales of the total social support, the illness and treatment index are also predictors of the HR-QOL. More specifically, in terms of social support, results suggest that significant people in the environment of the patient can play a major role considering the aspects of relationships, spirituality and the total QOL. Family also appears to interact with spirituality and the total QOL, and these results are in line with previous research showing there is an association of family support and HRQOL [16].

This study indicates friends as the most important and positive factor towards most of the HR-QOL's subsections: symptoms, relationships, well-being, spirituality and total QOL. Also friends were negatively correlated with both II and TI, which means that friends appear to play an important role in the way that a patient assesses his illness condition as well as the treatment. Overall, total social support was indicated by the results as an important factor associated with relationships, spirituality, well-being, total QOL, and negatively associated as the factor of friends, with II and TI. Perceived social support through the domain of friends is a predictor of HRQOL considering quality in relationships, well-being and the total QOL. Additionally, perception of the condition (II) and perception of the treatment (TI) were found as predictors of wellbeing, spirituality, symptoms, and total QOL verifying that the importance of illness and treatment beliefs, and perceived severity of symptoms in explaining distress in patients with a chronic illness such as diabetes during treatment [17].

Spirituality as part of HRQOL doesn't appear to have a social predictor. Also, the domain of functionality has found no predictor of social nature, agreeing with Aalto et al. [12] in the conclusion that social support is related to all HRQOL domains apart from pain and functioning. Also spirituality has no predictor neither in the II or TI domain. This could be due to the limitations of the present project and design. Because the distribution and the total procedure took place electronically, the mood of the participants could not be controlled. Spirituality appears to be a deeper aspect of HRQOL, and it might be implicated with variables that this study didn't include, e.g. cultural components, financial crisis of the country, and type of medication. Although family is an important component of the social variable specifically in Greek society, the results show that friends are even more associated and can predict almost every aspect of HRQOL. This could be explain by the fact that family relationships are often. Also, family relationships specially with a member that suffers with a chronic disease such as type 1 diabetes are characterized more by stringency and concern, while relationships with friends emphasizes in comfort, understanding and mutual interests in activities among friends, that usually increase the patient's overall mood and wellbeing.

The first limitation of the current study due to the design used are technical issues that the researcher is unable to control during the completion of the questionnaires by the participants. Technical difficulties might cause alterations in the responses, causing validity issues. Another limitation is lack of personal contact. Completing the questionnaires online is a design with many advantages however; lack of personal contact is one of the most important disadvantages. People respond differently when in personal, and tent to participate and answer with greater levels of consciousness and engagement. As third limitation of the present study, we could include the insecurity of participants towards internet safety and personal information. Because the questionnaires included some very sensitive and personal items, it is possible that responses can be altered and not candid across the total spectrum. Moreover, ethnicity was not taken under consideration in this study. Although all participants were Greece residence, the nationality of each individual was not specified or considered. Therefore, possible cultural differences of non-

Greeks, attitudes towards the variables that have been studied, HRQOL, social support, friends, family, perception of illness and treatment, are factors that might have affected the outcome of this study. The number of the sample could also be considered as a limitation, and future studies could investigate the association of the same variables in a bigger population.

5 Conclusions

This study demonstrates a significant association between type 1 diabetes, social support in terms of family, friends, total social support, and perception of the patient towards the illness and the treatment and HRQOL among Greek population. A positive and good total social support, positive perception towards the condition and the treatment can predict and increase an individual's HRQOL. Additionally this study demonstrates the reliability of both Illness and Treatment Index Scales among Greek population. Further research could determine whether these findings apply also to groups with different ethnicity living in Greece, and also the impact of more detailed cultural components such as religiosity.

Acknowledgements The author thanks patients, a friend, Kristina Xhaferaj, and the supervisor of this research, Dr. Paraskevi Theofilou.

Conflicts of Interest The author claims no conflict of interest. This project constitutes a thesis and no funding was received.

References

1. Guariguata, L., D. Whiting, C. Weil, and N. Unwin. 2011. The International Diabetes Federation Diabetes Atlas Methodology for Estimating Global and National Prevalence of Diabetes in Adults. *Diabetes Research and Clinical Practice* 94 (3): 322–332.
2. Hanberger, L. 2010. Quality of Care in Children and Adolescents with Type 1 Diabetes: Patients' and Healthcare Professionals' Perspectives. *Diabetes Care* 28 (1): 186–212.
3. The Diabetes Control and Complications Trial Research Group, D.M. Nathan, et al. 1993. The Effect of Intensive Treatment of Diabetes on the Development and Progression of Long-Term Complications and in Insulin-Dependent Diabetes Mellitus. *The New England Journal of Medicine* 329 (14): 977–986.
4. DiMatteo, M.R., K.B. Haskard, and S.L. Williams. 2007. Health Beliefs, Disease Severity, and Patient Adherence: A Meta-Analysis. *Medical Care 45* (6): 521–528.
5. Bendik, C.F., U. Keller, N. Moriconi, A. Gessler, C. Schindler, H. Zulewski, et al. 2009. Training in Flexible Intensive Insulin Therapy Improves Quality of Life, Decreases the Risk of Hypoglycaemia and Ameliorates Poor Metabolic Control in Patients with Type 1 Diabetes. *Diabetes Research and Clinical Practice* 83 (3): 327–333.
6. Engum, A., A. Mykletun, K. Midthjell, A. Holen, and A.A. Dahl. 2005. Depression and Diabetes A Large Population-Based Study of Sociodemographic, Lifestyle, and Clinical Factors Associated with Depression in Type 1 and Type 2 Diabetes. *Diabetes Care* 28 (8): 1904–1909.

7. Rousou, E., C. Kouta, and N. Middleton. 2016. Association of Social Support and Sociodemographic Characteristics with Poor Self-Rated Health and Depressive Symptomatology Among Single Mothers in Cyprus: A Descriptive Crosssectional Study. *BMC Nursing* 15 (1): 1.
8. Krantz, D.S., and D.C. Glass. 1984. Personality, Behavior Patterns, and Physical Illness: Conceptual and Methodological Issues. In *Handbook of Behavioral Medicine*, ed. W.D. Gentry, 38–86. New York: Guilford Press.
9. Shulman, R., J.D. Price, and J. Spinelli. 1989. Biopsychosocial Aspects of Long-Term Survival on End-Stage Renal Failure Therapy. *Psychological Medicine* 19 (04): 945–954.
10. Leventhal, H., R. Zimmerman, and M. Gutmann. 1984. Compliance: A Self-Regulation Perspective. In *Handbook of Behavioral Medicine*, ed. W.D. Gentry, 369–436. New York: Guilford Press.
11. Sacks, C.R., R.A. Peterson, and P.L. Kimmel. 1990. Perception of Illness and Depression in Chronic Renal Disease. *American Journal of Kidney Diseases* 15 (1): 31–39.
12. Aalto, A.M., A. Uutela, and A.R. Aro. 1997. Health Related Quality of Life Among Insulindependent Diabetics: Disease-Related and Psychosocial Correlates. *Patient Education and Counseling* 30 (3): 215–225.
13. Theofilou, P., A. Aroni, M. Ralli, M. Gouzou, and S. Zyga. 2013. Measuring Health: Related Quality of Life in Hemodialysis Patients. Psychometric Properties of the Missoula-VITAS. Quality of Life Index (MVQOLI-15) in Greece. *Health Psychology Research* 1 (2): 17.
14. Theofilou, P. 2015. Translation and Cultural Adaptation of the Multidimensional Scale of Perceived Social Support (MSPSS) for Greece. *Health Psychology Research* 3 (1): 45–47.
15. Greenberg, G.D., R.A. Peterson, and R. Heilbronner. 1997. *Illness Effects Questionnaire. Evaluating Stress: A Book of Resources*, 141–164. Lanham: Scarecrow.
16. Skinner, T.C., M. John, and S.E. Hampson. 2000. Social Support and Personal Models of Diabetes as Predictors of Self-Care and Well-Being: A Longitudinal Study of Adolescents with Diabetes. *Journal of Pediatric Psychology* 25 (4): 257–267.
17. Shrivastava, S.R., P.S. Shrivastava, and J. Ramasamy. 2013. Role of Self-Care in Management of Diabetes Mellitus. *Journal of Diabetes and Metabolic Disorders* 12 (1): 1.

Network Load Balancing Using Modular Arithmetic Computations

Stavros I. Souravlas and Angelo Sifaleras

Abstract Load-balanced routing has attracted considerable attention, especially in the recent years, where huge data volumes are carried over the computer networks. It is particularly important for non-all-to-all networks, where there is no direct communication between all the nodes of the network.

Telecommunication and network systems constitute complex dynamic systems with an ever-increasing number of users and network services. It has become apparent that, new routing demands can not be easily satisfied by conventional routing methods. Thus, intelligent optimization methods (e.g., nature-inspired methodologies) have arisen to improve network efficiency.

This paper presents a computational method that is based on modular arithmetic for achieving dynamic load balancing on data networks. The proposed algorithm organizes the overall communication into equal-sized packets, it divides the communication into a series of communication steps between the network nodes, and performs packet transfer. The last section includes discussion on the main costs each network routing operation inures: the data movement cost, the load information cost and the data reordering cost.

Keywords Networks • Routing • Load balancing

1 Introduction

Recently, telecommunication and network systems have become more complex and network problems have also been increased [1]. Perhaps, the most important problem that today's networks are facing is the efficiency improvement, in terms of minimizing the latencies, better utilizing the network sources like links, memory, etc and maximizing the overall throughput. To successfully deal with all these issues, researchers have developed several routing algorithms [2]. However, the increasing

S.I. Souravlas • A. Sifaleras (✉)
Department of Applied Informatics, University of Macedonia, 156 Egnatias Street, 54636 Thessaloniki, Greece
e-mail: sourstav@uom.gr; sifalera@uom.gr

P. Vlamos (ed.), *GeNeDis 2016*, Advances in Experimental Medicine and Biology 988, DOI 10.1007/978-3-319-56246-9_22

number of users and the continuously larger data volumes carried over all networks, necessitate the improvement of these techniques, so that many applications from different fields like scientific computations, medicine, biology [3, 4], etc. can be satisfied.

One of the most important methods towards increasing network efficiency in contemporary network systems is load-balancing. Load balancing is a technique used to keep the packets transmitted across the network as balanced as possible, such that all the communication links are equally loaded during a communication process and accordingly, the nodes involved receive almost equal data volumes at the end of the process. Due to its importance, load balancing has achieved considerable attention from many researchers. Several methods have been proposed in the literature to assure of balanced communication. These methods are divided into two main categories: *static* and *dynamic* [5–8]. In static methods, communication scheduling is predefined. On the contrary, a dynamic method takes into account the state of the system and the running application's parameters before taking any scheduling decision.

Generally, routing incurs three important costs: (1) *Data movement cost:* The cost of transferring units of data. Apparently, if good balancing is achieved, the data movement cost should be approximately the same for all network links at a time, (2) *Load information cost:* This cost incurs when a node requests information regarding the system's state. for example, availability of links, nodes with less load, etc., (3) *Data reordering cost:* This cost incurs when a node receives packets from different sources and has to combine all these packets in a single data packet. Then, a data reordering phase has to take place in the node's memory, increasing the total cost of routing, especially if this newly formed data packet has to be further transferred.

A number of network applications require that data elements from some nodes are packed together and form a bigger data packet delivered to a new node. An example is the Fast Fourier Transform (FFT), which necessitates that some network nodes receive and buffer packets of data samples while others process the data sets. Thus, a series of data samples may need to be combined into a single packet, which is then sent to a node for further processing. FFT has many applications, for example in Body Sensor Networks (BSN), a wireless sensing technology for healthcare applications.

In this work, we propose a load-balancing technique based on modular arithmetic, which applies in cases where larger data packets, formed by the data elements of some source network nodes, are subsequently transferred to a target node. The goal is to assure that, at any time, the data volumes carried over the network are equal between the links, thus the network performance is improved. The remaining of this paper is organized as follows: Sect. 2 briefly describes the related work. Section 3 presents the proposed load-balancing technique and performs computational analysis. Finally, Sect. 4 concludes the paper and offers aspects for future research.

2 Related Work

Numerous load-balancing techniques over data networks have been proposed in the literature. Some of them consider the load balancing problem as the problem of dividing a graph of all nodes to a finite number of partitions which are mapped then on different servers. To do this, some strategies take advantage of the geometric information of the nodes [9] or their graph connectivity [10]. Others investigate the possibility of controlling the stability of the load balancing system when the system scaling increases and propose techniques [11], while others focus on achieving the dynamic load balancing on MPI applications by creating new task scheduling techniques [12].

In [8], two load balancing algorithms are proposed; MELISA and LBA. A set of equations is defined, which is used to compute the loading parameters, like arrival and service rates. In MELISA, every node uses the model to estimates its arrival rate, service rate, and load, and then the load of its neighbors. These estimations are performed at predefined estimation time periods T_c. Then, each network node tries to distribute its load among its neighbors in such a way, that the processing of this load finishes almost simultaneously. For smaller networks, LBA (Load Balancing on Arrival) was developed. This algorithm achieves balancing by transferring data upon arrival. This approach is much faster, because it does not have to spend time for the estimation time period. All estimations are performed upon data arrival.

In [13], a novel set of load balancing protocols named ECLB (Earliest Completion Load Balancing) was proposed. ECLB includes five different protocols: (1) No Load Balancing, a strategy that does not balance the load, but transfers based on the total latency, (2) Basic Load Balancing, a strategy that transfers to the node with the lowest load, (3) Nearest Neighbor Load Balancing, a technique that transfers data from a source to its nearest neighbor, (4) Earliest Completion Load Balancing, a technique that estimates the time it takes to complete a transfer based on the neighbors' load and latency and selects the target node based on this estimation, and (5) Random Neighbor Load Balancing, a technique used to randomly chose a node from a list of available nodes. Randomized protocols were also proposed be Karger et al. [14].

In [15], the authors proposed a load-balancing technique that includes two basic load balancing operations: `NBRASJUST` and `REORDER`. The first is used to handle the sharing of the calling node with its neighbor, while the second moves a node to another location for sharing reasons. After these operations are performed, the network information regarding the load changes and routing is accordingly adjusted. An improved extension of this work was presented by Chawacheat et al. in [16], where for each data insertion or deletion in a node, the two balancing operations are called only once, as opposed to [15], where the calls are arbitrarily many.

Konstantinou et al. [17] presented a wave-based model to balance the load, after examining the load of the neighboring nodes. The idea is that a set of nodes transfers their load to a common neighbor and then share the load of the last node of the wave.

3 Load-Balancing Using Modular Arithmetic

In this section, the load-balancing routing algorithm will be described and afterward some performance issues will be discussed.

3.1 The Routing Strategy

To set up our problem, we initially define some variables. The network consists of N nodes and each node stores data in the form of data packets, where each packet consists of s elements (these packets will be referred to as initials), typically each element is a byte or multiple bytes. At any time, the nodes may need to build up larger data packets consisting of t elements (these packets will be referred to as finals), $s < t$. In other words, these larger data packets will contain data elements from a number of nodes.

Now, let us consider a big data set, with data packets equally distributed across the network nodes, say in a round robin fashion. For example newly produced data samples, stored in a round robin fashion throughout the network. Then, a data element indexed by i of a data packet indexed by l and local position inside the packet indexed by x, stored in node n_p, $p \in [0 \ldots N-1]$, can be described by $i = (lN + n_p)s + x$. Now, if we assume that larger data packets of size t have to be created to satisfy a specific application, then the data element just mentioned will become the data element indexed by j of a new data packet indexed by m and local position inside the packet indexed by y, stored in node n_q, $q \in [0 \ldots N-1]$. Thus, it will be described by $j = (mN + n_q)t + y$. Under these assumptions, we can describe this transfer by:

$$(lN + n_p)s + x = (mN + n_q)t + y \tag{1}$$

The number of variables (l, m, x, y) that solve Eq. (1) is the number of elements that need to be transferred between nodes n_p and n_q. Apparently, the data packets of size t will be formed by data elements from different sources. So, our problem is to define, for any values of N, s, and t, which data elements will be the initial ones in the final packets, which elements will follow, and which will be the last ones. Proposition 1 gives us the starting point:

Proposition 1 *The first elements (or bytes) of every final packet to be stored in n_q satisfies:*

$$-x \bmod N = \Phi, \tag{2}$$

where Φ is an integer mod N.

Proof Equation (1) can be rewritten as $lNs + n_p s + x = mNt + n_q t + y$, so it follows that $mNt - lNs = (x - y) + n_p s - n_q t$. However, N divides Ns and Nt, so $mNt - lNs$ is a multiple of N, or $mNt - lNs = \lambda N$, where λ is an integer. By setting $\phi = x - y$, Eq. (1) is rewritten as:

$$\lambda N - \phi = n_p s - n_q t \tag{3}$$

Now, if we divide (3) by N and get the modulo:

$$\begin{aligned} &(\lambda N - \phi) \bmod N = (n_p s - n_q t) \bmod N \\ &\Rightarrow (\lambda N) \bmod N - (\phi \bmod N) = (n_p s - n_q t) \bmod N \\ &\Rightarrow (-\phi) \bmod N = (n_p s - n_q t) \bmod N \\ &\Rightarrow (y - x) \bmod N = (n_p s - n_q t) \bmod N \end{aligned} \tag{4}$$

Because y is indexing final packets, it follows that a value of $y = 0$ is indexing the first bytes of a final data packet. By setting $y = 0$ and $\Phi = n_p s - n_q t$ to Eq. (4), we complete the proof. □

Thus, the first bytes of the final data packets to be stored to all n_qs will be found by the solutions of the following system of equations and constraints:

$$(y - x) \bmod N = (n_p s - n_q t) \bmod N, \text{ for all } (p, q) \in [0 \ldots N - 1] \tag{5}$$

$$x \leq s - 1 \text{ (initial packet size is } s) \tag{6}$$

$$y = 0 \tag{7}$$

The number of solutions for Eqs. (5)–(7) is given by the Initial Data Volume Transmitted *IDVT* function:

$$IDVT(n_p, n_q) = \{(l, m, x, y) : Eqs. (5)\text{–}(7)\} \tag{8}$$

Now, we can easily find the nodes that will contribute the remaining bytes of the final packets, as described by Proposition 2:

Proposition 2 *The remaining elements (or bytes) of every final packet to be stored in n_q are found by repeatedly solving a system similar to the one described by Eqs. (5)–(7), until no more solutions can be found.*

Proof Initially, let us divide a final packet of size t in three parts:

- Starting part t_{start}, that contains the first elements, as computed by Eqs. (5)–(7)
- Middle part t_{middle}, that contains all the data elements not in t_{start} and not in t_{end} (see below)
- Ending part t_{end}, the data elements at the very end of the packet

Now, for any node n_q, t_{start} has been found by solving Eqs. (5)–(7). The size of t_{start}, that is, the number of its bytes is given by IDVT [Eq. (8)]. So these elements will be found in positions indexed from 0, to $DVT(n_p, n_q) - 1$ of the final packet, and there remains to be found other $t - DVT(n_p, n_q)$ elements. It follows, that to find the middle part of a final packet for n_q, we have to solve a similar system like the one described in Eqs. (5)–(7), but with changes regarding y:

$$(y - x) \bmod N = (n_p s - n_q t) \bmod N, \text{ for all } (p, q) \in [0 \ldots N - 1] \tag{9}$$

$$x \leq s - 1 \tag{10}$$

$$y \in [IDVT(n_p, n_q) \ldots t - 1] \text{ (final packet size is } t) \tag{11}$$

Similarly, the Middle Data Volume Transmitted *MDVT* function computes the number of solutions for Eqs. (9)–(11):

$$MDVT(n_p, n_q) = \{(l, m, x, y) : Eqs.(9)\text{–}(11)\} \tag{12}$$

Now, let us express the middle part t_{middle} as the sum of $t_{M1} + t_{M2} + \ldots t_{Mk}$, where each of these summands is the number of the solutions (MDVT) for Eqs. (9)–(11), for a constant n_q and changing n_ps, from 0 to $N - 1$. Apparently, as we keep finding solutions, $t_{M1} + t_{M2} + \ldots t_{Mk}$ will reach $t - DVT(n_p, n_q)$, the number of elements apart from t_{start}, that need to be found for each final packet. Recall that the number of such solutions is finite, since x and y are limited by s and t, respectively. Thus, one of the solutions for Eqs. (9)–(11) will also be t_{end}, the ending part of the packet and this completes the proof. □

Now, it is easy to organize the data transmission in such a way, that we achieve load balancing between the network links. We need a definition beforehand.

Definition 1 A *packet set* PS_C is a set of packets that have C elements (bytes), where C is an integer, $C \leq t$.

Now, to achieve load balancing, we follow four phases described below:

`Phase 1:` Solve Eqs. (5)–(7) for all (n_p, n_q) that satisfy Eq. (4) to get all t_{start}s. These will have different IDVTs, so simply put all t_{start}s with equal IDVTs in the same packet set.

`Phase 2:` Transfer the packet sets found in `Phase 1`, in a series of communication steps, one packet set per step.

`Phase 3:` Solve Eqs. (9)–(11) for all (n_p, n_q), to get all t_{middle}s and t_{end}s. Again, these will have different MDVTs, so simply put all t_{middle}s with equal MDVTs in the same packet set. Do the same for all t_{end}s.

`Phase 4:` Repeat `Phase 2`, for the packet sets found in `Phase 3`.

From Steps 2 and 4, it is clear that same packet sizes are transferred between all the network nodes, thus we achieve load balancing in all network links. The next paragraph will give more details about the performance of our strategy, regarding load-balancing. Algorithm 1 gives the pseudocode of the proposed algorithm.

```
   input : Number of nodes N, initial packet size, s, final packet size, t
   output: Load balanced packet transmission, final packets of size t stored in the nodes of
           the network.
 1 Phase 1;
 2 Read N, s, and T ;
 3 for p ← 0 to N − 1 do
 4     // Keep n_p constant and run through all n_qs
 5     for q ← 1 to N do
 6         if the pair (n_p, n_q) satisfies Eq. (4) then
 7         Solve Eqs. (5)–(7) for (n_p, n_q)
 8         Find C = IDVT((n_p, n_q))
 9         Add communication (n_p, n_q) to packet set PS_C
10         else Move to next pair (n_p, n_q);
11         // All packet sets have been created
12     end
13 end
14
15 Phase 2;
16 //(Data Transmission, one communication step per packet set)
17 for C ← 1 to t do
18     if PS_C exists then Implement all communications added to it (see Line 8)
19     else Move to next packet set C;
20     // All nodes now have the first bytes of their final packets
21 end
22
23 Phase 3;
24 Read w = 0 //(w will keep the total number of solutions);
25 for p ← 0 to N − 1 do
26     // Keep n_p constant and run through all n_qs
27     for q ← 1 to N do
28         while w < t − DVT(n_p, n_q) do
29             Solve Eqs. (9)–(11) for (n_p, n_q)
30             Find C = MDVT((n_p, n_q))
31             Add communication (n_p, n_q) to packet set PS_C
32             w = w + MVDT
33         end
34         w ← 0 //(to start the next pair)
35     end
36     // All packet sets have been created
37 end
38
39 Phase 4;
40 Repeat Phase 2, for the packet sets found in Phase 3;
```

Algorithm 1: The four stages of the network load balancing algorithm

3.2 Discussion on Routing Costs and Other Properties

In this section we briefly discuss some cost issues and other properties of the proposed method. The proposed strategy has the following properties regarding the costs described in the introductory section:

1. It's *data movement cost* is at least equal to the cost of any other algorithm that handles the problem of transferring packets of different sizes over the network, like the one described in this paper, where different nodes have to contribute packets of different sizes to the formulation of larger packets. This cost is generally dictated by the number of communication steps multiplied by the number of bytes transferred per step. However, our strategy has two advantages that can be proven important: (a) Because each communication step includes equal-sized packets (load balancing), all communications per step are completed at approximately the same time, thus data link delays are avoided. These delays are present when packets of different sizes are transmitted across the network, so the "faster" links remain idle, waiting for the next transmission, and (b) Due to load balancing, it can work equally-well in cases when there are no direct connections between all the network nodes. The difference is that more communication steps through relay nodes would be required.
2. The system information requests can be handled like data packets, thus the *load information cost* can be restricted. There are two ways in which a node requests information regarding the system's state: (a) Through a centralized server, a case which is of no interest for our strategy, and (b) Through a request to the node it is interested. In the second case, even if the requests vary in terms of total cost, we can still use the proposed strategy to organize these submissions to equal-sized packets, especially in case where these requests are sent in a round-robin like fashion (for example, every node to its neighbor).
3. It's *data reordering cost* can be eliminated with minor changes. This is a big strength of the proposed scheme. We can organize the communication in such a way that the data packets are transmitted "in turn" so that the receiving nodes do not have to do any data reorganization upon receiving the packets. Attention must be paid, so that the selected packets that "take turns" to be transferred to n_qs must be further organized to packet subsets.

Two more important properties are worth to be mentioned [14, 16]:

1. The minimum data load transferred never decreases. Indeed, at any time there is a minimum load on the links, varied from 1 to $t - 1$ bytes (assuming that there are at least two contributing nodes n_p to each final packet).
2. In cases where all the nodes are both sources and targets, their load after transmission (the packets stored in their memories) also remains balanced.

4 Conclusions—Future Work

This work proposed a novel network load balancing algorithm in cases where different nodes have to unequally contribute to the formulation of larger packets, that must be transmitted across the network. This computational method uses modular arithmetic to compute the bytes transmitted and it is composed of four stages: In the first stage, the first bytes of each packet are found, in the second stage, these bytes are transmitted to their destination, in the third stage the middle and last bytes of these packets are found, and in the fourth and last stage, these bytes are transmitted. The first and third stage involve some type of grouping the packets into sets of equal sizes, thus achieving load balancing.

Our future work includes testing on real networks with real application data. Another challenge is to test the algorithm in real network topologies and make comparison. Computer networks which are not fully interconnected are of particular interest, because some type of relaying will be required. Finally, the development of other mathematical models on the same problems is also a subject that attracts attention.

References

1. Bhatia, N., R. Kundra, A. Chaurasia, and S. Chandra. 2012. *Load Balancing Using Hybrid ACO - Random Walk Approach*, 402–412. Berlin: Springer.
2. Sifaleras, A. 2015. *Classification of Network Optimization Software Packages*, 3rd ed., 7054–7062. Hershey, PA: IGI Global.
3. Dressler, F., and O.B. Akan. 2010. A survey on bio-inspired networking. *Computer Networks* 54(6): 881–900.
4. Meisel, M., V. Pappas, and L. Zhang. 2010. A taxonomy of biologically inspired research in computer networking. *Computer Networks* 54(6): 901–916.
5. Deng, Y., and R.W.H. Lau. 2012. On delay adjustment for dynamic load balancing in distributed virtual environments. *IEEE Transactions on Visualization and Computer Graphics* 18(4): 529–537.
6. El Kabbany, G.F., N.M. Wanas, N.H. Hegazi, and S.I. Shaheen. 2011. A dynamic load balancing framework for real-time applications in message passing systems. *International Journal of Parallel Programming* 39(2): 143–182.
7. Kim, S.S., E.A. Smith, and S.J. Hong. 2008. *Dynamic Load Balancing Using an Ant Colony Approach in Micro-Cellular Mobile Communications Systems*, 137–152. Berlin: Springer.
8. Shah, R., B. Veeravalli, and M. Misra. 2007. On the design of adaptive and decentralized load balancing algorithms with load estimation for computational grid environments. *IEEE Transactions on Parallel and Distributed Systems* 18(12): 1675–1686.
9. Steed, A., and R. Abou-Haidar. 2003. Partitioning crowded virtual environments. In *Proceedings of the ACM Symposium on Virtual Reality Software and Technology, VRST '03*, 7–14. New York: ACM.
10. Prasetya, K., and Z.D. Wu. 2008. Performance analysis of game world partitioning methods for multiplayer mobile gaming. In *Proceedings of the 7th ACM SIGCOMM Workshop on Network and System Support for Games, NetGames '08*, 72–77. New York: ACM.
11. Meng, Q., J. Qiao, J. Liu, and S. Lin. 2009. Research on the stability of load balancing algorithm for scalable parallel computing. In *Proceedings of the International Conference on Communication Software and Networks (ICCSN '09)*, 309–312.

12. Nian, S., and Guangmin, L. 2009. Dynamic load balancing algorithm for mpi parallel computing. In *Proceedings of the International Conference on New Trends in Information and Service Science, NISS '09*, 95–99. Washington, DC: IEEE Computer Society.
13. Haque, W., A. Toms, and A. Germuth. 2013. Dynamic load balancing in real-time distributed transaction processing. In *16th IEEE International Conference on Computational Science and Engineering*, 268–274.
14. D.R. Karger, and M. Ruhl. 2004. Simple efficient load balancing algorithms for peer-to-peer systems. In *Proceedings of the Sixteenth Annual ACM Symposium on Parallelism in Algorithms and Architectures, SPAA '04*, 36–43. New York: ACM.
15. Ganesan, P., M. Bawa, and H. Garcia-Molina. 2004. Online balancing of range-partitioned data with applications to peer-to-peer systems. In *Proceedings of the Thirtieth International Conference on Very Large Data Bases - Volume 30, VLDB '04*, 444–455. VLDB Endowment.
16. Chawachat, J., and J. Fakcharoenphol. 2015. A simpler load-balancing algorithm for range-partitioned data in peer-to-peer systems. *Networks* 66(3): 235–249.
17. Konstantinou, I., D. Tsoumakos, and N. Koziris. 2011. Fast and cost-effective online load-balancing in distributed range-queriable systems. *IEEE Transactions on Parallel and Distributed Systems* 22(8): 1350–1364.

A Quantum Inspired GVNS: Some Preliminary Results

Christos Papalitsas, Panayiotis Karakostas, and Kalliopi Kastampolidou

Abstract GVNS is a well known and widely used metaheuristic for solving efficiently many NP-Hard Combinatorial Optimization problems. In this paper, the qGVNS, which is a new quantum inspired variant of GVNS, is being introduced. This variant differs in terms of the perturbation phase because it achieves the shaking moves by adopting quantum computing principles. The functionality and efficiency of qGVNS have been tested using a comparative study (compared with the equivalent GVNS results) in selected TSPLib instances, both in first and best improvement.

Keywords Quantum inspired algorithms • qVNS • Variable neighborhood search • VNS • Metaheuristics • Optimization • TSP

1 Introduction

Many complex real world problems can be formulated as Combinatorial Optimization (CO) problems. From a technical point of view, CO problems involve finding a proper solution from a discrete finite set of solutions in order to achieve both the minimization (or maximization) of a cost function and the fulfillment of certain given constraints.

The Traveling Salesman Problem (TSP) is one of the most widely studied combinatorial optimization problems. Solving the TSP means finding the minimum cost route through which the traveler can start from an origin point and return to the same one after passing from all given nodes once. In graph representation, the problem is defined as a complete graph $G = (V, A)$, in which V is the set of the nodes $V = (v_1, v_2, \cdots)$ and $A = ((v_i, v_j), i, j \in V, i \neq j)$ the set of the arcs. Each arc is associated with a weight c_{ij} representing the cost (or the distance) of moving from

C. Papalitsas (✉) • K. Kastampolidou
Department of Informatics, Ionian University, Corfu, Greece
e-mail: c14papa@ionio.gr; p12kast@ionio.gr

P. Karakostas
Applied Informatics Department, University of Macedonia, Thessaloniki, Greece
e-mail: pkarakostas.tm@gmail.com

P. Vlamos (ed.), *GeNeDis 2016*, Advances in Experimental Medicine and Biology 988, DOI 10.1007/978-3-319-56246-9_23

node i to node j. If the c_{ij} is equal to the c_{ji}, the TSP is symmetric (sTSP), otherwise it is called asymmetric (aTSP). Also, the TSP is a NP-Complete CO problem and as it is known there is no any polynomial-time algorithm for finding an optimal solution regardless of the size of the problem instance [1].

For the acceleration of computational times, sacrificing some of the solution's quality by adopting heuristic and metaheuristic approaches is commonly accepted [2]. Heuristics are fast approximation computational methods divided into construction and improvement heuristics. Construction heuristics are used to build feasible initial solutions and improvement heuristics are applied to achieve better solutions. It is customary to apply improvement heuristics iteratively. Metaheuristics are general optimization frameworks which could be modified properly in order to generate heuristic approaches for solving specific optimization problems.

A well known issue on Computer Science is that Moore's law (a statement that computer power doubles every 2 years at the same cost) tends to reach its natural limitations. Moore's law, von Newmann machines etc. are parts of the classical computational model. Because of these limitations, the scientific community seems to have a tendency to revolve around new "alternative" computation ways and methods. These methods used to be called Unconventional methods. In general, unconventional computing is a wide range of new or unusual proposed computing models. A part of these computational models is Natural Computing.

We can define Natural computing as the procedure that make us to think computationally about nature and also to think "naturally" about the computational processes [3]. Nature inspired computing has emerged as an efficient paradigm to design and simulate innovative computational models inspired by natural phenomena to solve complex non linear and dynamic problems. Some of the well known computational systems and algorithms inspired by nature are:

1. Evolutionary algorithms inspired by biological systems.
2. Swarm intelligence algorithms inspired by the behavior of swarm/group of agents.
3. Social and Cultural Algorithms inspired by human interactions and beliefs in the society.
4. Quantum Inspired Algorithms, inspired by quantum physics [4].

2 Quantum Inspired Metaheuristics

2.1 Quantum Computing Principles

Quantum inspired procedures use the basic principles of Quantum Computing.Quantum computing, a subsection of Natural Computing and a field that was introduced relatively recently, in the 1980s, by Richard Feynman, who claimed that the efficient simulation of an actual quantum system using a classical computer is not possible, since there would be an exponential slowdown when simulating actual quantum processes [5, 6]. Quantum computing, a general notion that recognizes the computing process as a natural phenomenon, it could be an important addition to the existing standard computation processes.Quantum computing combines definitions, formalisms, and concepts from several natural sciences besides computer science. It mainly involves mathematics (e.g. linear algebra) and physics (quantum mechanics).

There are two types of quantum states: pure and mixed states. A pure state is a state represented by a single vector $|\chi\rangle$ over a complex Hilbert space. A mixed state is a statistical distribution of pure states expressed in the form of density matrices. A quantum register is the quantum analogue of a classical processor register. A mathematical description of a quantum register is achieved by using tensor products of qubit bra or ket vectors. More specifically, a n qubit quantum register is applied by the element: $|\chi\rangle = |\chi\rangle 1 \otimes |\chi\rangle 2 \otimes \cdots \otimes |\chi\rangle$n, in the tensor product Hilbert space: $H = H_1 \otimes H_2 \otimes \cdots \otimes H_n$, where n is the number of qubits. Our implementation is inspired by these quantum principles [7].

2.2 *Quantum Inspired Procedures in Literature*

The research community has shown great interest in the application of quantum inspired algorithms in metaheuristic procedures and in a type of algorithm, in general. Numerous studies have been published on Quantum Inspired procedures.

In 1996, a novel algorithm called quantum-inspired algorithm was proposed. It combined the quantum mechanism and the evolution computing characters which is why this Quantum-inspired algorithm is also called the quantum evolution algorithm (QEA) [8].

Sandip et al. proposed techniques which are Quantum Inspired Ant Colony Optimization, Quantum Inspired Differential Evolution and QuantumInspired Particle Swarm Optimization for Multi-level Colour Image Thresholding.These techniques find optimal threshold values at different levels of threshold-ing for colour images [4].

A quantum inspired Social Evolution algorithm is proposed by hybridizing Social evolution algorithm with the emerging quantum-inspired evolutionary algorithm. The proposed QSE algorithm is applied on a well-known 0–1 knapsack problem and the performance of the algorithm is compared to various evolutionary, swarm and quantum inspired evolutionary algorithm variants. Pavithr et al. claim that the performance of the QSE algorithm is better than or comparable to the different evolutionary algorithmic variants it is tested against [9].

Wei Fang et al. proposed a decentralized form of quantum-inspired particle swarm optimization with a cellular structured population for maintaining population diversity and balancing global and local search [10].

Zheng et al. introduced an interesting study where a novel Hybrid Quantum-Inspired Evolutionary Algorithm is applied to a permutation flow-shop scheduling problem. This new method initially proposes a simple representation method for the determination of job sequence in the PFSSP based on the probability amplitude of the information of qubits. Then the quantum chromosomes are encoded and decoded by using the quantum rotating angle [11].

Finally, Lu et al. presented a quantum inspired space search algorithm in order to solve numerical optimization problems. In the proposed algorithm, the feasible solution is decomposed into regions in terms of quantum representation. As the search progresses from one generation to the next, the quantum bits evolve gradually to increase the probability of region selection [12].

3 Quantum Variable Neighborhood Search: qVNS

3.1 Variable Neighborhood Search: VNS

Variable Neighborhood Search (VNS) is a metaheuristic proposed by Mladenovic and Hansen [13, 14]. The main idea of this framework is the systematic neighborhood change in order to achieve an optimal (or a close-to-optimal) solution [15]. Also, VNS and its extensions have been proven efficient in solving many combinatorial and global optimization problems [16, 17].

Each VNS heuristic consists of three parts. The first one is a shaking procedure (diversification phase) used to escape local optimal solutions. The next one is the neighborhood change move, in which the following neighborhood structure that will be searched is determined; during this part, an approval or rejection criterion is also applied on the last solution found. The third part is the improvement phase (intensification) achieved through the exploration of neighborhood structures through the application of different local search moves. Variable Neighborhood Descent (VND) is a method in which the neighborhood change procedure is performed deterministically [18]. General Variable Neighborhood Search (GVNS) is a VNS variant that VND method is used as improvement procedure. GVNS has been successfully applied in many applications [19, 20].

3.2 Description of the qGVNS

Like the classic GVNS, the quantum-inspired GVNS (qGVNS) consists of a VND local search, a shaking procedure and a neighborhood change step. More specifically, in our proposed method, the pipe-VND (exploration in the same neighborhood while improvements have also been made). The pipe-VND is comprised of two well-known neighborhood structures, the relocate local search operator (solutions obtained by node's relocation in tour) and 2-opt (solutions obtained by break and different reconnect of two tour arcs). The main difference between qGVNS and classic GVNS has been mentioned in the diversification phase. In our approach, perturbation has been achieved by adopting quantum computation principles.

In each shaking call a quantum register will generate a number of necessary qubits (for example n), and from these eigenvalues will be generated, at levels, which are equal or higher than the number of the nodes in tour. Then our algorithm will serially choose all the eigenvalues required in our problem's instance and put them in a 1×N vector. In addition, each one of them will be matched to each node of our current solution. Then, the eigenvalues will be used as a flag to each node of our current solution. Note, that eigenvalues can be $0 \leq C \leq 1$. Because of the matching between eigenvalues and nodes in tour, the sorting in the first vector will affect the node's order in the solution vector. So the ordered route will be the route occurring after the shaking move, driving our exploration effort in another search space.

At this point, it should be mentioned that the Nearest Neighbor heuristic is used in order to produce an initial feasible solution (as depot is set the first node). From an algorithmic perspective, qGVNS is summarized in the next pseudocode.

```
Data: an initial solution
Result: an optimized solution
Initialization of the feasibility distance matrix
begin
  X < − Nearest Neighbor heuristic;
  repeat
    X' < − Quantum-Perturbation(X)
    X" < − pipeVND(X')
    if X" is better than X' then
      X < − X"
    end
  until optimal solution is found or time limit is met;
end
```

Pseudocode of qGVNS

4 Experimental Results

The proposed qGVNS was implemented in Fortran and compiled by the Intel Fortran 64 Compiler XE v. 14.0.1.106. Also, it was applied on 16 benchmark TSPLib instances (9 sTSPs and 7 aTSPs). The experiments were carried out on a laptop PC running MS Windows 8 with an Intel Core i7-3610QM CPU at 2.30 GHz and 4 GB RAM. The finding of an optimal solution or a maximum 60-s limit were set as stopping criteria for the tests.

Furthermore, each instance was solved five times. Initially, a comparison between first and best improvement search strategies of both qGVNS and the equivalent GVNS (pipeVND: relocate/2-opt & shaking: relocate/2-opt) was presented. Then, the best results achieved by GVNS have been compared with the results of the most efficient search strategy of qGVNS.

As it is shown in Tables 1 and 2, the best improvement search strategy of qGVNS has been the most efficient strategy for both type of instances. On Table 3 it is evident that this qGVNS detects the optimal solution at seven of nine in total

Table 1 qGVNS on sTSP instances

	Best improvement			First improvement		
Problem	Av. value	AV. CPU time	Best value	Av. value	AV. CPU time	Best value
bayg29	1610	18.84 s	1610	1610	20.27 s	1610
bays29	2020	1.23 s	2020	2020	26.47 s	2020
fri26	937	0.27 s	937	937	0.59 s	937
gr17	2085	0.12 s	2085	2085	0.03 s	2085
gr24	1272	5.52 s	1272	1272	2.29 s	1272
ulysses16	6859	0.01 s	6859	6859	0.05 s	6859
ulysses22	7013	0.01 s	7013	7013	0.02 s	7013
gr48	5049.4	37.67 s	5046	5056.8	53 s	5046
hk48	11,508.6	43.3 s	11,461	11,519	60 s	11,470

Table 2 qGVNS on aTSP instances

	Best improvement			First improvement		
Problem	Av. value	AV. CPU time	Best value	Av. value	AV. CPU time	Best value
br17	39	0	39	39	0	39
ftv33	1357.8	60 s	1345	1444	60 s	1420
ftv35	1544.6	60 s	1516	1639.4	60 s	1615
ftv38	1616.6	60 s	1593	1749.4	60 s	1635
ftv44	1763.8	60 s	1739	1937	60 s	1895
p43	5654	60 s	5625	5661.4	60 s	5636
ry48p	14,698.2	60 s	14,651	14,991	60 s	14,824

Table 3 Optimal values of sTSP instances

Problem	Optimal	qGVNS
bayg29	1610	1610
bays29	2020	2020
fri26	937	937
gr17	2085	2085
gr24	1272	1272
ulysses16	6859	6859
ulysses22	7013	7013
gr48	5048	5049.4
hk48	11,461	11,508.6

benchmarks referring to symmetric TSP problems. However, this proposed qGVNS, applied to benchmarks containing asymmetric TSP's seems to produce improved solutions see: Table 4, but with some deviations from the optimal solution. These deviations as shown by the results on Table 4 are small. However, the most important results worth noting, are given in Table 7. Here, it is quite clear that qGVNS has achieved better results than the equivalent GVNS in terms of solution quality for

Table 4 Optimal values of aTSP instances

Problem	Optimal	qGVNS
br17	39	39
ftv33	1286	1357.8
ftv35	1473	1544.6
ftv38	1530	1616.6
ftv44	1613	1763.8
p43	5620	5654
ry48p	14,422	14,698.2

Table 5 GVNS on sTSP instances

	Best improvement		First improvement	
Problem	Best value	CPU time	Best value	CPU time
bayg29	1666	60s	1653	60s
bays29	2119	60s	2069	60s
fri26	971	60s	969	60s
gr17	2085	60s	2085	60s
gr24	1364	60s	1278	60s
ulysses16	6859	0.01s	6859	0.02s
ulysses22	7013	0.01s	7013	0.01s
gr48	5424	60s	5325	60s
hk48	11,884	60s	12,045	60s

Table 6 GVNS on aTSP instances

	Best improvement		First improvement	
Problem	Best value	CPU time	Best value	CPU time
br17	39	0	39	0
ftv33	1489	60s	1683	60s
ftv35	1791	60s	1791	60s
ftv38	1778	60s	1778	60s
ftv44	2014	60s	2014	60s
p43	5629	60s	5866	60s
ry48p	15,134	60s	16,757	60s

the small TSP instances (and CPU time in sTSP instances—see Tables 1 and 5). At the present point, it should be clarified that the best value achieved between first and best improvement search strategies is selected as the best result of each instance (sTSP and aTSP) with GVNS (Tables 6 and 7).

Table 7 Comparison between qGVNS and GVNS

Problem	Av. value (qGVNS)	Best (GVNS)	Problem	Av. value (qGVNS)	Best (GVNS)
bayg29	1610	1653	br17	39	39
bays29	2020	2069	ftv33	1357.8	1489
fri26	937	969	ftv35	1544.6	1791
gr17	2085	2085	ftv38	1616.6	1778
gr24	1272	1278	ftv44	1763.8	2014
ulysses16	6859	6859	p43	5554	5629
ulysses22	7013	7013	ry48p	14,698.2	15,134
gr48	5049.4	5325			
hk48	11,508.6	11,884			

5 Conclusion and Future Work

In this work, a qGVNS for solving the TSP has been introduced and it has been proven to outperform the equivalent classic GVNS in terms of solution quality for the small sTSP and aTSP instances from TSPLib.

Future work may include extended comparative study between classic GVNS and qGVNS or/and different neighborhood structures and neighborhood change moves in VND. Also, the current GVNS scheme may be applied on many TSP variants.

References

1. Rego, Csar, Dorabela Gamboa, Fred Glover, and Colin Osterman. 2011. Traveling salesman problem heuristics: Leading methods, implementations and latest advances. *European Journal of Operational Research* 211(3): 427–441.
2. Papalitsas, Ch., K. Giannakis, Th. Andronikos, D. Theotokis, and A. Sifaleras. 2015. Initialization methods for the TSP with time windows using variable neighborhood search. In *IEEE Proceedings of the 6th International Conference on Information, Intelligence, Systems and Applications (IISA 2015)*, 6–8 July, Corfu.
3. De Castro, Leandro Nunes. 2006. *Fundamentals of Natural Computing: Basic Concepts, Algorithms, and Applications*. London: Chapman & Hall/CRC.
4. Dey, Sandip, Siddhartha Bhattacharyya, and Ujjwal Maulik. 2016. New quantum inspired meta-heuristic techniques for multi-level colour image thresholding. *Applied Soft Computing* 46: 677–702.
5. Feynman, Richard P. 1982. Simulating physics with computers. *International Journal of Theoretical Physics* 21(6–7): 467–488.
6. Feynman, Richard Phillips, J.G. Hey, and Robin W. Allen. 1998. Feynman lectures on computation.
7. Hagouel, Paul Isaac, and Ioannis G. Karafyllidis. 2012. Quantum computers: Registers, gates and algorithms. In *2012 28th International Conference on Microelectronics Proceedings*, May. Institute of Electrical and Electronics Engineers (IEEE).
8. Narayanan, A., and M. Moore. Quantum-inspired genetic algorithms. In *Proceedings of IEEE International Conference on Evolutionary Computation*.

9. Pavithr, R.S., and Gursaran. 2016. Quantum inspired social evolution (QSE) algorithm for 0–1 knapsack problem. *Swarm and Evolutionary Computation* 29: 33–46.
10. Fang, Wei, Jun Sun, Huanhuan Chen, and Xiaojun Wu. 2016. A decentralized quantum-inspired particle swarm optimization algorithm with cellular structured population. *Information Sciences* 330: 19–48.
11. Zheng, Tianmin, and Mitsuo Yamashiro. 2009. A novel hybrid quantum-inspired evolutionary algorithm for permutation flow-shop scheduling. *Journal of Statistics and Management Systems* 12(6): 1165–1182.
12. Lu, Tzyy-Chyang, and Jyh-Ching Juang. 2011. Quantum-inspired space search algorithm (QSSA) for global numerical optimization. *Applied Mathematics and Computation* 218(6): 2516–2532.
13. Mladenovic, N., and P. Hansen. 1997. Variable neighborhood search. *Computers & Operations Research* 24(11): 1097–1100.
14. Hansen, Pierre, Nenad Mladenovic, Raca Todosijevic, and Sad Hanafi. 2016. Variable neighborhood search: Basics and variants. *EURO Journal on Computational Optimization* 1–32. doi:10.1007/s13675-016-0075-x.
15. Mladenovic, Nenad, Raca Todosijevic, and Dragan Uroevic. 2016. Less is more: Basic variable neighborhood search for minimum differential dispersion problem. Inf. Sci. 326: 160–171.
16. Da Silva, Rodrigo Ferreira, and Sebastin Urrutia. 2010. A general VNS heuristic for the traveling salesman problem with time windows. *Discrete Optimization* 7(4): 203–211.
17. Jarboui, Bassem, Houda Derbel, Sad Hanafi, and Nenad Mladenovic. 2013. Variable neighborhood search for location routing. *Computers & Operations Research* 40(1): 47–57.
18. Sifaleras, Angelo, and Ioannis Konstantaras. 2017. Variable neighborhood descent heuristic for solving reverse logistics multi-item dynamic lot-sizing problems. *Computers & Operations Research* 78: 385–392.
19. Sifaleras, A., I. Konstantaras, and N. Mladenović. 2015. Variable neighborhood search for the economic lot sizing problem with product returns and recovery. *International Journal of Production Economics* 160: 133–143.
20. Sifaleras, A., and I. Konstantaras. 2015. General variable neighborhood search for the multi-product dynamic lot sizing problem in closed-loop supply chain. *Electronic Notes in Discrete Mathematics* 47: 69–76.

Self-Optimized Computational Method Calculating Robust b Values for Earthquake Catalogs Containing Small Number of Events

Konstantinos Arvanitakis, Romanos Kalamatianos, and Markos Avlonitis

Abstract In the Gutenberg-Richter relation that describes the frequency-magnitude distribution of earthquakes, the b value represents the distribution's slope. Since b values can be used for mapping the dynamic response of earthquake source, methodologies for calculating robust b values are of great importance. Although nowadays software which is meant for statistical analysis of earthquake data can determine b values with high accuracy, in occasions where catalogs that contain small number of earthquake events, the produced results are not satisfactory. In this paper we present a new self-optimized algorithm for a more efficient calculation of the b value. The algorithm's results are compared with two widely known software for statistical analysis of earthquake data, showing a better performance in evaluating b values for earthquake catalogs containing small number of events.

Keywords b Value • Statistical seismology • Correlation coefficient

1 Introduction

Classical seismology is a part of geophysics science, dealing with the physics of earthquakes and the propagations of seismic waves. In order to understand in a better way the earthquake phenomenon, geophysicists introduced statistical methods in seismology, forming what it is known as Statistical Seismology. This field's interest focuses on the results and knowledge that can be gained from the application of statistical theories on earthquake events. The statistical analysis uses earthquake signals describing single earthquakes or earthquake swarms (patterns of seismicity) and their propagation in time and space.

In order to monitor the earthquake activity and earth's surface behavior, vast networks of stations have been installed, equipped with modern instruments to record wave signals from earth's interior and deformations (slip) on the surface. These stations produce large amounts of data in real time. Due to the importance of

K. Arvanitakis (✉) • R. Kalamatianos • M. Avlonitis
Department of Informatics, Ionian University, Corfu, Greece 49100,
e-mail: c14arva@ionio.gr; c14kala@ionio.gr; avlon@ionio.gr

P. Vlamos (ed.), *GeNeDis 2016*, Advances in Experimental Medicine and Biology 988, DOI 10.1007/978-3-319-56246-9_24

the wanted information and the large amount of data, the need of fast and automated real time analysis emerged. Modern Information and Communication Technologies (ICT) which occupy the ability of massive real time computing power can fulfill this need. Therefore in the last decades a lot of software applications, meant for statistical analysis of earthquake data, were introduced.

The existing software contain a variety of functions such as catalog declustering, creating synthetic catalogs, calculating the Magnitude of completeness, estimating the a and b values of the Gutenberg-Richter law, etc. These software have been proven useful, by producing robust results in short amount of time. In earlier works such software were used to calculate the b value of the frequency magnitude distribution. After numerous experiments it was demonstrated that the implemented methods on these software, were not capable of producing robust results, in cases of catalogs with small number of events. All these results had to be revised manually, which is a time consuming process. Thus a new method of calculating the b value for catalogs with small number of events is needed.

In this work a method for calculating robust b values on catalogs with small number of events is presented. This method does not take into consideration the Magnitude of Completeness, and tries to identify in a self-optimized way the most linear part of the Guttenberg-Richter distribution of earthquakes, in order to produce reliable results for the b value of the corresponding distribution's slope.

2 General Methodology for Calculating b Values

A very well-known relation among seismologists is the Gutenberg Richter law. This law describes the relation between earthquake magnitudes and their frequency of occurrence [1]. This correlation is attributed from (1), where N(M) is the number of earthquakes with magnitude greater than M, a and b values are constants that describe a region's seismicity.

$$\log_{10} N(M) = a - bM \tag{1}$$

The plot of the cumulative frequency–magnitude distribution of a region's earthquakes appears to be a power law distribution. By logging the Y-axis of this plot, the distribution transforms in a straight line (Fig. 1).

For magnitudes smaller than approximate 3.2 as shown in Fig. 1 the plot doesn't follow a power law distribution and a straight line respectively on the distributions. This magnitude is a catalog's magnitude of completeness (Mc) [2]. The Mc separates the catalog in the complete and incomplete part. The complete part of the catalog contains all the earthquake events with magnitude > = Mc. All the

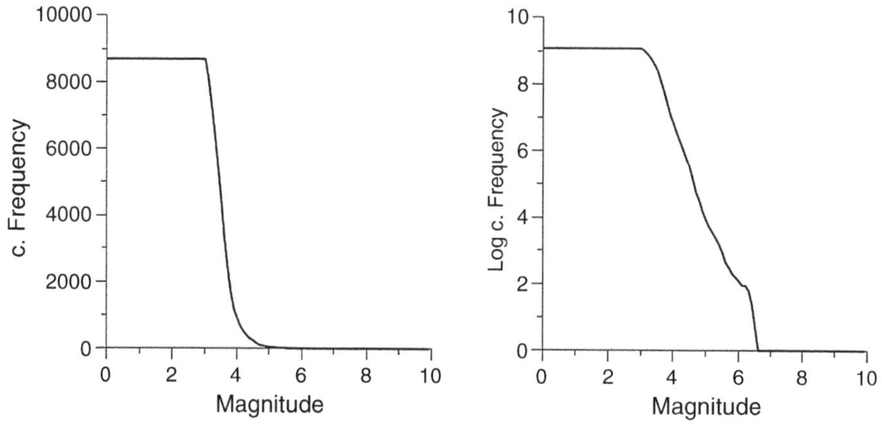

Fig. 1 Cumulative frequency-magnitude distribution and the corresponding logged distribution

remaining earthquake events compose the incomplete part of the catalog. The data of the incomplete part of the catalog are not considered robust enough for further experimentation.

Due to the incompetence of modern seismographs to record small magnitude earthquakes in great distance, the noise caused from human activity on earth's surface and the noise caused from natural phenomena of the planet, small magnitude earthquakes are not recorded sufficiently thus leading to the incomplete part of the catalog.

There are many methods to estimate a catalog's Mc, such as the Goodness of fit test [3] where synthetic power law distributions are compared to the cumulative frequency-magnitude distribution and the Day and night noise modulation method [4] which compares the seismic activity during day time with that of the night, to short out the noise that human activity creates on the data. The most popular and simplest method among seismologist to determine the Mc, is the Maximum Curvature technique. This method identifies a catalog's Mc as the magnitude with the biggest frequency of occurrence.

The most important parameter, for this study, is the b value of the Gutenberg Richter law as it is described from (1), which is the distribution's slope. This attribute describes the ratio of small and big earthquakes in an earthquake catalog. Furthermore, in literature [5, 6] it has been proposed that the variations of b value through time can produce useful information for the imminent seismic activity. Also, in more recent studies [7, 8] the spatial variations of b value has been used in locating asperities and the results were very promising.The most common method of calculating the b value, is the Maximum Likelihood Estimate of b [9] created by Keiiti Aki, were the b value is given from (2), where n is the total number of earthquake in the catalog, Mi is the magnitude of each event in the catalog and Mc the catalog's magnitude of completeness.

$$b = \frac{n \log(e)}{\sum_{i=1}^{n} (Mi) - nMc} \tag{2}$$

Due to the fact that (2) contains the attribute of Mc, the efficiency of this method to estimate the b value depends on the accuracy in which the Mc was calculated. In catalogs with small number of earthquake events, where the data are not capable of producing a smooth power law distribution, because of their own low population, the Maximum Curvature technique produces unreliable results. Most of the times the calculated Mc has a lower magnitude than it should.

As it is obvious, the misscalculated Mc leads on inefficient b value estimation. From the plots in Fig. 3 it is shown that even the bigger earthquakes of the catalog distort the distribution in a way that does not fit in a power law distribution. In their study [10] Stirling, Wesnousky and Shimazaki proposed that earthquakes of great magnitude, do not adhere to the Gutenberg-Richter law. Thus, it should be expected that earthquakes of great magnitude shall increase the error of the b value estimation.

As it is obvious, the accuracy of the estimation is depended on the magnitude range of the used data. Therefore a method that ensures a data range of high credibility in the b value calculation should be introduced.

3 Robust b Values for Earthquake Catalogs Containing Small Number of Events

The proposed method, presented in this paper, uses a self-optimized approach to determine the proper magnitude range on the G-R Distribution, for the b value calculation. The least squares line method is used to determine the b value.

In more details the algorithm starts by creating the frequency magnitude distribution and the cumulative log frequency magnitude distribution. The Magnitude of completeness (Mc) of the catalog is determined with the Maximum curvature technique. As it was mentioned earlier, this method may not be accurate in catalogs with small number of events. However, we can safely assume, that magnitudes smaller than the Mc don't belong in the complete part of the catalog, therefore they can't be used for the b value estimation. The catalogs Mc serves as a lower bound of magnitudes that will be tested by the algorithm to determine the optimum magnitude range for the b value calculation.

As it was mentioned in the previous section, large earthquakes don't follow the G-R Law and thus those earthquakes shall not participate in the b value calculation. The magnitude that excludes those big earthquakes will serve as the upper bound of the wanted magnitude range. The upper bound is identified as the first largest earthquake, followed by three magnitudes in a row (by step of 0.1) with 0 events.

Inside that range formed from the lower and upper bound, the algorithm determines the range of magnitudes greater than 1, which appears to have the smallest variability of data. To do so the correlation coefficient (3) is tested on all the possible combinations of ranges on the cumulative log distribution.

$$CV = \left(\frac{n\sum xy - \sum x \sum y}{\sqrt{\left[n\sum \left(x^2 - \left(\sum x\right)^2\right]\left[n\sum y^2 - \left(\sum y\right)^2\right]}} \right)^2 \tag{3}$$

The correlation coefficient is close to 1 for data with low variation and gets closer to 0 as the variation gets bigger. Thus, the magnitude range with a plot closer to a straight line will have the biggest correlation coefficient score. This magnitude range is selected as the optimum range for the b value calculation.

Finally, the b value is determined as the slope of the least square line fitting on the data of the selected magnitude range. The slope of the least square line is given by (4).

$$l = \frac{\Sigma xy - \frac{\Sigma x \Sigma y}{n}}{\Sigma x^2 - \frac{(\Sigma x)^2}{n}} \tag{4}$$

4 Results

Experiments were conducted on an earthquake catalog of Corinth Gulf in Greece. The data used in this study were taken from the Greek National Seismological Network and were compiled from the bulletins of the Central Seismological Station of Geophysics Department of the Aristotle University of Thessaloniki.[1] For every earthquake event recorded, the catalog contains the time (year/month/day/hour/minute), location (latitude/longitude/depth) and magnitude of the earthquake. The area's catalog was separated in a grid by 0.1 latitude and longitude degrees. For the purpose of this study 17 cells of the grid were selected and for each cell a sub-catalog, containing 50 earthquake events, was created. The coordinates of the corresponding areas can be seen in Table 1.

[1] http://geophysics.geo.auth.gr/ss/.

Table 1 Latitude and Longitude degrees of the examined areas

Area ID	Latitude	Longitude
1	37.9–38.0	21.8–21.9
2	37.9–38.0	21.9–22.0
3	37.9–38.0	22.0–22.1
4	37.9–38.0	22.5–22.6
5	38.0–38.1	21.7–21.8
6	38.0–38.1	22.5–22.6
7	38.1–38.2	21.4–21.5
8	38.1–38.2	21.7–21.8
9	38.2–38.3	22.4–22.5
10	38.2–38.3	22.5–22.6
11	38.3–38.4	22.4–22.5
12	38.3–38.4	22.5–22.6
13	38.3–38.4	22.6–22.7
14	38.4–38.5	21.4–21.5
15	38.4–38.5	22.4–22.5
16	38.5–38.6	21.8–21.9
17	38.5–38.6	22.3–22.4

For every catalog the b value was calculated with the ZMAP [11], the SSHParameters of the AH-TCS web platform (tcs.ah-epos.eu) and the proposed software, here after called Self-cor, from the self-optimized algorithm for the corresponding linear correlation coefficient. According to the b value that was calculated, the fitting line was estimated for the magnitude range that each software used for the b value calculation (Fig. 2).

To measure each software's efficiency the R-squared test was used to determine the "Goodness of Fit" between the G-R Distribution and the produced fitting lines. The R-squared is given by (5) where n is the total number of observations, y the observed values, ŷ the expected values and ȳ the average of the observed value s.

$$R^2 = 1 - \frac{\sum_{i=1}^{n} (y_i - \widehat{y}_i)^2}{\sum_{i=1}^{n} (y_i - \overline{y})^2} \quad (5)$$

When the observed values are exactly the same as the expected values then R^2 equals 1. As the distance between the observed and expected values gets bigger, the R^2 gets a value close to 0. In occasions that the fit between the observed and expected values is really bad, R^2 can even get negative values. In Table 2 and Fig. 3 the Goodness of fit of the software's fitting lines on the corresponding catalogs are presented.

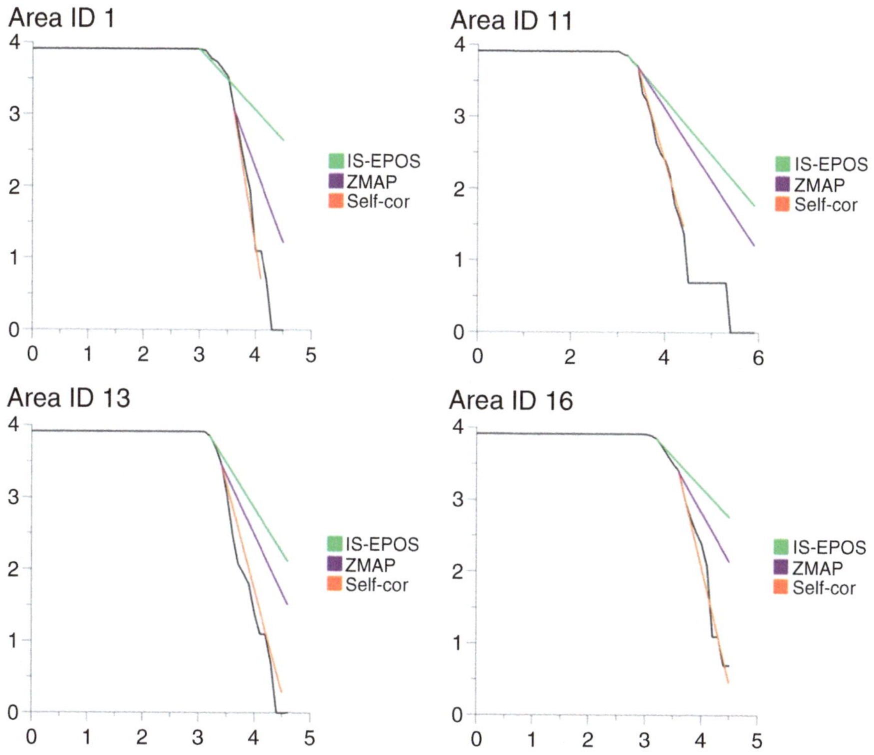

Fig. 2 The produced fits on the G-R Distribution for selected areas

Table 2 Software's R^2 score on the corresponding areas

Area ID	SSHParameters	ZMAP	Self-cor
1	−0.075315	0.158235	0.919752
2	−0.379112	−0.303105	0.963504
3	−0.235471	−0.139031	0.765141
4	0.484036	0.662816	0.982772
5	0.540506	0.312656	0.960971
6	−0.024116	−0.021550	0.988267
7	−0.232720	−0.172142	0.913819
8	−0.646892	−0.533475	0.738784
9	−0.178427	−0.107809	0.987873
10	0.353916	0.518650	0.934868
11	−0.352246	−0.170917	0.978815
12	−0.432339	−0.300654	0.978604
13	−0.811956	−0.097049	0.882917
14	−0.070514	−0.039331	0.957291
15	−0.304725	−0.146794	0.968649
16	0.217044	0.331485	0.911172
17	0.510571	0.505562	0.825544

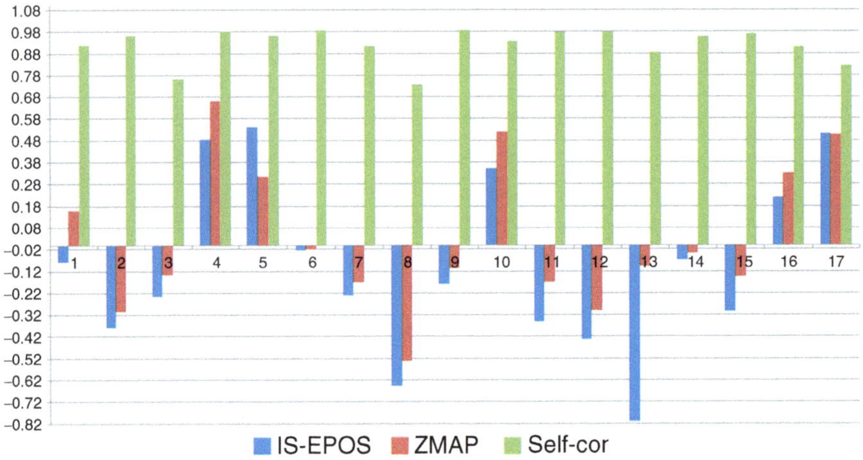

Fig. 3 R^2 score on the corresponding areas, for each software

As it can be seen, in all cases the Self-cor software has a better fit than the other two software applications. Also in 13 cases the R^2 score is higher than 0.9. By average the R^2 score, for each software, is as follows **SSHParameters**: −0.096338824, **ZMAP**: 0.026914529 and **Self-cor**: 0.921102529.

References

1. Gutenberg, B., and C.F. Richter. 1944. Frequency of Earthquakes in California. *Bulletin of the Seismological Society of America* 34: 185–188.
2. Woessner, J., and S. Wiemer. 2005. Assessing the Quality of Earthquake Catalogues: Estimating the Magnitude of Completeness and Its Uncertainty. *Bulletin of the Seismological Society of America* 95: 984–698.
3. Wiemer, S., and M. Wyss. 2000. Minimum Magnitude of Completeness in Earthquake Catalogs: Examples from Alaska, the Western United States, and Japan. *Bulletin of the Seismological Society of America* 90: 859–869.
4. Rydelek, P.A., and I.A. Sacks. 1989. Testing the Completeness of Earthquake Catalogs and the Hypothesis of Self-Similarity. *Nature* 337: 251–253.
5. Papadopoulos, G.A., M. Charalampakis, A. Fokaefs, and G. Minadakis. 2010. Strong Foreshock Signal Preceding the L'Aquila (Italy) Earthquake (Mw 6.3) of April 2009. *Natural Hazards and Earth System Sxiences* 10: 19–24.
6. Avlonitis, M., and S. Tassos. n.d. *Towards Realistic Self-Organized Critical Models for Earthquakes, ESC 2010*. France: Monpellier.
7. Arvanitakis, K., C. Gounaropoulos, and M. Avlonitis. 2016. Locating Asperities by Means of Stochastic Analysis of Seismic Catalogs. Bulletin of the Geological Society of Greece, vol. L. Thessaloniki.
8. Arvanitakis, K., and M. Avlonitis. 2016. Identifying Asperity Patterns via Machine Learning Algorithms. *Artificial Intelligence Applications and Innovations*: 87–93.

9. Aki, K. 1965. Maximum Likelihood Estimate of b in the Formula logN=a-bM and its Confidence Limits. *Bulletin of the Earthquake Research Institute* 43: 237–239.
10. Stirling, M.W., S.G. Wesnousky, and K. Shimazaki. 1996. Fault Trace Complexity, Cumulative Slip and the Shape of the Magnitude-Frequency Distribution for Strike-Slip Faults: A Global Survey. *Geophysical Journal International* 124: 833–868.
11. Wiemer, S. 2001. A Software Package to Analyze Seismicity: ZMAP. *Seismological Research Letters* 72: 373–382.

The Use of Translational Research Platforms in Clinical and Biomedical Data Exploration

Konstantina Skolariki and Antigoni Avramouli

Abstract The rise of precision medicine combined with the variety of biomedical data sources and their heterogeneous nature make the integration and exploration of information that they retain more complicated. In light of these issues, translational research platforms were developed as a promising solution. Research centers have used translational tools for the study of integrated data for hypothesis development and validation, cohort discovery and data-exploration. For this article, we reviewed the literature in order to determine the use of translational research platforms in precision medicine. These tools are used to support scientists in various domains regarding precision medicine research. We identified eight platforms: BRISK, iCOD, iDASH, tranSMART, the recently developed OncDRS, as well as caTRIP, cBio Cancer Portal and G-DOC. The last four platforms explore multidimensional data specifically for cancer research. We focused on tranSMART, for it is the most broadly used platform, since its development in 2012.

Keywords Precision medicine • Translational research platforms • Biomedical research • Clinical data • High-throughput technologies

1 Introduction

The rise of personalized medicine and the generation of high-throughput data has lead to an increasing need for sufficient translational research tools. We identified eight platforms, that enable researchers to explore and analyze data. These are: BRISK (Biology-Related Information Storage Kit), iCOD (integrated clinical omics database), iDASH (integrating data for analysis, anonymization and sharing), tranSMART, and OncDRS (Oncology Data Retrieval Systems). OncDRS is an extended 'Integrated Informatics from Biology to the Bedside' (i2b2) project. There

K. Skolariki
University of Surrey, Guildford, Surrey, GU2 7XH, UK
e-mail: ks00430@surrey.ac.uk

A. Avramouli (✉)
Department of Informatics, Ionian University, Corfu 49100, Greece
e-mail: c15avra@ionio.gr

P. Vlamos (ed.), *GeNeDis 2016*, Advances in Experimental Medicine and Biology 988, DOI 10.1007/978-3-319-56246-9_25

are also three more research analysis platforms created specifically to integrate oncogenomics data: caTRIP (Cancer Translational Research Informatics Platform), G-DOC (Georgetown Database of Cancer), cBio Cancer Genomics Portal.

The background knowledge and the basic functionalities available for these translational research platforms has been described by Canuel et al. [1]. These platforms provide: (1) the storage and integration of big data, (2) an analysis context, allowing scientists to integrate and explore the data and (3) additional information from external sources.

The variety of data sources and their heterogeneous nature make the unification and exploration of the data more composite. These tools were created as a solution to the growing amount of -omics data. They use informatics methods to link molecular and clinical data. Research centers have used translational tools for the study of integrated data for hypothesis development and validation, cohort discovery and data-exploration. In this review we decided to focus on already published papers that describe the application of these tools in the integration of large data sets highlighting their crucial role in defining certain biomarkers for precision medicine.

2 Materials and Methods

We searched PubMed for scientific literature and identified articles possibly describing translational platforms. We assessed the articles to categorize the systems based on (1) their ability to integrate data and (2) the data analysis functionalities that they provide to scientists. Translational research platforms are publicly available solutions and are used to integrate data for analysis, anonymization and sharing. Exploiting openly accessible resources (i.e. studies published before November 2016) defining these tools, we established the key features and uses of each platform. They were selected according to the validation of their use in biomedical big data processing and their applications in exploring new biomarkers for certain diseases.

3 Overview of Translational Research Platforms

3.1 Cancer Translational Research Informatics Platform (caTRIP)

caTRIP was developed in 2006 by the Duke Comprehensive Cancer Center (DCCC), in collaboration with SemanticBits LLC. caTRIP was a translational research system used in the 'cancer Biomedical Informatics GridTM' (caBIG) project, which was initiated by the U.S. National Cancer Institute [2, 3]. It was a data-integration project with the goal to develop an open-source information network across the United States for secure data exchange on cancer research.

The purpose of caTRIP was to allow physicians to find patients with similar characteristics, analyze their clinical outcomes and find information about the successful treatments that were administered across the caBIG data network. The most promising application of caTRIP was that oncologists could access a rich live data network that could provide statistically significant facts in mere minutes. This could benefit clinicians to query data from a cohort of preexisting patients to help guide treatment of another patient, rather than relying on single cancer centers or limited facts published in the literature. In addition, caTRIP was adept at facilitating research, since scientists could have access to a large bio-specimen repository with samples available for further investigation.

Nevertheless, caBIG project was cancelled in 2011, as the working group discovered that only few cancer centers were using caBIG's clinical data management tools and its cloud computing infrastructure. Also, caBIG applications costed millions of dollars to build, while the majority of the life sciences research tools developed for caBIG have had limited use or impact, according to the report [4]. Even though, it had admirable goals and seemed workable in theory, in the end it turned out to be overly complicated to use. Essentially, caBIG relied on standardized data formats, and its one-size-fits-all approach suited nearly nobody.

Despite the lack of recent updates in the caTRIP platform we included it in this review because of its substantial contribution to the progress of translational research tools, through the facilitation of its data sharing and storing processes. Besides, in 2012 the U.S. National Cancer Institute established the National Cancer Informatics Program (NCIP) as a successor to caBIG. NCIP is funding and developing projects related to cancer genomics data management, creating a software sharing hub, and releasing code under the open-source model.

3.2 cBioPortal for Cancer Genomics

The cBioPortal for Cancer Genomics is developed and maintained by the Memorial Sloan-Kettering Cancer Center (MSKCC) [5, 6]. This web resource provides visualization, analysis, and download of large-scale oncogenomics datasets data sets. It was particularly designed to address the unique data integration issues posed by large-scale cancer genomics projects and to make the raw data they generated more easily and directly available to the entire cancer research community.

3.3 G-DOC Georgetown Database of Cancer (G-DOC)

G-DOC was created by the Lombardi Comprehensive Cancer Center at Georgetown University, in 2011 [7]. It uses a cloud computing translational informatics infrastructure to facilitate translational and systems-based approaches in cancer research.

G-DOC Web portal was designed specifically to allow researchers with a variety of backgrounds to surpass the difficulties of using powerful bioinformatics tools for biomedical research. G-DOC offers examination and imagining of four major "omics" types (DNA, mRNA, miRNA, and metabolites) in a unified environment. The platform facilitates a broad assembly of systems biology and a range of bioinformatics tools. It also contains a wide variety of analytic tools and capabilities, integrated views of detailed molecular and genomic characteristics, as well as three-dimensional drug-target complex structures, enabling the discovery of new candidate molecules for cancer-targeted therapies. G-DOC also allows researchers to securely share data.

Madhavan et al. [8], used G-DOC and developed a minimum number of multi-omics features that allow the best classification accuracy of relapse phenotype in colorectal cancer (CRC) patients. Researchers integrated the results of molecular profiling of several omics data types to determine the most reliable molecular prognostic biomarkers for relapse in CRC. The data types were analyzed with a multi-step approach with sequential exclusion of redundant molecular features. That approach enabled the identification of 31 multi-omics features that highly correlate with relapse. The biomarkers detected in urine and blood samples of patients with cancer at the time of surgery, point to a strong association between immune system processes and biomarkers that indicate the possibility of future relapse. Thus, highlighting the way towards precise cancer treatment.

Recently Luo et al. [9], used G-DOC platform in order to study the significance of lymphocyte antigen-6 (Ly6) family members, a group of alloantigens, in healthy and cancer-infected tissue. Specifically, they used Oncomine (Invitrogen) [10] for gene expression micro array analysis, and G-DOC tools to form the clinical outcome information. With these tools, they analyzed 130 gene expression omnibus (GEO) datasets and realised that four different Ly6 gene family members (Ly6D, Ly6E, Ly6H, Ly6K) show increased gene expression levels in cancerous tissue. Their analysis established that these genes could act as novel prognostic markers as well as novel candidates for the development of targeted therapies. CBioPortal for Cancer Genomics was also used to investigate genetic variations across Ly6 family members in different cancer types.

G-DOC Plus platform, an enhanced web platform enabling Next Generation Sequencing (NGS) data and medical imaging was presented earlier this year by the University of Georgetown [11]. This platform enables the analysis of -omics and clinical information in order to create new hypothesis for precision medicine. Moreover, G-DOC Plus expanded further its utilities. In addition to cancer datasets collections, the platform now includes data on several non-cancer diseases (e.g. Alzheimer's and Duchene Muscular Dystrophy Disease). G-DOC Plus currently holds data from over 10,000 patients selected from public and private resources while the system hosts huge records of clinical and other omics data, processed NGS data, medical images and meta-data. In addition to the translational research module, G- DOC Plus contains two new modules—Precision medicine and Population genetics workflows.

3.4 Biology-Related Information Storage Kit (BRISK)

BRISK is an open source data management tool developed by the University of British Columbia, Vancouver, Canada at 2011 [12]. The demand of a collaborative data sharing tool for investigators in the AllerGen (The Allergy, Genes and Environment Network) consortium lead to the platforms' creation.

BRISK has many service tools and it uses a central hub that supports efficient data sharing and information deployment to aid researchers in organizing, mining and analyzing data collected from various studies. These data include clinical phenotype descriptions and single nucleotide polymorphisms information and can be analyzed through genome-wide association studies (GWAS). Until now there is no literature published on BRISK, except from one research paper which showed no associations between the risk of non-Hodgkin lymphoma and genes controlling lymphocyte development [13].

3.5 Integrated Clinical Omics Database (iCOD)

iCOD was developed in 2010 at the Information Center for Medical Sciences in Tokyo, Japan [14]. It is one of the first platforms to provide broad pathological, clinical and life-style information, additional to the molecular omics data for a patient's saved medical record. This platform also enables scientists to estimate the interrelation among the above parameters in order to comprehensively clarify the omics basis of a disease for the creation of plausible disease pathways.

Because of the fact that this database contains data from patients who were medically treated at Medical and Dental University Hospital in Tokyo and other collaborating institutions, the data sources are of limited amount.

Nonetheless, iCOD, amongst other databases, was used by Dropmann et al. [15] in the investigation of TGF-β1 and TGF-β2 expression in liver diseases. The researchers compared their own data with those obtained from other open source databases. Their study demonstrated an upregulation of TGF-β2 expression in liver diseases which suggested that TGF-β2 could serve as a promising therapeutic target for hepatocellular carcinoma. Their work may lead the way to a new era for fibrosis and liver precise therapeutics.

Correspondingly Itzel et al. [16] identified two novel oncogenes, CDCA3 and KIF18B, and demonstrated that there is a highly significant upregulation of both genes in many tumors compared to normal tissue.

3.6 Integrated Clinical Omics Integrating Data for Analysis, Anonymization and Sharing (iDASH)

iDASH was developed in 2011 by the National Center for Biomedical Computing, United States [17]. iDASH is a multi-institutional team of quantitative scientists (mathematicians, information and computer scientist, software engineers and biostatisticians) that have created a powerful computational infrastructure required for data integration and data analysis. This scientific team leverages tools and algorithms focused on sharing data in a privacy-preserving manner. iDASH provides a cyber-infrastructure that can be used by biomedical and behavioral researchers. The platform enables them to access data, software and a high performance computing environment. Thus, qualifying them to generate and test new hypothesis and therefore, reducing barriers in data sharing.

3.7 Oncology Data Retrieval Systems (OncDRS)

OncDRS is an enterprise translational informatics system developed at Dana-Farber Cancer Institute (DFCI) in the United States in 2014 [18]. This suite, effectively and seamlessly integrates clinical and genomic data and also extracts related genomic results from heterogeneous sources.

OncDRS is a self-service application, providing researchers with easier access to the above-mentioned data, for hypothesis validation, cohort identification, and thorough data analysis. This informatics framework extends Integrated Informatics from Biology to the Bedside (i2b2) project with the following features: (1) by allowing researchers to associate data of cancer disease and episodes, (2) by developing specific data hierarchies for oncology data, like histology, disease stage, and relapse, (3) by providing facts about the complete hierarchical path to a search hypothesis query, (4) by offering correlative studies and (5) by enabling cohort identification and recruitment for clinical trials.

Since its launch in 2014, OncDRS has facilitated up to 1500 research queries and has delivered data for more than 50 research studies. The success of this system in such a short period of time can be attributed to the ease of use, its self-service capability, as well as to fact that, investigators are able to evaluate easily the estimated cohort amount before investing time in initiating research projects.

4 tranSMART

The lack of translatability of preclinical models into meaningful biological knowledge as was identified by Johnson & Johnson (J&J), lead to a partnership between J&J and the Recombinant Data (R&D) Corporation in 2008. The result of that partnership was a knowledge management and high-content analysis platform that would provide access to all R&D data. That platform is known as tranSMART.

tranSMART has been developed, concentrating on medical research requirements and clinical analysis purposes. The first iteration of tranSMART was arranged in 2010. It was initially developed as a precompetitive collaboration platform for pharmaceutical companies, nonprofit organizations as well as academic institutions. It became available as an open source in 2012, with 32 instances in operation. tranSMART ties the productive world of basic science and clinical practice data by merging numerous kinds of data from different sources into a shared environment. It offers exploration and analysis tools. It is based on the open-source i2b2 collaborative data warehouse [19].

4.1 tranSMART Applications

tranSMART is used as a research tool in association with numerous diseases. University of Michigan (U-M) utilizes tranSMART to support the NIH-funded ColoRectal Transdisciplinary Study (CORECT) [20], while additional uses are planned for diabetes and renal testbeds. One Mind for Research's Knowledge Integration Network (KIN), that focuses on traumatic brain injury (TBI) and post-traumatic stress disorder (PTSD) is building upon existing tranSMART capabilities.

Another tranSMART application example is the analysis of the gut microbiome. The members of sysINFLAME, a multi-centre consortium funded by the German Ministry of Education and Research as part of their systems medicine initiative 'e:Med' aims at jointly investigating inflammatory diseases from a systems perspective, with an emphasis on chronic inflammatory diseases of the gut, the joints and the skin. Key phenotypes of interest to the sysINFLAME consortium are inflammatory bowel diseases (Crohn disease and ulcerative colitis). They illustrate the functionalities of the tranSMART toolbox by relating the gut microbiome to genotypic information based on previously published data.

4.2 tranSMART Datathon

tranSMART Foundation hosted its first datathon on June 30th through July 2nd 2015. The goal of this Datathon was to evaluate various data science approaches to biomarker discovery, pathway analysis to distinguish similarities and differences across different neurodegenerative diseases, specifically Alzheimer's disease and Parkinson's disease, and to ascertain new insights into the diagnosis and treatment for these diseases.

According to Barash et al. [21], the findings of datathon include a: (1) single nucleotide polymorphism (SNP) that with Multiple Sclerosis severity, (2) SNP that is linked with brain cancer, (3) SNP associated with both Parkinson's Disease (PD) and Alzheimer disease and potential new biomarkers shared by both diseases, (4) the monocyte level in PD patients is lower than prodromal and healthy controls. Furthermore, analysis of Alzheimer's Disease Neuroimaging Initiative (ADNI) data showed that converters display high cerebral spinal fluid (CSF) T-Tau levels and lower precuneus thickness.

4.3 tranSMART Advantages

From a precision medicine point of view, integrating heterogeneous data is needed to provide a unified view for analysis. tranSMART's extensile data model, its corresponding integration processes, the prompt data analysis features, and its open source nature make it a vital tool in translational and clinical research.

Translational tools, provide researchers a comprehensive source to develop, test, refine, form and validate research hypotheses as well as the necessary resources to search various data sets for probable drug targets, pathways and biomarkers. Such was the case with integrated colorectal cancer (CRC) data. tranSMART assisted researchers in developing new hypotheses and in validating an existing hypothesis. In the absence of access to integrated CRC data via tools like tranSMART, analyses would take from weeks to months. Thus, said platforms can potentially accelerate research translation to cures [22].

Bio-specimen and cohort discovery has been made accessible with the added advantage of being able to request access to the bio-samples of patients from the subsequently identified cohort. Hence, habilitating the reuse valuable existing data.

tranSMART allows prompt exploration of research trials data by permitting users to identify cohorts, and complete pre-construed statistical analyses with the simplicity of a drag and drop interface. Complex queries can be constructed without any programming experience.

tranSMART promotes data sharing, by easily sharing data broadly. This way tranSMART allows its user community to benefit from experts globally. It also enables collaboration, not only at the institutional level by offering a setting where different departments can access research data from the entire institution, but also across academic and corporate research sectors [23].

tranSMART enables scientists to (1) contrast data from proteomics, genomics and other "omics" studies, (2) compare patterns of gene expression in healthy and diseased individuals and tissue samples, (3) assess connections between genotype and phenotype in clinical trial figures, (4) extract pre-clinical data for understanding the biology of human disease, (5) examine genetic and environmental factors correlated with human diseases, (6) present data visually using a graphical interface and (7) arrange clinical data into molecular subtypes of a specific disease [23].

Finally, tranSMART allows access to externally available resources, such as PubMed. The minimizing cost and increasing research effectiveness, also adds on their perks.

The application of research findings into clinical data has been a slow and challenging process, fraught with barriers. Research tools and data handling practices have been lacking. They showed limited success in standardizing data integration processes. There still are numerous concerns in the data management workflow in a typical research setting.

4.4 tranSMART Limitations

Translational research data derived from varied data sources, is stored using different formats, thus making it difficult to integrate and analyze them. In addition, an overlay between some queries was caused since there is no maintenance for temporal data in tranSMART.

The implementation of translational research platforms in healthcare institutions has had several obstacles from the technical perspective (e.g. there is often no common identifier and no shared data model). According to Firnkorn et al. [24], data import and export tools of the biomedical data warehouse system i2b2 are limited to a set schema and hence offer unsatisfactory flexibility. Statistical analyses inside i2b2 and tranSMART are feasible, but constrained to the applied functions. Statistical software packages like SPSS or R as well as for data mining tools, typically expect all information for a patient in one single row. That creates a problem because the set of key-value pairs for a patient in i2b2 and tranSMART are contained in a table called OBSERVATION_FACT. Thus, the exported data cannot externally analysed.

Although, external databases (e.g. metabolic pathways or annotation tracks of genome browsers) should be integrated immediately into the analysis track, this is not the case. For example, it would be highly efficient if simple enrichment computations would be accessible directly from tranSMART environment.

Lastly, tranSMART allows secondary use of healthcare data for researchers and clinicians but at the same time it raises new technical and ethical enquiries. Typically, researchers do not have unimpeded access to the required data because of poor obtainability.

5 Conclusion

The rise of precision medicine combined with the need for reduced cost of high throughput molecular techniques has opened unprecedented opportunities for the application of genomics in the treatment of patients. Over the past decade, several translational research platforms have been developed, in order to integrate clinical data and support life science research. In this review, we described the functionalities of the main platforms providing both integration and analysis features for clinical and omics data. Notwithstanding the various implementation challenges of these tools, such as data integration, sharing models and policies, we strongly believe that these networks will enhance global research collaborations and assist in a deeper understanding of molecular disease mechanisms. Therefore, transforming large datasets into actionable knowledge.

References

1. Canuel, V., B. Rance, P. Avillach, P. Degoulet, and A. Burgun. 2014. Translational Research Platforms Integrating Clinical and Omics Data: A Review Of Publicly Available Solutions. *Briefings in Bioinformatics* 16 (2): 280–290.
2. Fenstermacher, D., C. Street, T. McSherry, V. Nayak, C. Overby, and M. Feldman. 2005. The cancer biomedical informatics grid (caBIG™). *Conference Proceedings IEEE Engineering in Medicine and Biology Society* 1: 743–746.
3. McConnell, P., R.C. Dash, R. Chilukuri, R. Pietrobon, K. Johnson, R. Annechiarico, and A.J. Cuticchia. 2008. The Cancer Translational Research Informatics Platform. *BMC Medical Informatics and Decision Making* 8 (1): 60.
4. NIH Report. 2011. An Assessment of the Impact of the NCI Cancer Biomedical Informatics Grid (caBIG). Available at http://deainfo.nci.nih.gov/advisory/bsa/archive/
5. Cerami, E., J. Gao, U. Dogrusoz, B.E. Gross, S.O. Sumer, B.A. Aksoy, A. Jacobsen, C.J. Byrne, M.L. Heuer, E. Larsson, Y. Antipin, B. Reva, A.P. Goldberg, C. Sander, and N. Schultz. 2012. The cBio Cancer Genomics Portal: An Open Platform for Exploring Multidimensional Cancer Genomics Data. *Cancer Discovery* 2 (5): 401–404.
6. Gao, J., B.A. Aksoy, U. Dogrusoz, G. Dresdner, B. Gross, S.O. Sumer, Y. Sun, A. Jacobsen, R. Sinha, E. Larsson, E. Cerami, C. Sander, and N. Schultz. 2013. Integrative Analysis of Complex Cancer Genomics and Clinical Profiles Using the cBioPortal. *Science Signaling* 6 (269): pl1.
7. Madhavan, S., Y. Gusev, M. Harris, D.M. Tanenbaum, R. Gauba, K. Bhuvaneshwar, A. Shinohara, K. Rosso, L.A. Carabet, L. Song, R.B. Riggins, S. Dakshanamurthy, Y. Wang, S.W. Byers, R. Clarke, and L.M. Weiner. 2011. G-DOC: A Systems Medicine Platform for Personalized Oncology. *Neoplasia* 13 (9): 771–783.
8. Madhavan, S., Y. Gusev, T.G. Natarajan, L. Song, K. Bhuvaneshwar, R. Gauba, A. Pandey, B.R. Haddad, D. Goerlitz, A.K. Cheema, H. Juhl, B. Kallakury, J.L. Marshall, S.W. Byers, and L.M. Weiner. 2013. Genome-Wide Multi-Omics Profiling of Colorectal Cancer Identifies Immune Determinants Strongly Associated with Relapse. *Frontiers in Genetics* 4: 236.
9. Luo, L., P. McGarvey, S. Madhavan, R. Kumar, Y. Gusev, and G. Upadhyay. 2016. Distinct Lymphocyte Antigens 6 (Ly6) Family Members Ly6D, Ly6E, Ly6K and Ly6H Drive Tumorigenesis and Clinical Outcome. *Oncotarget* 7 (10): 11165–11193.
10. Rhodes, D.R., J. Yu, K. Shanker, N. Deshpande, R. Varambally, D. Ghosh, T. Barrette, A. Pander, and A.M. Chinnaiyan. 2004. ONCOMINE: A Cancer Microarray Database and Integrated Data-Mining Platform. *Neoplasia* 6 (1): 1–6.
11. Bhuvaneshwar, K., A. Belouali, V. Singh, R.M. Johnson, L. Song, A. Alaoui, M.A. Harris, R. Clarke, L.M. Weiner, Y. Gusev, and S. Madhavan. 2016. G-DOC Plus – An Integrative Bioinformatics Platform for Precision Medicine. *BMC Bioinformatics* 17 (1): 193.
12. Tan, A., B. Tripp, and D. Daley. 2011. BRISK—Research-Oriented Storage Kit for Biology-Related Data. *Bioinformatics* 27 (17): 2422–2425.
13. Schuetz, J.M., D. Daley, S. Leach, L. Conde, B.R. Berry, R.P. Gallagher, J.M. Connors, R.D. Gascoyne, P.M. Bracci, C.F. Skibola, J.J. Spinelli, and A.R. Brooks-Wilson. 2013. Non-Hodgkin Lymphoma Risk and Variants in Genes Controlling Lymphocyte Development. *PLoS ONE* 8 (9): e75170.

14. Shimokawa, K., K. Mogushi, S. Shoji, A. Hiraishi, K. Ido, H. Mizushima, and H. Tanaka. 2010. ICOD: An Integrated Clinical Omics Database Based on the Systems-Pathology View of Disease. *BMC Genomics* 11: S19.
15. Dropmann, A., T. Dediulia, K. Breitkopf-Heinlein, H. Korhonen, M. Janicot, S.N. Weber, M. Thomas, A. Piiper, E. Bertran, I. Fabregat, K. Abshagen, J. Hess, P. Angel, C. Coulouarn, S. Dooley, and N.M. Meindl-Beinker. 2016. TGF-β1 and TGF-β2 Abundance in Liver Diseases of Mice and Men. *Oncotarget* 7 (15): 19499–19518.
16. Itzel, T., P. Scholz, T. Maass, M. Krupp, J.U. Marquardt, S. Strand, D. Becker, F. Staib, H. Binder, S. Roessler, X.W. Wang, S. Thorgeirsson, M. Muller, P.R. Galle, and A. Teufel. 2014. Translating Bioinformatics in Oncology: Guilt-by-Profiling Analysis and Identification of KIF18B and CDCA3 as Novel Driver Genes in Carcinogenesis. *Bioinformatics* 31 (2): 216–224.
17. Ohno-Machado, L., V. Bafna, A.A. Boxwala, B.E. Chapman, W.W. Chapman, K. Chaudhuri, M.E. Day, C. Farcas, N.D. Heintzman, X. Jiang, H. Kim, J. Kim, M.E. Matheny, F.S. Resnic, and S.A. Vinterbo. 2012. IDASH: Integrating Data for Analysis, Anonymization, and Sharing. *Journal of the American Medical Informatics Association* 19 (2): 196–201.
18. Orechia, J., A. Pathak, Y. Shi, A. Nawani, A. Belozerov, C. Fontes, C. Lakhiani, C. Jawale, C. Patel, D. Quinn, D. Botvinnik, E. Mei, E. Cotter, J. Byleckie, M. Ullman-Cullere, P. Chhetri, P. Chalasani, P. Karnam, R. Beaudoin, S. Sahu, Y. Belozerova, and J.P. Mathew. 2015. OncDRS: An Integrative Clinical and Genomic Data Platform for Enabling Translational Research and Precision Medicine. *Applied & Translational Genomics* 6: 18–25.
19. Bierkens, M., W. van der Linden, K. van Bochove, W. Weistra, R.J.A. Fijneman, R. Azevedo, J.-W. Boiten, J. Beliën, and G.A. Meijer. 2015. TranSMART. *Journal of Clinical Bioinformatics* 5 (Suppl 1): S9.
20. Guo, Y., M. Braxenthaler, B.D. Athey, and M. Haas. 2013. tranSMART: An Open Source and Community-Driven Informatics and Data Sharing Platform for Clinical and Translational Research. *AMIA* 2013: 6–8.
21. Barash, C.I., K. Elliston, and R. Potenzone. 2015. TranSMART Foundation Datathon 1.0: The Cross Neurodegenerative Diseases Challenge. *Applied & Translational Genomics* 6: 42–44.
22. Jonnagaddala, J., J.L. Croucher, T.R. Jue, N.S. Meagher, L. Caruso, R. Ward, and N.J. Hawkins. 2016. Integration and Analysis of Heterogenous Colorectal Cancer Data for Translational Research. *Nursing Informatics*: 387–391.
23. Schumacher, A., T. Rujan, and J. Hoefkens. 2014. A Collaborative Approach to Develop a Multi-Omics Data Analytics Platform for Translational Research. *Applied & Translational Genomics* 3 (4): 105–108.
24. Firnkorn, D., S. Merker, M. Ganzinger, T. Muley, and P. Knaup. 2011. Unlocking Data for Statistical Analyses and Data Mining: Generic case Extraction of Clinical Items from i2b2 and tranSMART. *Studies in Health Technology and Informatics* 228: 567–571.

EEG Analysis of the Neurofeedback Training Effect in Algorithmic Thinking

Antonia Plerou, Panayiotis Vlamos, and Aikaterini Margetaki

Abstract Although significant advances have been made in understanding several cognitive states, the algorithmic thinking ability is yet to be analyzed in terms of neuroscience and brain imaging techniques. Studies on the effects of neurofeedback on learning disabilities especially mathematics disorders are limited. The objective of the present study is to evaluate the brain activity and activation differences between neurofeedback trained participants and controls, during the overall EEG analysis during continuous algorithmic tasks performance. A study of 182 children of upper education is proposed to assess the efficacy of two protocols of neurofeedback training as means of algorithmic thinking ability evaluation. Results suggest statistical significant variation in the mean SD values in terms of several brain waves ratios during algorithmic task solving epochs.

Keywords Neurofeedback training • Learning disabilities • Alpha waves • Beta waves • Algorithmic thinking • EEG

1 Introduction

This study focuses on the algorithmic thinking ability evaluation with the use of neurofeedback training approach. The purpose of the research is to evaluate the different human mental behavior through Electroencephalogram (EEG) signal with time-frequency analysis by receiving information from the internal changes of brain state. Several EEG signals have been collected for these states and analyzed using the Acqknowledge software. Participants were asked to deal with interactively presented algorithmic procedures in order to assess their ability in algorithmic processing stimuli. While engaging in the given tasks, EEG signals of the brain are recorded by means of a sensor system in order to study and analyze their

A. Plerou (✉) • P. Vlamos
Department of Informatics, Bioinformatics and Human Electrophysiology Laboratory, Ionian University, Corfu, Greece
e-mail: tplerou@ionio.gr; vlamos@ionio.gr

A. Margetaki
Medical School, University of Crete, Heraklion, Greece

P. Vlamos (ed.), *GeNeDis 2016*, Advances in Experimental Medicine and Biology 988, DOI 10.1007/978-3-319-56246-9_26

brain activity. This is carried out by protocols based on neuroscience and mainly in neurofeedback methods. In specific two protocols based on neurofeedback were employed namely the Alpha-Theta and the SMR-low Beta protocol in order to reinforce Alpha and Beta frequencies respectively during the various stages of the neuroeducational approach. Statistical measures, like mean values and standard deviation, in each sub-band, are chosen in order to analyze different mental states of participants while dealing with an algorithmic thinking training evaluation test.

2 Related Work

Jacob's study in 2006 on two children with learning disabilities using neurofeedback treatment showed that neurofeedback is a successful treatment for this disease [1]. Fernandez in 2007 showed that neurofeedback is an effective treatment for children with learning disabilities, with a high abnormal ratio of Alpha-Theta [2]. Becerra in follow-up study performed on children with learning disorders showed that neurofeedback is an effective treatment for a long period [3]. According to Hashemian et al. study of 28 third grade primary school children, the 14 received neurofeedback treatment and the 14 non-real neurofeedback treatment. This approach was based on enhancement of Beta-Theta ratio in CZ region and was conducted with 20 sessions that lasted 30 min for 10–12 weeks. In this case, the comparison between real and sham groups showed that the effect of real neurofeedback therapy was significant versus sham group [4].

3 Participants

The sample of 182 participants was randomly selected (convenience sample) while the participants were voluntarily evaluated and two equal groups were formulated. 91 subjects were allocated to neurofeedback group and 91 participates were allocated to control group. Participants of this case study are adults graduate students of the Department of Informatics of the Ionian University in Corfu. An IQ pre-test had conducted in order to match two groups according to the degree of intelligence. In addition, the two groups were matched for age, sex, and learning disabilities. Participant's brain activity and their ability of algorithmic lesion stimuli were evaluated with engaging with the given problems. The training was conducted within 10 sessions that lasted from 60–80 minutes for approximately 9 months. Ten algorithmic tasks were provided, mainly derived from the computer science, graph and game theory. Both control group and neurofeedback trained participants brain activity were recorded while dealing with the above given algorithmic tasks. Therefore ten epochs were created for each participant of both control and neurofeedback group and they were further analyzed statistically with ANOVA analysis [5].

4 Material

The recording of the brain's activity obtained by using electrodes is called electroencephalogram or EEG. The EEG records the signal of the specific brain region where the electrodes are placed. Five major brain waves can be distinguished by their frequency ranges, namely Delta (δ) 0.5–4 Hz, Theta (θ) 4–8 Hz, Alpha (α) 8–12 Hz, Beta (β) 12–30 Hz and Gamma (γ) 30–128 Hz [6]. In this research, the BIOPAC data acquisition unit (MP150) and AcqKnowledge 4.3 software from Biopac Systems Inc. are used for data acquisition, analysis, storage, and retrieval. Silver chloride electrodes were applied following the 10–20 system. The EEG is recorded at 500 samples/sec with a resolution of 12 bits/sample. The data is digitally filtered using 1–50 Hz band pass filter. Each of the participants underwent continuous electroencephalographic record of their brain wave activity, at rest with eyes closed for 3 min, and during a continuous performance task of approximately from 60 to 80 min long. The neurofeedback training sessions took place in a quiet, dimly lit room in the Bioinformatics and Human Electrophysiology Laboratory in Corfu to ensure reduced distraction [7].

5 Methods

Neurofeedback treatment was performed based on enhancement of Alpha-Theta ratio in the C4 region and SMR-low Beta ratio enhancement in the P4 region. The reference electrode was placed in the right earlobe. EEG biofeedback training is an operant conditioning technique used to reinforce or inhibit specific forms of EEG activity. In the Alpha-Theta protocol employed by the Peniston studies, low-frequency EEG activity was reinforced. The efficacy of Alpha-Theta EEG biofeedback may lie in its ability to allow participants to better tolerate stress, anxiety, and anxiety-eliciting situations, which are particularly evident during the initial phases of recovery [8].

The Alpha-Theta state is believed to promote self-awareness, as well as a spiritual and intuitive enhancement [9]. During the neurofeedback training and while participants are dealing with the algorithmic tasks, a visual guidance is provided in order to keep stress effect and anxiety at low levels. Recurrent audiovisual reminders of encouragement are provided to the participants in order to remain calm and stay focused. In addition, they were frequently repeatedly to check their breathing and breathe tranquility with the use of audiovisual guidance which is related to performance expectancy effect.

At the initial phase of neurofeedback training, Alpha-Theta ratio is enhanced while participants relax with their eyes closed while hearing pleasing sounds, such as waves gently crashing on the beach or a babbling brook. In this stage, the low frequencies are reinforced, namely Alpha ratio (8–12 Hz) and Theta ratio (4–8 Hz) respectively while Delta ratio (0.5–4 Hz) and Gamma ratio (30–128 Hz) were suspended [6]. The participants remain for a time interval in a relaxing state, which is essential in order to feel calm, to minimize the stress effect and enhance the quality of learning during the neurofeedback training. Participants are guided in relaxing situations, to reinforce the intuitive perception and to come into deeper levels of consciousness [10].

After this stage, the participants are considered to be free of stress and in the state of mental clarity in order to deal with the algorithmic tasks in order their performance to be evaluated. During their engagement with the given processes, the SMR- low Beta protocol is applied and signals of their brain are recorded. According to the SMR-low Beta training protocol the low Beta ratio (12–15 Hz) is enhanced and this is related to high alertness, concentration and focused attention. The SMR-low Beta protocol is often used for treating ADHD, and other disorders [10].

The analysis of the electrical signals obtained from the brain to assess cognitive effects is used in order the learning ability of the brain on the algorithmic thinking to be understood and to provide suggestions for novel and improved learning methods and learning approach. An essential element of the research is the use techniques for the reduction of artifact effects and the elimination potential conflicts and limitations of the above process [11].

6 EEG Analysis

In the case of this study, identical features of the EEG signals such as mean and standard deviation have been extracted using statistical analysis to detect the predetermined mental states. Mean computes the mean amplitude value of the collected EEG data samples between the endpoints of the selected area. Equation (1) is used to extract the mean value of EEG signal, where, n_s represents starting point and n_e represents the ending point of the sample of data and the total number of samples and i represents the values of points at horizontal axis and $X_{i\ EEG}$ is the values of points of a curve on the vertical axis [12].

$$mean = \frac{1}{n_e - n_s} \sum_{i=n_s}^{n_e - n_s} x_{i_{EEG}} \quad (1)$$

Standard deviation measures the amount of variation or dispersion from the average of the selected EEG data Standard deviation computes a standard deviated value from the mean value of the EEG data samples between the endpoints of the selected area. The formula used to compute standard deviation is shown in Eq. (2) where X_{EEG} is the mean value of the EEG data set [13].

$$Stddev = \sqrt{\frac{1}{n_e - n_s - 1} \sum_{i=n_s}^{n_e - n_s} \left(x_{i_{EEG}} - \overline{x}_{EEG}\right)^2} \quad (2)$$

7 ANOVA Analysis Results: Mean Values of EEG Signal

The comparisons were conducted with analyses of variance (ANOVA), alpha was set at P (0.05). The obtained data were analyzed using STATA software, version 13.0 [14]. A post hoc analyses were conducted for each epoch [15]. The epochs that were compared in this analysis concerns the time period that participants were dealing with the algorithmic tasks. The difference between the control and neurofeedback group is also evaluated by determining the mean values of the each epoch respectively. In the case of the overall EEG signal evaluation, the p-value of the mean value was computed to be 0.8025 suggesting no significant difference between the groups. No significant difference was found is the case of the mean Alpha ratio in the same epochs (p-value = 0.1600). The mean Beta wave comparison between the two groups suggest no significant difference in this ratio as well (p-value = 0.1968). In the case of the Theta and Delta, mean value results reveal no significant difference (p-values 0.5357and 0.5130 respectively). The overall differences are not considered to be significant, therefore no further analysis is conducted.

8 ANOVA Analysis Results: Mean SD Values of EEG Signal

A post hoc analyses were conducted for significant interactions for the task solving phase [15]. In the case of computing the Standard Deviation (SD) of the overall EEG wave the mean values of SD for the control and neurofeedback group are 871959.2 and 1.2e + 06 while the standard deviation for these values is 765448.7 and 963976.2. This difference is considered to be highly statistical significant whereas the p-value is 0.0450. As noticed in Fig. 1, in the 3rd epoch the EEG Standard Deviation for the CG and NFB group differ significantly (p-value = 0.0636). In particular, the mean value of the SD for the CG and the NFB group is 761042.6 and 1,918,752 respectively, while the standard deviation between these values is 409981.8 and 2,107,168 respectively. The difference between the mean values of SD for the overall EEG recording for the two groups is statistically high significant for the 5th epoch where the p-value was computed to be 0.0170. The mean of the SD value is 644995.8 and 1,259,369 for the control and neurofeedback group respectively. In reference to the 6th epoch the difference is significant as well (p-value = 0.0553) and the corresponding values for the mean of SD values for the CG and NFB group respectively are 900777.2 and 1,478,930. Additionally, the difference of the SD between the two groups is additionally significant (p-value = 0.0463) in the 8th epoch (mean SD values 791376.8 and 1,363,176 for the CG and NFB group respectively). The p-value is computed to be 0.0255 for the 8th epoch and the mean SD values are 647943.2 and 1,247,700 respectively for the two groups). A highly significant difference is also noticed in the 10th epoch (p-value = 0.0057) where the

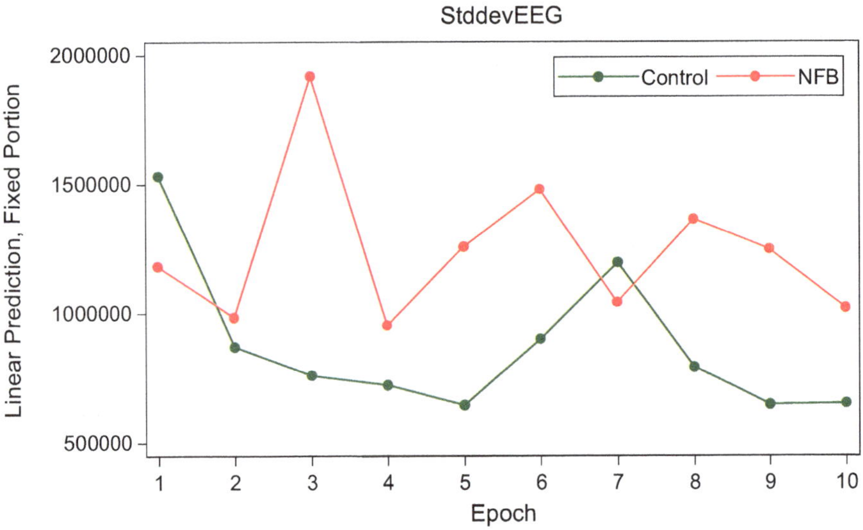

Fig. 1 Mean SD values of EEG signal and epochs (p-value = 0.0450)

mean of the SD values were 651702.2 and 1,020,758 in respect with the two groups. In Fig. 1 the mean SD values of EEG signal for each epoch for neurofeedback and control group is presented.

9 ANOVA Analysis Results: Mean SD Values of Alpha Ratio

A post hoc analyses were conducted for each epoch in the case of the Alpha ratio for the tasks solving phase as well [15]. A highly significant difference is noticed in the case of evaluating the SD value in reference to the group Alpha wave recording. In particular, the p-value was computed to be 0.0164 and the mean of the SD values of the overall Alpha wave was 276621.9 to 390334.2 and the standard deviation 239091.2 to 235885.3 for the CG and NFB group respectively. In Fig. 2 the significant differences in several epochs between the two groups are presented. Specifically, for the 3rd epoch, the p-value is 0.0493 and the SD mean values 238317.6 and 461278.7 for the control and neurofeedback group. In the 4th epoch, the difference was not highly significant (p-value = 0.1949) while the mean of the SD values for the CG and NFB group was 225396.4 and 299390.8 respectively. Highly significant was the difference in the Alpha SD values in the 5th epoch (p-value = 0.0046), namely the mean SD value is 200000.6 (control group) and 420857.9 (NFB group). In the case of the SD values in the 6th epoch, the difference is considered to be statistically significant (p-value = 0.0431) while the SD mean values are 288489.7 and 499927.3 for the CG and neurofeedback group respectively.

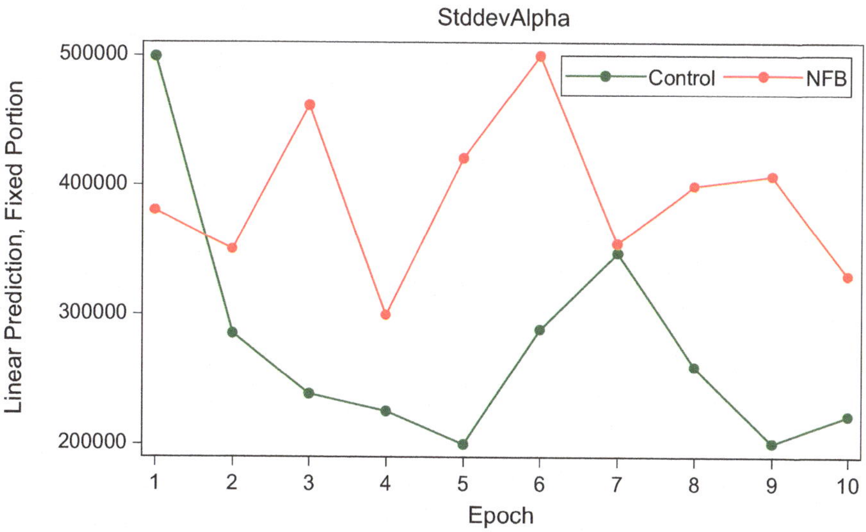

Fig. 2 Mean SD values of Alpha ratio and epochs (p-value = 0.0164)

A significant difference is noticed as well in the 8th epoch where the p-value was computed to be 0.0960. The mean value of the standard deviation of the Alpha wave is 259401.5 for the control and 399353.1 for the NFB group. A statistical high difference comparing the Alpha ratio standard deviation is noticed in the 9th epoch (p-value = 0.0039). More specifically the mean value of the control group is 200599.1 and 407196.4 for the neurofeedback trained group. For the 10th epoch, the difference is statistically significant as well (p-value = 0.0083). The SD mean values were computed from 221756.3 to 330295.5 for the control and NFB group respectively. In Fig. 2 the mean SD values of Alpha ratio of each epoch in respect with the neurofeedback and control group is presented.

10 ANOVA Analysis Results: Mean SD Values of Beta Ratio

In reference to the overall standard deviation for the Beta ratio the mean SD values for each epoch are evaluated. A post hoc analyses were conducted for each of the ten epoch [15]. The overall difference between the mean SD values is presented in Fig. 3 and is considered to be statistically significant (p-value = 0.00184). The Beta ratio standard deviation mean values were measured to differ significantly, namely 306497.4 and 418220.8 for the SD values of the control group and the neurofeedback group respectively. While checking specific epochs of interest interesting findings are noticed. Namely for the 4th epoch, the p-value was computed to be 0.0656 suggesting that there is a significant difference for the analyzing

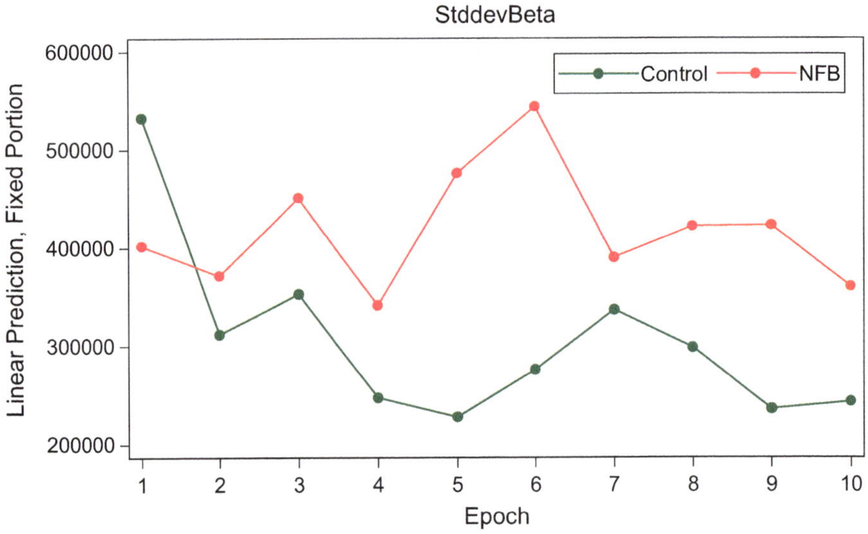

Fig. 3 Mean SD values of Beta ratio and Epochs (p-value = 0.0084)

variable. Respectively the control group SD means value was 247671.9 while for the Neurofeedback group was 342117.8. The p-value for the 5th epoch was 0.0043 suggesting the strong difference between the control and NFB group SD value of Beta rhythm. In particular, the mean SD value for the two group respectively was 227943.1 and 476191.8 for the CG and NFB group. In the 6th epoch, a significant difference occurs where the p-value was 0.0101. The mean SD values 276166.7 and 544131.2 for control and neurofeedback trained group respectively. The difference noticed in the 8th epoch is a result of the difference of the control group mean value in reference to the standard deviation of Beta rhythm (298722.1) and the NFB group mean value 421896.5 (p-value is 0.0792). The p-value (0.0029) for the 9th epoch suggest a statistically significant difference, namely, mean SD values 236203.8 and 422940.5 for the CG and NFB group respectively. Finally, the p-value computed to be 0.0007 in the 10th epoch suggest the statistical strong difference between the two groups in terms of mean SD values (243514.8 and 360507.1). In Fig. 3 the mean SD values of Beta ratio of each epoch and both groups is presented.

11 ANOVA Analysis Results: Mean SD Values of Theta Ratio

A post hoc analyses were conducted for each epoch of the Theta wave as well [15]. While referring to the difference noticed between the overall standard deviation values of the Theta ration the p-value was computed to be 0.1025 while the mean SD value of the control group was 354919.2 and mean SD value of the neurofeedback

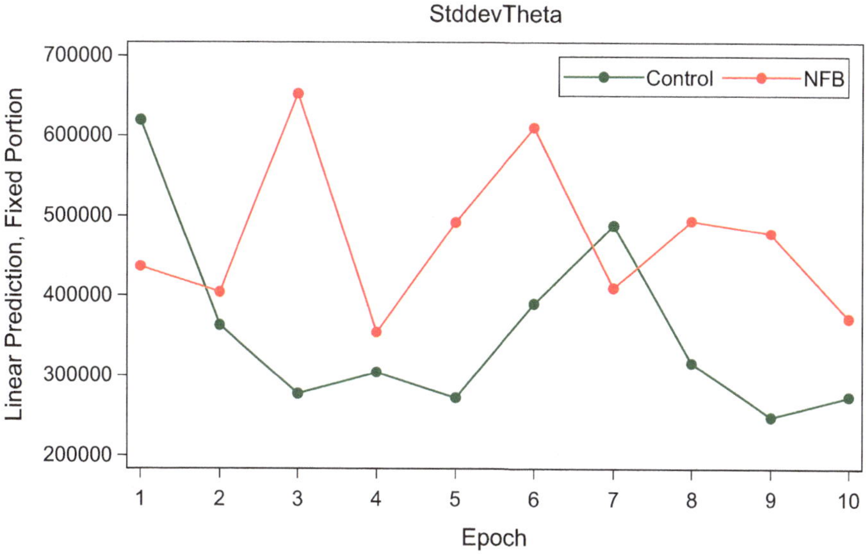

Fig. 4 Mean SD values of Theta ratio and epochs (p-value = 0.1025)

group was 343866.9. For the 3rd epoch in the case of the Theta overall standard deviation, the p-value was computed 0.0445 suggesting a significant difference between the two groups and the mean SD values, were computed 277031.2 and 652867.1 for the CG and NFB group respectively. For the 5th epoch, the overall p-value of the Theta ratio mean SD value was 0.0298 suggesting that the difference is statistically significant while the mean SD value were 271783.1 for the control group and the 492101.2 for the NFB group. In the case of the 9th epoch, the p-value = 0.0250 suggests a significant difference between the mean SD values (247834.8 and 478178.7) of the two group respectively. The difference in the 10th epoch is less significant (p-value = 0.0989) thought the mean SD values differ for 273075.5–372330.4 for CG and NFB group respectively. In Fig. 4 the mean SD values of Theta ratio of each epoch and both groups is presented.

12 ANOVA Analysis Results: Mean SD Values of Delta Ratio

A post hoc analyses were conducted for significant interactions for the specified epochs for Delta ratio [15]. In the case of the overall Delta wave standard deviation evaluation, the p-value is 0.1047 suggesting no significant difference between the two groups. Specifically, the mean SD values vary from 625,898 for the control group to 898491.3 for the NFB group respectively. In the 3rd epoch, the p-value was 0.1093 suggesting a difference between two group, namely from 552416.8 to 1,452,849 and respectively for the CG and NFB group. In the 5th epoch the

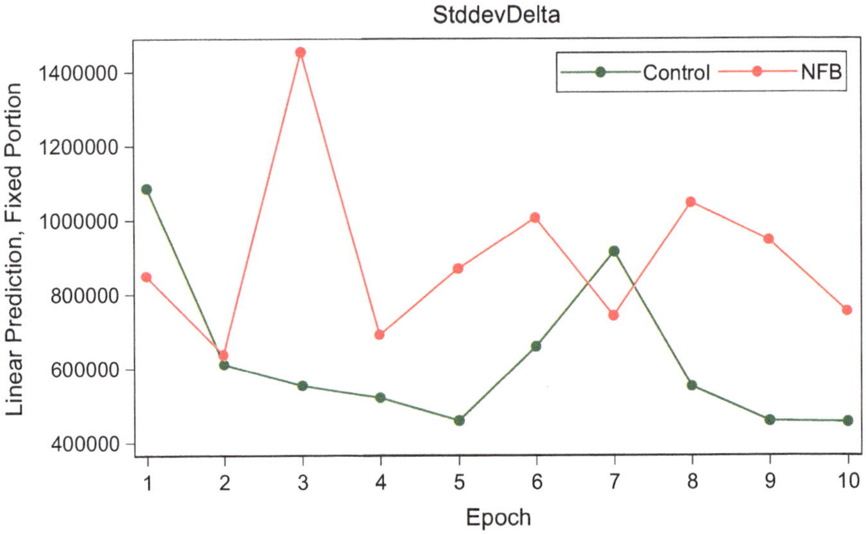

Fig. 5 Mean SD values of Delta ratio and epochs (p-value = 0.1047)

difference is even more significant (p- value = 0.0645) and the difference between the mean SD values are from 458259.9 to 868494.4 for the CG and NFB groups respectively. The significance of the two groups is even higher in the 8th epoch (p-value = 0.0490) and the original differences between the two groups in reference with the mean values are from 550,625 to 1,046,021. The difference is statistically important as well in the case of 9th epoch where the p-value is computed to be 0.0594. The difference in respect the mean SD values vary from 456918.7 to 945125.6 for the CG and the NFB trained group respectively. For the 10th epoch, the difference is statically significant (p-value = 0.0133) and the mean SD values are 454835.6 and 751535.4 for the control and neurofeedback group respectively. In Fig. 5, the mean SD values of Delta ratio for each epoch of neurofeedback and control group is presented.

13 Discussion and Conclusions

The analysis of electrical signals obtained from the brain by means of a sensor system is conducted in order to assess the cognitive functions associated with algorithmic thinking in order to enhance the educational process. The comparison between neurofeedback and control groups showed that the effect of real neurofeedback therapy was significant versus the control group. The difference between the control and neurofeedback group was evaluated by analyzing the variations of the mean and standard deviation values of each ratio and epoch respectively. In the case of the overall EEG signal evaluation, the p-value of the standard deviation value was computed to be 0.0450 suggesting a highly significant difference between the groups with the NFB values to be higher. A highly significant difference was found is the case of the mean SD values of the Alpha ratio for each epoch (p-value = 0.0164) while the NFB group values were significantly greater. The mean SD values of the Beta wave comparison between the two groups suggest a really high significant difference in this ratio with the NFB group values to be superior as well (p-value = 0.0084). In the case of the Theta and Delta, standard deviation value results reveal a significant difference (p-values 0.1025 and 0.1047 respectively) with the NFB trained group values to be larger. These findings suggest that the variations of the brain activity in the case of the neurofeedback trained group are statistically significant. These findings are also related to the enhanced performance that the neurofeedback group posed while dealing with the given algorithmic problems. Not only the Mean SD values were noticed to differ significantly, moreover, the mean SD values of the NFB trained group are higher. Mean values in respect to the different recorded EEG sub-bands were noticed to vary as well, and the mean SD values are noticed to higher in the case of the neurofeedback trained group. Nevertheless, this difference was not considered to be significant, there for no further analysis was conducted. Therefore the brain activation differences between neurofeedback trained participants and controls, suggest the enhanced activation of the brain activity during continuous algorithmic tasks performance. Author's future directions are addressed towards a follow-up study to confirm the overall signal variance stability over time.

References

1. Jacobs, E.H. 2006. Neurofeedback Treatment of Two Children with Learning, Attention, Mood, Social, and Developmental Deficits. *Journal of Neurotherapy* 9 (4): 55–70.
2. Fernández, T., et al. 2007. Changes in EEG Current Sources Induced by Neurofeedback in Learning Disabled Children. An Exploratory Study. *Applied Psychophysiology and Biofeedback* 32 (3–4): 169–183.
3. Becerra, J., et al. 2006. Follow-Up Study of Learning-Disabled Children Treated with Neurofeedback or Placebo. *Clinical EEG and Neuroscience* 37 (3): 198–203.
4. Hashemian, P., and P. Hashemian. 2015. Effectiveness of Neuro-Feedback on Mathematics Disorder. *Journal of Psychiatry* 18 (2).
5. Cohen, Y., and J.Y. Cohen. 2008. Analysis of Variance. In *Statistics and Data with R*, 463–509. Chichester, UK, John Wiley & Sons.
6. Basar, E., et al. 1995. Time and Frequency Analysis of the Brain's Distributed Gamma-Band System. *IEEE Engineering in Medicine and Biology Magazine* 14 (4): 400–410.
7. Howells, F.M., V.L. Ives-Deliperi, N.R. Horn, and D.J. Stein. 2012. Mindfulness Based Cognitive Therapy Improves Frontal Control in Bipolar Disorder: A Pilot EEG Study. *BMC Psychiatry* 12 (1): 15.
8. Scott, W.C.W., et al. 2005. Effects of an EEG Biofeedback Protocol on a Mixed Substance Abusing Population. *The American Journal of Drug and Alcohol Abuse* 31: 455–469.
9. Gruzelier, J. 2009. A Theory of Alpha/Theta Neurofeedback, Creative Performance Enhancement, Long Distance Functional Connectivity and Psychological Integration. *Cognitive Processing* 10 (Suppl 1): S101–S109.
10. Holten, V.. 2010. Bio- and neurofeedback applications in stress regulation. Master's thesis.

11. Delorme, A., Sejnowski, T. and Makeig, S., 2007. Enhanced detection of artifacts in EEG data using higher-order statistics and independent component analysis.
12. Islam, M., et al. 2015. Cognitive State Estimation by Effective Feature Extraction and Proper Channel Selection of EEG Signal. *Journal of Circuits Systems and Computers* 24 (2): 1540005.
13. Islam, M., T. Ahmed, S.S. Mostafa, M.S.U. Yusuf, and M. Ahmad. 2013. Human Emotion Recognition Using Frequency & Statistical Measures of EEG Signal. 2013 International Conference on Informatics, Electronics & Vision (ICIEV), pp. 1–6. IEEE.
14. StataCorp. 2007. *Stata Statistical Software: Release 10*. College Station, TX: StataCorp LP.
15. Holmbeck, G.N. 2002. Post-hoc Probing of Significant Moderational and Mediational Effects in Studies of Pediatric Populations. *Journal of Pediatric Psychology* 27 (1): 87–96.

Formal Models of Biological Systems

Georgia Theocharopoulou, Catherine Bobori, and Panayiotis Vlamos

Abstract Recent biomedical research studies are focused in the mechanisms by which misfolded proteins lead to the generation of oxidative stress in the form of reactive oxygen species (ROS), often implicated in neurodegenerative diseases and aging. Moreover, biological experiments are designed to investigate how proteostasis depends on the balance between the folding capacity of chaperone networks and the continuous flux of potentially nonnative proteins. Nevertheless, biological experimental methods can examine the protein folding quality control mechanisms only in individual cells, but not in a multicellular level. Formal models offer a dynamic form of modelling, which allows to explore various parameter values in an integrated time-dependent system. This paper aims to present a formal approach of a mathematical descriptive model using as example a representation of a known molecular chaperone system and its relation to diseases associated to protein misfolding and neurodegeneration.

Keywords Molecular chaperones • Neurodegenerative diseases • Alzheimer disease • Huntington's disease • Misfolded proteins • UPR • Protein folding

1 Introduction

The term proteostasis describes this state of healthy proteome balance, whereas proteostasis network (PN) refers to the cellular components that participate in proteostasis maintenance [1]. Failure of proteostasis is implicated in disease and the deleterious effects of aging [2].

Maintaining protein homeostasis is a challenging task, since the concentration and subcellular localization of each individual protein species needs to be carefully

G. Theocharopoulou (✉) • C. Bobori
Department of Informatics, Ionian University, Corfu, Greece
e-mail: zeta.theo@ionio.gr; p12bobo@ionio.gr

P. Vlamos
Department of Informatics, Bioinformatics and Human Electrophysiology Laboratory, Ionian University, Corfu, Greece
e-mail: vlamos@ionio.gr

P. Vlamos (ed.), *GeNeDis 2016*, Advances in Experimental Medicine and Biology 988, DOI 10.1007/978-3-319-56246-9_27

controlled [3]. Recent research studies suggest an emerging proteostasis network (PN), which is comprised by a plethora of factors, including molecular chaperones. The concept of this network is that these components function as a coordinator in proteome integrity, in order to prevent the pathologic accumulation of protein aggregates, which potentially form toxic species. Protein aggregation is implicated in neurodegenerative diseases and other medical disorders, ranging from Alzheimer's disease (AD) to type 2 diabetes [4].

An understanding of protein folding is important for the analysis of the resulting interplay of many events involved in cellular regulation and the development of novel therapeutic strategies for human diseases that are associated with the failure of proteins to fold correctly [5].

Because the number of possible conformations a protein chain can adopt is very large, folding reactions are highly complex and heterogeneous [6]. Although aggregation primarily leads to amorphous structures largely driven by hydrophobic forces, it may also lead to the formation of amyloid-like fibrils, which are often reported in a wide range of neurodegenerative diseases [7]. A protein aggregate is described as any association of two or more protein molecules in a non-native conformation [3]. Aggregates cover a range of structures, from amorphous assemblies to highly ordered fibrils (amyloid) with cross-β-structure [3].

While there are different misfolding mechanisms, they all converge on common conformational changes, which may promote either gain of a toxic activity, or the lack of a biological function of the natively folded protein. Since there is no single folding mechanism, the folding rate maybe enhanced by interactions between the domains of proteins, while the topology of the protein is also important [8]. Chaperones may bind to nascent polypeptide chain and facilitate proteins folding, or are ATP-driven machines to assist unfolding of misfolded proteins [9].

Protein homeostasis (proteostasis) is essential for maintaining cell' s functionality, therefore imbalance of proteostasis, is considered to contribute to aberrantly folded proteins that typically lose their function [10]. Molecular chaperones are capable of suppressing protein aggregation and promote efficient refolding or degradation. Therefore, recent advances are targeting chaperone network as a therapeutic approach [11]. In this paper, we focus on the importance of molecular chaperones in neurodegenerative diseases, and especially of Hsp70, a Heat Shock Protein which is used in our formal model in order to understand their protective mechanisms.

1.1 The Role of Chaperones in Protein Folding

Chaperones play a major role in modulating de novo folding and refolding (i.e., the chaperonins, Hsp70s, and Hsp90s) [6]. ATP binding and hydrolysis are essential for the chaperone activity [12]. Studies have shown that for efficient folding chaperones act like a catalyst, in oder the rate of folding to be faster than the rates of aggregation, or chaperone rebinding. Otherwise, these polypeptide chains fail to effectively reach their native state and they are transferred to the degradation machinery

[13]. Partially folded, or misfolded states often tend to aggregate, because they expose hydrophobic amino acid residues and regions of unstructured polypeptide backbone [14].

Accumulation of misfolded proteins in the endoplasmic reticulum (ER) can disrupt ER function resulting in ER stress [15]. The ER responds by triggering specific signaling pathways, including the UPR, which increases the abundance of stress-regulated chaperones [16]. Moreover, retro-translocation and degradation of ER-localized proteins occurs, which results minimization of the accumulation and aggregation of misfolded proteins, increasing the capacity of the ER machinery for folding and degradation [17, 18].

The general function of chaperones in regulating protein folding is significant in neuronal proteostasis. Besides assisting the formation of nonnative proteins and the refolding of misfolded proteins, chaperones also regulate critical cellular processes, such as protein trafficking, protein degradation and macromolecular complex assembly [10].

1.2 Proteostasis Network

Although research has made major advances which unveiled the importance of HSP-mediated quality control in normal cellular function, we are far from understanding the mechanics of the chaperone system machinery which in conjunction with the protein transport and degradation machineries to ensure proteome integrity [19].

The physiological regulation of the PN is complex and involves several interconnected stress response pathways, including the cytosolic heat shock response (HSR) [20], the unfolded protein response (UPR) in the endoplasmic reticulum [21], and the mitochondrial UPR [22]. Many research studies review the link between protein aggregation and the function of the PN [3, 23, 24] and suggest that the failure of cells to maintain proteostasis contributes to the load of misfolded proteins which the toxic effects of protein aggregates leading to neuronal death in numerous diseases and is also involved in the aging process.

Protein homeostasis is highly interconnected to molecular chaperone network which triggers the pathways of protein degradation in order to remove nonfunctional, misfolded, or aggregated proteins [25].

Molecular chaperones guide the conformation of proteins from their initial folding and assembly and throughout their lifetime [26]. Neurons are especially susceptible to degeneration because of their long lifetime. Thus, research on neurodegenerative diseases in recent years has been focused on the neuroprotective properties of chaperones [27].

Specific proteins may interact with as many as 25 different types of chaperones throughout their lifetime, as shown in yeast [28]. Failure of conformational maintenance is particularly relevant to the onset of age-related degenerative disorders, which typically involve protein aggregation [29].

Aggregated proteins that cannot be unfolded for proteasomal degradation may be removed by autophagy and lysosomal/vacuolar degradation [6]. Loss of autophagy has been implicated in the formation of inclusion bodies in neurodegeneration, even in the absence of additional stress, demonstrating the importance of this maintaining proteostasis [30]. Additional pathways of autophagy include chaperone-mediated autophagy (CMA) [31] and chaperone-assisted selective autophagy (CASA) [32].

The ability of cells and tissues to maintain proteostasis declines during aging [6]. The gradual decline in proteostasis capacity during aging results in the accumulation of misfolded (or oxidized) proteins, leading to the deposition of aggregates, cellular toxicity, and cell death [4]. Thus, age is an additional risk factor for a range of degenerative diseases associated with protein misfolding and aggregate deposition. Diseases in which the misfolded protein forms aggregates or toxic dominant negative protein species are usually categorized as a gain-of-function pathogenesis and include type 2 diabetes and major neurodegenerative diseases, such as Parkinson's, Huntington's, and Alzheimer's disease, as well as amyotrophic lateral sclerosis (ALS) [6, 33].

Furthermore, toxic protein aggregation also compromises the cellular response to stress stimuli (221).

Our paper is structured as follows: After the necessary introduction, where related literature is referred, in the next section, we present the description of our model. In Sect. 3 we present our simulations results and indicative figures. Finally, in Sect. 4 there is the discussion.

2 Description of the Model

The Hsp70 system plays a housekeeping role in quality control of the cell, as it acts as a folding catalyst. More specifically, it is involved in a wide range of folding processes, including the folding and assembly of newly synthesized proteins, refolding of misfolded and aggregated proteins, membrane translocation of organellar and secretory proteins, and control of the activity of regulatory proteins [12].

Spontaneous protein folding occurs on the timescale of milliseconds, while Hsp70-dependent protein folding in vitro occurs typically on the time scale of minutes or longer. A fraction of the unfolded molecules often misfolds and aggregates [34]. Misfolded and aggregated proteins may be trapped in these states for minutes or even hours, until they reach their native state via the chaperone-activity cycle (Fig. 1). Two alternative modes of action are referred in literature. In the first mechanism Hsp70s sequesters through repetitive substrate binding and release cycles in order to prevent aggregation, while allowing free molecules to fold to the native state [12, 35]. In the second mechanism, Hsp70 chaperones are considered as unfoldases that use free energy from ATP binding and/or hydrolysis to unfold or pull apart misfolded and aggregated proteins [12, 35].

Hsp70 consists of an ATP-binding domain, which binds and hydrolizes ATP and a C-terminal substrate binding domain, which binds extended polypeptides [36, 37]. Since Hsp70s perform many different tasks within the same cellular compartment, regulation of the interplay between these cellular functions needs to be examined.

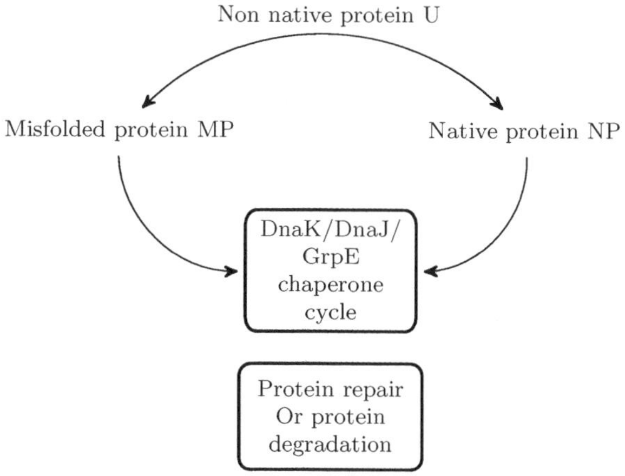

Fig. 1 The chaperone activity cycle

Many co-chaperones, including J proteins, nucleotide exchange factors (NEFs), and tetratricopeptide repeat (TPR) domain-containing proteins [38, 39] cooperate with Hsp70s, in order to guide the diverse cellular activities. The main Hsp70 chaperone of E. coli is DnaK which is highly expressed under all conditions but even more expressed under heat shock conditions [40]. ATP binding and hydrolysis is absolutely essential for the chaperone activity. In order to complete a functional chaperone cycle, the NEFs facilitate ADP-ATP exchange [26]. Considering the fact that there has been identified a broad range of localized functions that the chaperone machinery is involved, a major question is how targeted chaperone action is selectively triggered [41].

Our model represents the role for DnaK in the folding of nascent polypeptide chains, as it has been described in biological studies [12, 42, 43]. The functional versatility of Hsp70 molecules is feasible because of the involvement of a single Hsp70 with multiple cochaperones, mainly consisting of various JDPs. ATP hydrolysis in Hsp70 is thought to be a major determinant of chaperone function [39]. In the DnaK-ATPase cycle Hsp70 chaperone DnaK from Escherichia coli interacts with its co-chaperone DnaJ and the nucleotide exchange factor GrpE, in order these co-chaperones to raise the overall ATPase rate [44]. Many denatured or misfolded substrate proteins are first captured by DnaJ and subsequently transferred to the DnaK-ATP complex, with DnaK [45]. Substrate and DnaJ trigger DnaK - ATPase activity, which leads to locking of the substrate in the DnaK-ADP complex [45]. Driven by the following GrpE-catalyzed ADP-ATP exchange, the DnaK-substrate complex is released [46].

For our modelling purposes we use the systems biology markup language (SBML) [47]. SBML is a computer-readable format for representing models of biochemical reaction networks. A stochastic approach of the mechanisms of molecular chaperons can be found in [48].

2.1 Describing Model's Interactions

For simplicity in this model we assume that proteins may be in their non native state (U) and either fold in their native form, (NP), or are misfolded (MP). The reactions of the model are shown in Table 1.

Native proteins are synthesized at rate r1. NonNative $\xrightarrow{r1}$ Native

DnaJ captures non native proteins and forms a complex at rate r2.

DnaK in its ATP bound form interacts with the substrate.

DnaJ associated with the substrate interacts with DnaK ATP and through rapid stimulation of ATP Hydrolysis by DnaK mediates formation Dnak-DnaJ- substrate complexes containing ADP. This reaction is performed at rate r3.

$$\text{JUcom} \xrightarrow[r3]{ATP} \text{KJUcom+ADP}$$

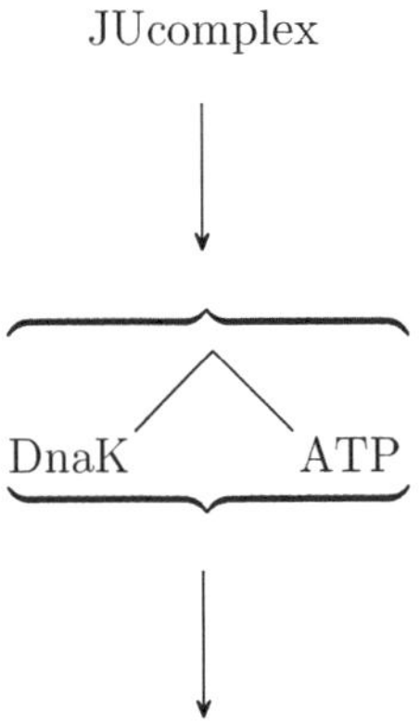

KJUcomplex + ADP

DnaJ might dissociate from these complexes in rate r4.

$$\text{KJUcomplex} \xrightarrow{\text{DnaJ r4}} \text{KUcomplex}$$

GrpE (E) associates with DnaK complexed, triggering nucleotide release and forming GrpE-DnaK-DnaJ-substrate complexes at rate r5.

Dnak dissociates with binding ATP at rate r6.

GrpE facilitates the exchange of ADP for ATP at rate r7. Then, the dissociated substrate will undergo further folding to its native state or misfold.

We assume that unsuccessful folding occurs at rate r8.

Unsuccessful refolding by Hsp70 directs the misfolded proteins to polyubiquitylation and degradation by the proteasome [49] (Fig. 2).

2.2 Reactions

This model contains 23 Reactions. All reactions are listed in the following table.

Table 1 Overview of all reactions

№	ID	Name	Reaction equation
1	NativeProtSynthesis	ProtSynthesis	$\mathtt{U\,source} \longrightarrow \mathtt{NP}$
2	misfolding	misfolding	$\mathtt{NP} + \mathtt{ROS} \longrightarrow \mathtt{MP} + \mathtt{ROS}$
3	DnaJUBinding	DnaJUBinding	$\mathtt{U} + \mathtt{J} \longrightarrow \mathtt{JCom}$
4	unsuccessfulRefolding	unsuccessfulRefolding	$\mathtt{JCom} \longrightarrow \mathtt{MisP} + \mathtt{DnaJ}$
5	refolding	refolding	$\mathtt{MCom} + \mathtt{ATP} \longrightarrow \mathtt{Hsp90} + \mathtt{NatP} + \mathtt{ADP}$
6	proteinDegradation	proteinDegradation	$\mathtt{MisP} + \mathtt{ATP} \longrightarrow \mathtt{ADP}$
7	proteinAggregation1		$2\ \mathtt{MP} \longrightarrow \mathtt{AggP}$
8	proteinAggregation2		$\mathtt{MisP} + \mathtt{AggP} \longrightarrow 2\ \mathtt{AggP}$
9	DnaKInteraction	DnaKinteraction	$\mathtt{K} + \mathtt{ATP} \longrightarrow \mathtt{J} - \mathtt{K} - \mathtt{Com}$
10	DnaJDissociation	DnaJDissociation	$\mathtt{J} - \mathtt{K} - \mathtt{Com} \longrightarrow \mathtt{J} + \mathtt{K} - \mathtt{Com}$
15	GrpEAssociation	GrpEAssociation	$\mathtt{E} + \mathtt{J} - \mathtt{K} - \mathtt{Com} \longrightarrow \mathtt{J} - \mathtt{K} - \mathtt{E} - \mathtt{Com}$
16	GrpERelease	GrpERelease	$\mathtt{J} - \mathtt{K} - \mathtt{E} - \mathtt{Com} + \mathtt{ATP} \longrightarrow \mathtt{J} - \mathtt{K} - \mathtt{Com}$
19	countTime	countTime	$0\ \mathtt{source} \longrightarrow \mathtt{X}$
20	ATPformation	ATPformation	$\mathtt{ADP} \longrightarrow \mathtt{ATP}$
21	ATPconsumption	ATPconsumption	$\mathtt{ATP} \longrightarrow \mathtt{ADP}$
22	radicalFormation	radicalFormation	$0\ \mathtt{source} \longrightarrow \mathtt{ROS}$
23	radicalScavenging	radicalScavenging	$\mathtt{ROS} \longrightarrow \emptyset$

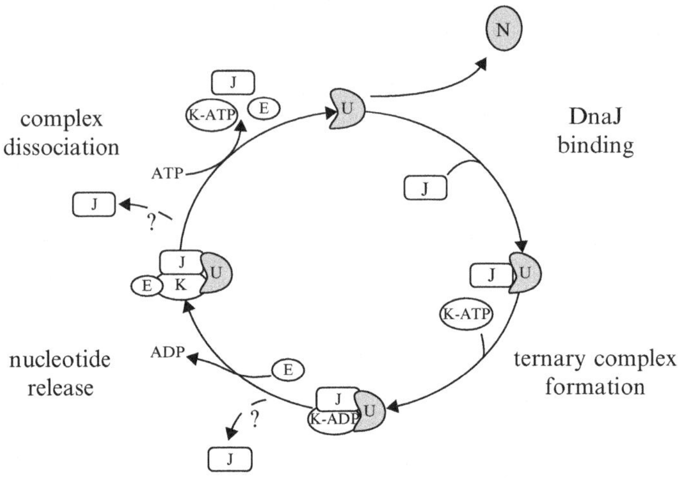

Fig. 2 Schematic representation of the biological cycle of the DnaK chaperone system, as proposed in [42]

3 Simulation Tests Results

There is experimental evidence that chaperone systems are dramatically upregulated during periods of high stress [50, 51]. Decreased protein degradation and repair has been associated to an increased susceptibility to oxidative stress [52]. Since endoplasmic reticulum (ER) stress triggers Reactive oxygen species (ROS) signalling, ROS may provide a common link between cellular stress and the initiation of the cell's antioxidant defense system, including chaperone machinery. Based on these research results, we tried to incorporate into our model the response of the DnaK molecular chaperone system to the effects of ROS. Both the "resting state" and stress-induced levels of activity needs to be examined in order to identify the reaction of cells, when they are subject to high levels of stress over long time periods. This will elucidate the role of chaperone machinery. Moreover, the ability of maintaining cellular proteostasis declines with aging, which results in the accumulation of misfolded proteins, deposition of aggregates, cellular toxicity and eventually cell death [53–55]. This deterioration of proteostasis is a characteristic risk factor of many human pathologies and represent a hallmark which is considered to contribute to the aging process [56] (Figs. 3 and 4).

Oxidized proteins due to increased level of ROS are either degraded or repaired depending on whether the damage is reversible or not. Thus, there is a critical balance between the capacity of Hsp70 chaperones to bind to the misfolded proteins and the aggregated proteins that tend to accumulate (Fig. 5).

Moreover, the age-related accumulation of oxidized protein has been proposed to be due to either, or both, increased protein oxidative damage and decreased oxidized protein degradation and repair [52].

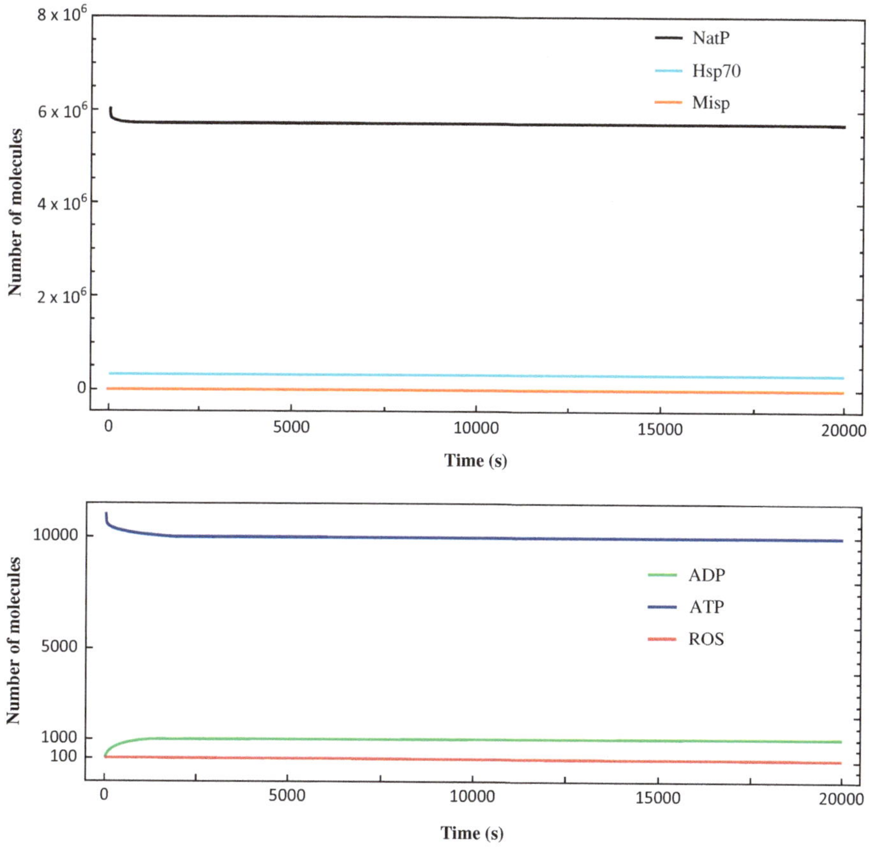

Fig. 3 Simulation results for normal conditions (unstressed cell). (**a**) Levels of native protein (NatP), misfolded protein (MisP) and unbound Hsp70. (**b**) Levels of ATP, ADP and ROS in initial conditions (ROS was scaled for clearer visualisation)

The role of chaperone function and the simultaneous protein oxidation, misfolding and aggregation in aged organisms, suggests that preservation of protein homeostasis and long- range protein organization can be major determinants in longevity [57–59].

When the accumulation of misfolded or unfolded proteins occurs within the ER, this disturbs ER homeostasis, giving rise to ER stress [60]. ER stress results in activation of the unfolded protein response (UPR) which aims to alleviate the stress [60]. The UPR involves up-regulation of protein chaperones to promote protein folding, translational attenuation to reduce the load of proteins within the ER to prevent further accumulation of misfolded proteins, and up-regulation of ER-associated protein degradation (ERAD) and autophagy to promote degradation of misfolded proteins [61]. Therefore, ER stress plays a pivotal role in cell survival by maintaining proteostasis and is implicated as a key mechanism relevant to pathogen-

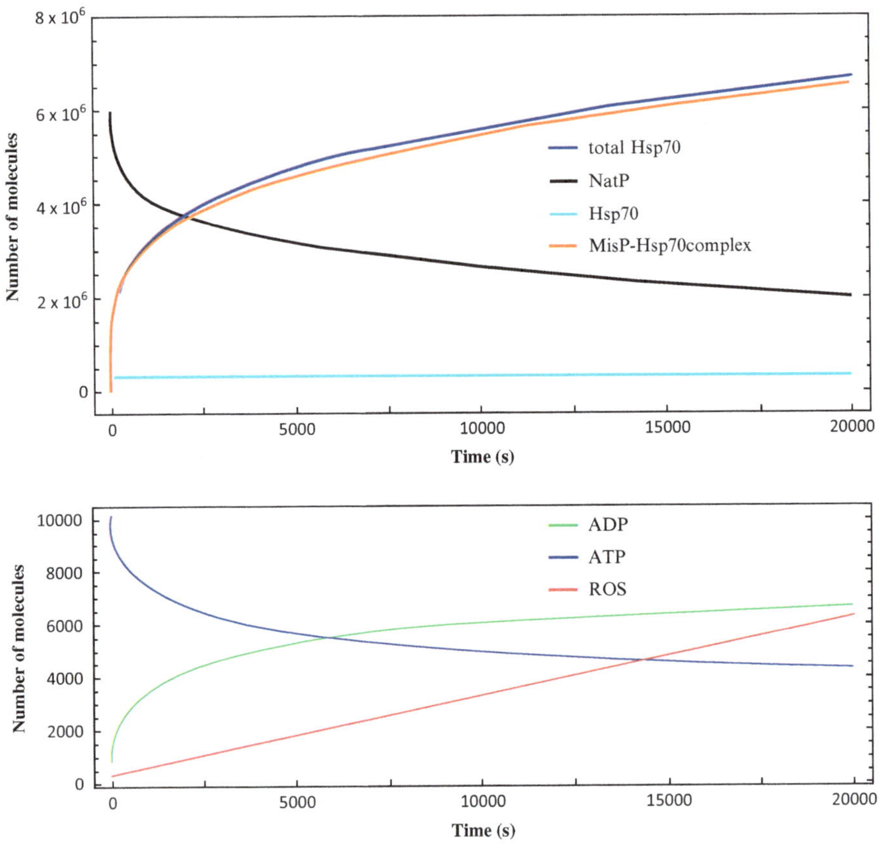

Fig. 4 Simulation results for increased level of ROS (with respect to time). (**a**) Levels of native protein (NatP), misfolded protein (MisP), misfolded protein complexed with Hsp70, and unbound Hsp70. Total Hsp70 increases in step with misfolded protein complexed to Hsp70, native protein corresponding decreases. (**b**) Levels of ATP, ADP and ROS

esis in neurodegenerative diseases. Since chaperones promote the correct folding of proteins into their native conformations and prevent protein misfolding, novel therapeutic strategies may be beneficial in disorders involving protein misfolding.

4 Discussion

Studies over the past two decades have elucidated the important role of the chaperone machinery in assisting protein folding. Nevertheless, understanding how the pathways of misfolding are connected to the neurodegenerative disorders, is still elusive. One of the obstacles is, that in vitro experiments are not easily transferable

Fig. 5 The critical balance between protein folding and degradation activity

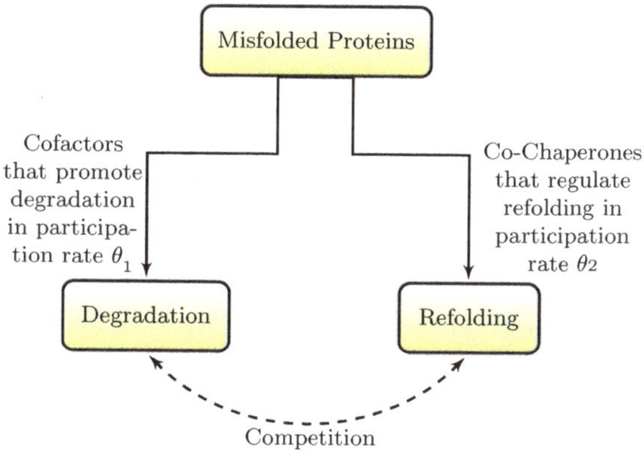

to the in vivo. The development of formal mathematical models allow us to conduct simulation experiments in order to observe the conformational changes in a single polypeptide chain and incorporate the various aspects which are involved in the complex chaperone machinery. This will help us better understand the role of molecular chaperones as a key element in the network of proteostasis. Unravelling the complex signalling pathways of this network, will contribute in solving the problem of how cells react to conformational stress, or proteostasis deficiency. The critical balance of aberrantly folding proteins which aggregate into toxic species, whereas others are degraded and how these pathways change during ageing, will reveal the important relationship between proteostasis and longevity.

References

1. Muntau, A.C., J. Leandro, M. Staudigl, F. Mayer, S.W. Gersting. 2014. Innovative strategies to treat protein misfolding in inborn errors of metabolism: Pharmacological chaperones and proteostasis regulators. *Journal of Inherited Metabolic Disease* 37(4): 505–523.
2. Morimoto, R.I., and A.M. Cuervo. 2014. Proteostasis and the aging proteome in health and disease. *The Journals of Gerontology Series A: Biological Sciences and Medical Sciences* 69(Suppl 1): S33–S38.
3. Hipp, M.S., S.H. Park, and F.U. Hartl. 2014. Proteostasis impairment in protein-misfolding and-aggregation diseases. *Trends in Cell Biology* 24(9): 506–514.
4. Hartl, F.U., A. Bracher, and M. Hayer-Hartl. 2011. Molecular chaperones in protein folding and proteostasis. *Nature* 475(7356): 324–332.
5. Dobson, C.M., A. Šali, M. Karplus. 1998. Protein folding: a perspective from theory and experiment. *Angewandte Chemie International Edition* 37(7): 868–893.
6. Kim, Y.E., M.S. Hipp, A. Bracher, M. Hayer-Hartl, and F. Ulrich Hartl. 2013. Molecular chaperone functions in protein folding and proteostasis. *Annual Review of Biochemistry* 82: 323–355.
7. Witt, S.N. (ed). 2011. *Protein Chaperones and Protection from Neurodegenerative Diseases*, pp. 1–427. Hoboken: Wiley.
8. Kessel, A., and N. Ben-Tal. 2010. *Introduction to Proteins: Structure, Function, and Motion*. Boca Raton: CRC Press.

9. Garrett, R.H., and C.M. Grisham. 2002. *Principles of Biochemistry: With a Human Focus*. Pacific Grove: Brooks/Cole/Thomson Learning.
10. Smith, H.L., W. Li, and M.E. Cheetham. 2015. Molecular chaperones and neuronal proteostasis. *Seminars in Cell & Developmental Biology* 40: 142–152. doi:10.1016/j.semcdb.2015.03.003.
11. Ou, J.R., M.S. Tan, A.M. Xie, J.T. Yu, and L. Tan. 2014. Heat shock protein 90 in Alzheimer's disease. *BioMed Research International* 2014: Article ID 796869, 7 p. doi:10.1155/2014/796869.
12. Mayer, M., and B. Bukau. 2005. Hsp70 chaperones: cellular functions and molecular mechanism. *Cellular and Molecular Life Sciences* 62(6): 670–684.
13. Balchin, D., M. Hayer-Hartl, and F.U. Hartl. 2016. In vivo aspects of protein folding and quality control. *Science* 353(6294): aac4354.
14. Vabulas, R.M., S. Raychaudhuri, M. Hayer-Hartl, and F.U. Hartl. 2010. Protein folding in the cytoplasm and the heat shock response. *Cold Spring Harbor Perspectives in Biology* 2(12): a004390.
15. Kopito, R.R., and D. Ron. 2000. Conformational disease. *Nature Cell Biology* 2(11): E207–E209.
16. Rao, R.V., and D.E. Bredesen. 2004. Misfolded proteins, endoplasmic reticulum stress and neurodegeneration. *Current Opinion in Cell Biology* 16(6): 653–662.
17. Bertolotti, A., Y. Zhang, L.M. Hendershot, H.P. Harding, and D. Ron. 2000. Dynamic interaction of BiP and ER stress transducers in the unfolded-protein response. *Nature Cell Biology* 2(6): 326–332.
18. Araki, K., and K. Nagata. 2012. Protein folding and quality control in the ER. *Cold Spring Harbor Perspectives in Biology* 4(8): a015438.
19. Labbadia, J., and R.I. Morimoto. 2015. The biology of proteostasis in aging and disease. *Annual Review of Biochemistry* 84: 435.
20. Anckar, J., and L. Sistonen. 2011. Regulation of hsf1 function in the heat stress response: Implications in aging and disease. *Annual Review of Biochemistry* 80: 1089–1115.
21. Walter, P., and D. Ron. 2011. The unfolded protein response: from stress pathway to homeostatic regulation. *Science* 334(6059): 1081–1086.
22. Haynes, C.M., and D. Ron. 2010. The mitochondrial UPR–protecting organelle protein homeostasis. *Journal of Cell Science* 123(22): 3849–3855.
23. Saez, I., and D. Vilchez. 2014. The mechanistic links between proteasome activity, aging and agerelated diseases. *Current Genomics* 15(1): 38–51.
24. Vilchez, D., I. Saez, and A. Dillin. 2014. The role of protein clearance mechanisms in organismal ageing and age-related diseases. *Nature Communications* 5.
25. Morimoto, R.I. 2008. Proteotoxic stress and inducible chaperone networks in neurodegenerative disease and aging. *Genes & Development* 22(11): 1427–1438.
26. Kampinga, H.H., and E.A. Craig. 2010. The hsp70 chaperone machinery: J proteins as drivers of functional specificity. *Nature Reviews Molecular Cell Biology* 11(8): 579–592.
27. Ali, Y.O., B.M. Kitay, and R.G.Zhai. 2010. Dealing with misfolded proteins: Examining the neuroprotective role of molecular chaperones in neurodegeneration. Molecules 15(10): 6859–6887.
28. Gong, Y., Y. Kakihara, N. Krogan, J. Greenblatt, A. Emili, Z. Zhang, and W.A. Houry. 2009. An atlas of chaperone–protein interactions in saccharomyces cerevisiae: implications to protein folding pathways in the cell. *Molecular Systems Biology* 5(1): 275.
29. Chiti, F., and C.M. Dobson. 2006. Protein misfolding, functional amyloid, and human disease. *Annual Review of Biochemistry* 75: 333–366.
30. Calabrese, V., C. Cornelius, A.T. Dinkova-Kostova, E.J. Calabrese, and M.P. Mattson. 2010. Cellular stress responses, the hormesis paradigm, and vitagenes: novel targets for therapeutic intervention in neurodegenerative disorders. *Antioxidants & Redox Signaling* 13(11): 1763–1811.
31. Kaushik, S., and A.M. Cuervo. 2012. Chaperone-mediated autophagy: A unique way to enter the lysosome world. *Trends in Cell Biology* 22(8): 407–417.

32. Arndt, V., N. Dick, R. Tawo, M. Dreiseidler, D. Wenzel, M. Hesse, D.O. Fürst, P. Saftig, R. Saint, B.K. Fleischmann, et al. 2010. Chaperone-assisted selective autophagy is essential for muscle maintenance. *Current Biology* 20(2): 143–148.
33. Gregersen, N., P. Bross, M.M. Jorgensen. 2005. Protein folding and misfolding: the role of cellular protein quality control systems in inherited disorders. In: *The Online Metabolic and Molecular Bases of Inherited Disease (OMMBID)*. ed. Valle, D., A.L. Beaudet, B. Vogelstein, K.W. Kinzler, S.E. Antonarakis, A. Ballabio. General Themes. Mc-Graw Hill.
34. Hartl, F.U., and M. Hayer-Hartl. Molecular chaperones in the cytosol: From nascent chain to folded protein. Science 295(5561): 1852–1858 (2002).
35. Tapley, T.L., T.M. Franzmann, S. Chakraborty, U. Jakob, and J.C. Bardwell. 2010. Protein refolding by pH-triggered chaperone binding and release. *Proceedings of the National Academy of Sciences* 107(3): 1071–1076.
36. Flaherty, K.M., D.B. McKay, W. Kabsch, K.C. Holmes. 1991. Similarity of the three-dimensional structures of actin and the atpase fragment of a 70-kda heat shock cognate protein. *Proceedings of the National Academy of Sciences* 88(11): 5041–5045.
37. Flaherty, K.M., C. DeLuca-Flaherty, D.B. McKay, et al. 1990. Three-dimensional structure of the atpase fragment of a 70 k heat-shock cognate protein. Nature 346(6285): 623–628.
38. Meimaridou, E., S.B. Gooljar, J.P. Chapple. 2009. From hatching to dispatching: The multiple cellular roles of the hsp70 molecular chaperone machinery. *Journal of Molecular Endocrinology* 42(1): 1–9.
39. Assimon, V.A., A.T. Gillies, J.N. Rauch, and J.E Gestwicki. 2013. Hsp70 protein complexes as drug targets. *Current Pharmaceutical Design* 19(3): 404–417.
40. Calloni, G., T. Chen, S.M. Schermann, H.C. Chang, P. Genevaux, F. Agostini, G.G. Tartaglia, M. Hayer-Hartl, F.U. Hartl. 2012. Dnak functions as a central hub in the E. coli chaperone network. *Cell Reports* 1(3): 251–264.
41. Siegenthaler, R.K., and P. Christen. 2006. Tuning of dnak chaperone action by nonnative protein sensor DnaJ and thermosensor GrpE. *Journal of Biological Chemistry* 281(45): 34448–34456.
42. Gething, M.J. 1997. Guidebook to molecular chaperones and protein-folding catalysts. Oxford: Oxford University Press.
43. Mayer, M.P. 2013. Hsp70 chaperone dynamics and molecular mechanism. *Trends in Biochemical Sciences* 38(10): 507–514.
44. Young, J.C. 2010. Mechanisms of the hsp70 chaperone system this paper is one of a selection of papers published in this special issue entitled "Canadian Society of Biochemistry, Molecular & Cellular Biology 52nd Annual Meeting-Protein Folding: Principles and Diseases" and has undergone the journal's usual peer review process. *Biochemistry and Cell Biology* 88(2): 291–300.
45. Kellner, R., H. Hofmann, A. Barducci, B. Wunderlich, D. Nettels, and B. Schuler. 2014. Single-molecule spectroscopy reveals chaperone-mediated expansion of substrate protein. *Proceedings of the National Academy of Sciences* 111(37): 13355–13360. doi:10.1073/pnas.1407086111.
46. Siegenthaler, R.K., J. Grimshaw, and P. Christen. 2004. Immediate response of the Dnak molecular chaperone system to heat shock. *FEBS Letters* 562(1–3): 105–110.
47. Hucka, M., A. Finney, H.M. Sauro, H. Bolouri, J.C. Doyle, H. Kitano, A.P. Arkin, B.J. Bornstein, D. Bray, A. Cornish-Bowden, et al. 2003. The systems biology markup language (SBML): A medium for representation and exchange of biochemical network models. *Bioinformatics* 19(4): 524–531.
48. Proctor, C.J., C. Sőti, R.J. Boys, C.S. Gillespie, D.P. Shanley, D.J. Wilkinson, T.B. Kirkwood. 2005. Modelling the actions of chaperones and their role in ageing. *Mechanisms of Ageing and Development* 126(1): 119–131.
49. Goldberg, A.L. 2003. Protein degradation and protection against misfolded or damaged proteins. *Nature* 426(6968): 895–899.
50. Hartl, F.U. 2015. Molecular chaperones: Guardians of the proteome. *Journal of Neurochemistry* 134: PL02, 2–2.

51. Fulda, S., A.M. Gorman, O. Hori, and A. Samali. 2010. Cellular stress responses: Cell survival and cell death. *International Journal of Cell Biology* 10: 1–23.
52. Friguet, B. 2006. Oxidized protein degradation and repair in ageing and oxidative stress. *FEBS Letters* 580(12): 2910–2916.
53. Morley, J.F., H.R. Brignull, J.J. Weyers, R.I. Morimoto. 2002. The threshold for polyglutamine-expansion protein aggregation and cellular toxicity is dynamic and influenced by aging in Caenorhabditis elegans. *Proceedings of the National Academy of Sciences* 99(16): 10417–10422.
54. David, D.C., N. Ollikainen, J.C. Trinidad, M.P. Cary, A.L. Burlingame, and C. Kenyon. 2010. Widespread protein aggregation as an inherent part of aging in C. elegans. *PLoS Biology* 8(8): e1000450.
55. Olzscha, H., S.M. Schermann, A.C. Woerner, S. Pinkert, M.H. Hecht, G.G. Tartaglia, M. Vendruscolo, M. Hayer-Hartl, F.U. Hartl. 2011. Vabulas, R.M.: Amyloid-like aggregates sequester numerous metastable proteins with essential cellular functions. *Cell* 144(1): 67–78.
56. López-Otín, C., M.A. Blasco, L. Partridge, M. Serrano, and G. Kroemer. 2013. The hallmarks of aging. *Cell* 153(6): 1194–1217.
57. Söti, C., and P. Csermely. 2002. Chaperones and aging: role in neurodegeneration and in other civilizational diseases. *Neurochemistry International* 41(6): 383–389.
58. Verbeke, P., J. Fonager, B.F. Clark, and S.I. Rattan. 2001. Heat shock response and ageing: mechanisms and applications. *Cell Biology International* 25(9): 845–857.
59. Leak, R.K. 2014. Heat shock proteins in neurodegenerative disorders and aging. *Journal of Cell Communication and Signaling* 8(4): 293–310.
60. Perri, E.R., C.J. Thomas, S. Parakh, D.M. Spencer, J.D. Atkin. 2016. The unfolded protein response and the role of protein disulfide isomerase in neurodegeneration. *Frontiers in Cell and Developmental Biology* 3: 80.
61. Tsai, Y.C., and A.M. Weissman. 2010. The unfolded protein response, degradation from the endoplasmic reticulum, and cancer. *Genes & Cancer* 1(7): 764–778.

Printed by Printforce, the Netherlands